THE FLUID ENVELOPE OF OUR PLANET
How the Study of Ocean Currents Became a Science

THE FLUID ENVELOPE OF OUR PLANET

How the Study of Ocean Currents Became a Science

ERIC L. MILLS

UNIVERSITY OF TORONTO PRESS
Toronto Buffalo London

© University of Toronto Press Incorporated 2009
Toronto Buffalo London
www.utppublishing.com
Printed in Canada

ISBN 978-0-8020-9697-5

Printed on acid-free paper

Library and Archives Canada Cataloguing in Publication

Mills, Eric L., 1936–
  The fluid envelope of our planet : how the study of ocean currents
became a science / Eric L. Mills.

  Includes bibliographical references and index.
  ISBN 978-0-8020-9697-5

  1. Oceanography – History.   I. Title.

  GC29.M54 2009   551.4609   C2009-902728-3

University of Toronto Press acknowledges the financial assistance to its
publishing program of the Canada Council for the Arts and the Ontario
Arts Council.

University of Toronto Press acknowledges the financial support for its
publishing activities of the Government of Canada through the Book
Publishing Industry Development Program (BPIDP).

This book has been published with the help of a grant from the Canadian
Federation for the Humanities and Social Sciences, through the Aid to
Scholarly Publications Program, using funds provided by the Social Sciences
and Humanities Research Council of Canada.

Thy way is in the sea, and Thy path in the great waters, and Thy footsteps are not known.

Psalm lxxvii, quoted on the title page of James Rennell's *Currents of the Atlantic Ocean*, 1832

The ocean currents are brought about by continuously blowing winds, differences in specific gravity which depend on heat or salt content of the water, variations in barometric pressure, through accumulation of water (as in the Mexican Current) or disturbance of the level, through strong evaporation (as in the Mediterranean) ... The directions of the currents are variously modified through the configuration of coasts, through the rotation of the Earth, as a water particle progressing toward the equator or toward a pole can assume the rotational velocity of each degree of latitude only gradually, and through winds and countercurrents. It is the task of physicists to determine, and continuously improve ... the numerical proportions of those elements (the prime causes of motion and its disruption) upon which the ocean and the atmosphere depend; though following the model of the astronomical sciences is admittedly unattainable it will at least lead to a knowledge of some of the eternal laws which bring about climatic changes in the construction of currents in the fluid envelope of our planet.

Alexander von Humboldt, ca. 1830, quoted by Peterson et al. 1996, p. 69

CONTENTS

viii   Contents

ACKNOWLEDGMENTS

I owe special debts to five people who, in a diversity of ways, contributed to the genesis and completion of this book. Most important of all, they remain my friends and colleagues in our continuing studies of the history of the marine sciences. They are Jacqueline Carpine-Lancre, retired Librarian and Archivist of the Musée océanographique de Monaco; Deborah Day, Archivist of the Scripps Institution of Oceanography; Jennifer Hubbard, historian of science at Ryerson University, Toronto; Helen Rozwadowski, Coordinator of the Maritime Studies Program, University of Connecticut at Avery Point; and Vera Schwach of NIFU STEP, Oslo. We have batted around ideas, shared resources and publications, and helped each other's research for many years. No one could have had – or have – better colleagues.

The germ that grew into this book was the fine scholarship of Dr Harold Burstyn, now a lawyer in New York State. His publications on ideas about oceanic and atmospheric circulation from the Middle Ages into the nineteenth century stimulated my teaching of the history of oceanography and eventually influenced a new research project when I began to speculate what happened in oceanography after the *Challenger* expedition that turned it toward quantitative studies of ocean circulation.

Chronologically, the direct research for this book began in Germany, as a result of Professor Bernt Zeitzschel's hospitality in the then Institut für Meereskunde of the University of Kiel, supported by the resources of its library, and with the interest and help of the then Kustos, Professor Gerhard Kortum. Bettina Probst and the collections of the Museum für Verkehr und Technik, Berlin, and Dr Walter Lenz and Professor Hjalmar Thiel of the University of Hamburg, contributed to my knowledge of the marine sciences in Germany since the late nineteenth century.

In Norway, a wellspring of the marine sciences, I am indebted for help and information to the archivists of the Staatsarchiv i Bergen and the Manuscripts Section of the University of Oslo Library; also to Professors Hans Brattström and William Helland-Hansen (University of Bergen), Professor Karin Pittman and Helge Bottnen (of the then Department of Fisheries and Marine Biology, University of Bergen), Professor Odd Saelen (Geophysical Institute, University of Bergen), and Vera Schwach (NIFU STEP, Oslo). Mrs Mette Haugen of Straumsgrend kindly provided biographical information on her uncle Olav Mosby. In Sweden, Dr Artur Svansson read chapter 3, provided one of the few known pictures of Johan Sandström, sent me his book on Otto Pettersson, and generously shared his knowledge of the history of Scandinavian oceanography.

In Monaco, my research was advanced thanks to a *bourse d'études* from the Musée océanographique, the better part of a year in the wonderful library of the Musée, and the remarkable opportunity to learn from Jacqueline Carpine-Lancre and Dr Christian Carpine. With typical generosity, Mme Carpine-Lancre not only found us an outstanding place to live (with a view of Corsica on the clearest winter days), but also provided working space, all the resources of the Musée's library and archives, and her unparalleled knowledge of the history of French, Monégasque, and non-francophone European marine sciences.

My debts to the Scripps Institution of Oceanography are long-standing, including a William E. and Mary B. Ritter Memorial Fellowship, two Visiting Scholar Positions, and a number of less formal visits to work in the SIO Archives and Library. Deborah Day provided access to the Archives, all kinds of material and personal help, a wealth of information about the history of SIO, and lots of hands-on advice about retrieving information. George and Betty Shor were my hosts and friends, with a wealth of first-hand knowledge of SIO and La Jolla, and Robert Marc Friedman provided Anne Mills and me with accommodation during a critical year of study for both of us. Professors Robert Arthur, Walter Munk, and Joe Reid sympathetically listened to my ideas and provided me with information about early physical oceanography at SIO. Elsewhere in the United States, Helen Rozwadowski (University of Connecticut at Avery Point) provided the enthusiastic support that kept this project afloat.

In Scotland, the huge but little-used archive of D'Arcy Wentworth Thompson in the Manuscripts Section of the University of St Andrews Library was opened to me by Robert Smart, Keeper of Manuscripts.

To my deep regret, H.B. Hachey of the then Atlantic Oceanographic Group at the Atlantic Biological Station, St Andrews, New Brunswick, died when the research for this book was barely in embryo stage. But since Harry Hachey's time, Marilynn Rudi and Charlotte McAdam, successive librarians of the St Andrews Biological Station, along with past Director Dr Robert E. Cook and its current Director, Dr Rob Stephenson, have helped me obtain access to the archival collections at St Andrews and have provided a welcoming research environment in that beautiful and historic location.

On the West Coast of Canada, Gordon Miller, librarian of the Pacific Biological Station, Fisheries and Oceans Canada, in Nanaimo, gave me access at all hours of the day to the PBS library and its manuscript collections, allowing me to work *ad libitum* with one of the most historically important marine science collections in Canada. At DFO's Institute of Ocean Sciences in Sidney, Sharon Thomson and Susan Johnson helped me with document access. And in the University of British Columbia's Archives, Chris Hives and his staff helped me with access to information on the origins of oceanography in that university. I am grateful too for interviews giving me personal views of the development of Canadian oceanography in British Columbia and Canada as a whole with the late Dr William M. Cameron (West Vancouver), the late A.J. Dodimead (Nanaimo), Dr W.N. English (Victoria), the late Richard H. Herlinveaux (Sidney), Professor Tim Parsons (Vancouver), the late Professor George Pickard (Vancouver), Les Spearing (IOS, Sidney), Mrs Lorraine Tully (Nanaimo), and the late Dr Michael Waldichuk (West Vancouver).

Graeme Durkin and Joyce Kennedy of the Department of Fisheries and Oceans Library in Ottawa gave me access to information on the organization of governmental marine science in Canada. That is a subject that Jennifer Hubbard (Ryerson University) and I have talked about enthusiastically for years. I am grateful to her for that, and for her groundbreaking work on the history of marine sciences in Canada.

My colleague in the Department of Oceanography at Dalhousie University, Dr Dan Kelley, helped me out with a review of a critical chapter – and, of equal importance, was always enthusiastic about my forays into his field. His scholarly generosity is typical of Oceanography at Dalhousie, which was and is tolerant of tangential interests: quality of scholarship has always trumped disciplinarity in that admirable department.

Working with University of Toronto Press was a pleasure. For their help, I thank especially Len Husband, Ken Lewis, and Frances Mundy.

Finally, I must thank the Social Sciences and Humanities Research Council of Canada, which supported the beginning of my research on this project with two research grants – although its completion, including a good deal of travel, depended completely on funding from what I some-times whimsically call 'The Anne and Eric Mills Charitable Foundation.'

Late one night early in 1996, Anne Mills and I were bundled in sleeping bags in the deep cold and stillness of the Anza-Borrego Desert of Southern California. The stars were brilliant. Comet Hyakutake blazed overhead, and a few coyotes yapped. I was on leave from inten-sive reading of the German literature that gave rise to chapter 5 of this book. Anne had been researching the biology and history of the desert plant *Ocotillo*. Is this relevant? Of course. I was deep into a newly oriented career as a historian of science, and Anne was beginning a period of self-training as a botanist. We were sharing the uncertainties of the new. The desert at night seemed like a reward for what we shared then and continue to share.

Lower Rose Bay, Nova Scotia
April 2008

THE FLUID ENVELOPE OF OUR PLANET
How the Study of Ocean Currents Became a Science

Introduction

The Fluid Envelope of Our Planet

If philosophers of old, who visited
So many lands to study their secrets,
Had witnessed the marvels I witnessed,
Spreading my sail to such different winds,
What great writings they would have left us!
What revelations about the heavens,
What marvellous testimonies to Nature's youth!
And all without hyperbole. Plain truth!

Vasco da Gama, as imagined by Luis Vaz de Camões in *The Lusíads*, Canto
Five (translated by Landeg White, 1997)

This book is about plain truth – the plain scientific truths about the
oceans constructed by those who changed the ocean sciences from matters of observation and common-sense report to a branch of mathematical geophysics. But upon inspection, scientific truths – or better, the
introduction of influential changes into science – are painted in shades
of gray, not absolute black and white, and depend upon contingencies of
time, place, and personal interaction that defy any kind of deterministic
historiography. Such is the case with the developments that took place in
the study of ocean circulation between the 1890s and the middle of the
twentieth century. It might be possible to regard them as the triumph of
a particular approach to the physics of the atmosphere and the ocean,
but as I intend to show, the development of a mathematical approach
to the oceans was not so much a march of progress as an evolution and

a knitting together of ideas about ocean circulation in a variety of personal, national, and scientific contexts.

For historians, the time scale is short, a mere sixty or seventy years – but the changes that I describe were built upon use and study of the oceans going back to the early Middle Ages and earlier. When Vasco da Gama and his companions rounded the southern tip of Africa and sailed down the Southwest Monsoon winds to Calicut in 1497/98, finally achieving India by sea from Europe, the sea was full of the wonders reported by Camões – waterspouts and torrential downpours, St Elmo's fire and violent storms. And so it is today, as the long tradition of adventure writing about the sea attests. But beginning mainly with the German-speaking geographers of the nineteenth century, who, as the century went on, passed their mantle increasingly to physicists, significant changes occurred in how the oceans were viewed. First the systematic approach of geology was applied as physical geography. Then, just before the end of the nineteenth century, the precision, and the abstraction, of mathematical analysis were applied to the oceans by a few physicists in an attempt to bring the attributes of water in motion under control, at least intellectually. In the chapters that follow, I show, from the vantage point of physical scientists, how the oceans came to be seen not as trackless wastes (the view of ancient and classical authors), nor as part of a great interlinked cosmic machine (Humboldt's view in the early nineteenth century), but as physical phenomena subject to mathematical analysis.[1]

Throughout the Renaissance voyages of exploration and commerce mentioned in chapter 1, it is striking how infrequently any currents of the oceans were depicted, despite the accumulation of seaman's lore about winds and surface circulation. Even though Arab traders and pilots seem to have had detailed knowledge of the monsoonal winds and currents of the Indian Ocean by the ninth century, and the Portuguese, and quickly thereafter the Spanish, navigators of the Age of Exploration found the equatorial currents and the Gulf Stream, it seems surprising that large-scale depictions of currents are absent from maps until the last half of the seventeenth century. It was then too, at the time of the Scientific Revolution, that cloistered scholars, not unlettered mariners, suggested, based upon the increasing knowledge of currents, that there was a continuous *system* of circulation, for otherwise the flowing water would leave empty spaces, which manifestly was not the case. The idea of *continuity* in ocean circulation had been born.

Mariners had known for centuries that there was a link between the surface winds and the direction of ocean currents. The winds themselves, especially the Trades north and south of the equator, appeared to blow just as they should – either following the westward motion of the stars, as Aristotle had said, or as a result of the eastward rotation of the Earth.[2] Whatever the cause, the Trades seemed to produce the equatorial currents, girdling the globe in Atlantic, Pacific, and Indian Oceans. It required no great insight to suggest that these west-flowing currents, when they met obstructions (the northern coast of South America was an example) would be deflected to the north or south, resulting in the meridional currents and eastwardly directed currents that became better and better known during the sixteenth and seventeenth centuries. By the end of James Cook's explorations of the Pacific, late in the eighteenth century, nearly all the major surface current systems of the world were known, and their causes were presumed to be the global wind systems.[3] This was the basis of James Rennell's great compilation in 1832, *The Currents of the Atlantic Ocean*, which gave an authoritative seal of approval, based upon large amounts of information, to the prevailing view that it was the winds alone that moved the water.

Alexander von Humboldt, viewing the Earth as a synthesis of interlocked physical mechanisms, was not so sure. Searching for what he called 'knowledge of the chain of connection, by which all natural forces are linked together, and made mutually dependent upon each other,'[4] he claimed that nearly any physical mechanism could be proposed to move the water, including, in addition to the wind, the effect of changes in density. The use of self-registering thermometers since the first decades of the nineteenth century had shown the presence of deep cold water. How could the great depths of the ocean be filled with frigid water if it had not flowed there from polar latitudes, where it had sunk below the surface? Humboldt's suggestion, not original to him, but certainly made credible by his authority, lay fallow for fifty years, until explorations of the deep sea showed the ubiquity of deep cold water, apparently of polar origin, and called for an explanation.

The late-nineteenth-century debate about the mechanism of ocean circulation – was the driving mechanism the wind alone or did it include polar cooling as well? – was the last great debate about ocean circulation that took place almost completely outside a mathematical framework. Neither James Croll nor W.B. Carpenter, the antagonists in one of the nineteenth century's bitterest and least-known scientific debates, had sufficient physical training or insight to see how their differences over the causes of ocean circulation could be decided. The Carpenter-Croll debate during the 1870s, described in chapter 2, is the great divide between common-sense explanations of how the oceans move and a new approach based upon mathematical simplification, analysis, and the hope of prediction, far removed from the unifying intentions of Humboldt. *Mathematical analysis* became the underlying theme of the new science of physical oceanography, having closer relations with mathematical geophysics than with the geographical description of the physical world that predominated throughout the middle decades of the nineteenth century.[5]

This is not to say that the analytical approach (wrapped up in the term 'physical oceanography') had a trouble-free birth. Old habits do die hard, and the use of a special kind of fluid dynamics, originating in Scandinavia in the late decades of the nineteenth century, was foreign to most physicists and certainly beyond the competence, and usually the comprehension, of the biologically trained marine scientists of the early 1900s who faced the need to explain the distribution of marine animals (especially fish) and their food in a medium that was in constant motion. Much of this book deals with the varieties of ways that marine scientists reacted to a new physical framework proposed for their work and the variety of ways that their working habits and habitats were changed – or in some cases not changed – by the wave of innovation resulting from the increasing use of dynamic and other mathematical methods in oceanography. Modern-day physical oceanographers may find my approach here paradoxical or even perverse, because the names of most of the protagonists have disappeared from physical oceanography and even the main technique that they used is now little more than a footnote. But it was these nearly forgotten early physical oceanographers who, haltingly and for their own reasons, contributed to the self-sufficient science of today. They are the main subjects of this book.[6]

From its roots in Scandinavia, the mathematical analysis of ocean currents ('dynamic oceanography') was first exported to Canada as part of a very practical project – to find and develop fishery resources for the young Dominion. Canadian marine scientists were ill-prepared to use it, or even accept the utility of dynamic physical oceanography, so that, as I show in chapters 4 and 8, there was a long delay between the first attempt to transplant these methods from Europe and their eventual use in a totally new context. It was only when new challenges arose – major problems in managing marine habitats and, soon thereafter, the submarine threat during the Second World War – that the orientation of Canadian marine science was changed in ways that were unimaginable a generation earlier.

Although we are accustomed to seeing Europe as the wellspring of scientific innovation in the early twentieth century, reactions in Europe to the mathematical approach in physical oceanography were strikingly varied. In Germany, which had a long and distinguished tradition of physical geography, academic oceanography arose from this stock, especially in the University of Berlin. But the adoption of quantitative techniques, especially dynamic oceanography, was halting, and took place very slowly at first, mainly because of lack of training, as chapter 5 shows. When it became clear that mathematical physical oceanography could promote the models of oceanic circulation upon which younger German oceanographers had built their careers, the method was embraced, and even verified, by comparing direct measurement of currents with mathematical calculation based upon data from the same area, the Gulf Stream off Florida.

No greater patron of oceanography existed in the late nineteenth century than Albert Ier Prince of Monaco. His fortune, his ships, his example, and his Institut océanographique in Paris set the pattern of French research on the oceans for decades. In this, physics played a role mainly through adaptation of physical measurement to a wet, salty environment, the oceans, rather than as a new mode of analysis. French oceanography until the 1930s was profoundly conservative, the conclusion of chapter 6, and little aided by the eccentricities of Albert's widely versed one-time protégé, Julien Thoulet, who dabbled in mathematical oceanography without fully understanding it, thereby delaying the introduction of dynamic physical oceanography into France for several decades.

Despite the sterile soil for mathematical physical oceanography in Canada early in the twentieth century, the story in the United States was different. George McEwen, a young physicist hired at the laboratory of the Marine Biological Association of San Diego (which became the Scripps Institution of Oceanography), was certainly among the first physicists anywhere to understand and use Scandinavian mathematical oceanography. But it offended his scientific scruples, which insisted upon causal analysis[7] rather than upon what he considered expedients. As a result, his kind of oceanography fell behind the times, and it was a young U.S. Coast Guard officer, Edward H. Smith, a student at Harvard of the multi-talented (but unmathematical) H.B. Bigelow, who went to Scandinavia to learn modern mathematical oceanography at its source and who, in effect, brought it to North America for good. Thereafter, especially with the stimulus for new knowledge of the oceans during the Second World War, simple mathematical techniques, like dynamic oceanography, became part of the tool kit of a new profession, physical oceanography, which concerned itself more and more with physical modelling of the oceans to aid analysis and give the prospect of prediction.[8] It is then, in the 1960s, that this book ends.

In a recent review of the history of oceanography, Margaret Deacon and Colin Summerhayes write tellingly of the past century:

> In spite of the great changes that have taken place in oceanography over the past century, close links still exist between the thoughts and aspirations of scientists in the past and those working at the present day, and the underlying requirements of successful ocean science remain essentially the same: enthusiastic and persistent individuals, challenged by ideas, able to gain access to financial and other resources, and in possession of technology capable of working in a difficult environment to gain the necessary data.[9]

But, as I show, the 'necessary data' have not been the same for all who attempted to understand ocean circulation. It has been only for a few decades that any unanimity has existed concerning what kinds of information about the oceans are relevant and important, how they may be gathered,[10] and what kind of analysis is appropriate. In the middle of the nineteenth century, as earlier, there was no agreed-upon approach to ocean circulation. Even when dynamic oceanography began to evolve as the outcome of related work in meteorology there was nothing inherent in it that resulted in its widespread application. Instead, it was the utility of the technique to several groups of physicists working separately and on different projects that ensured its use.

Elsewhere, Margaret Deacon has shown us how haltingly the ocean sciences have developed since at least the time of the Scientific Revolution of the seventeenth century.[11] My work reinforces that insight, showing in detail for one branch of the ocean sciences, physical oceanography, how dependent scientific innovation is upon the contingencies of time, place, and personality. A seemingly simple change of practice, from observation of winds and surface drift of the waters to the use of mathematical techniques to determine the direction and velocity of currents, was not the same change everywhere, for the conditions were different from one scientific environment to another. Now, early in the twenty-first century, one can use the term 'physical oceanography' and expect to evoke similar responses and expectations in marine scientists. One hundred years ago, the term was little known and its boundaries were not fixed. In the chapters following, I aim to show in detail the variety of ways that a branch of ocean science developed, until divergent approaches converged on ideals and practices common to all physical scientists of the oceans. The way to plain truth was complex and multifarious five hundred years ago, was still so a century ago, and is so even today.

1

The Way of the Sea: Knowledge of Oceanic Circulation before the Nineteenth Century

All the seas of the world are one.

J.H. Parry, *The Discovery of the Sea*, 1974, p. xi

The 'Wonderful Phenomena' of the Oceans

For at least three centuries, maps of the Earth have depicted the circulation of the oceans – ocean currents – by arrows filling the virtually unknown areas of our planet's surface. A particularly beautiful and influential example is the set of maps prepared by the German cartographer Heinrich Berghaus (1797–1884)[1] under the influence of the great traveller and polymath Alexander von Humboldt (1769–1859) and published in 1845. These were modified for English-speaking readers by the Scottish geographer and cartographer Alexander Keith Johnston (1804–1871) and published in his *Physical Atlas of Natural Phenomena* in 1848, making them available to a wide audience.[2]

These maps summarized centuries of tacit and recorded knowledge of the surface currents of the Atlantic, Pacific, and Indian Oceans (fig. 1.1), and the text accompanying the Berghaus/Johnston charts of the oceans quotes Humboldt:[3]

The currents of the ocean supply new and most abundant matter of research for the elucidation of the physical phenomena of the earth. While they carry the temperature of one zone into the other, they sometimes promote and sometimes retard the distribution of the race of men, and the commercial intercourse of civilized nations ... It is the duty of philosophers to determine and adjust their various elements, according to the sublime

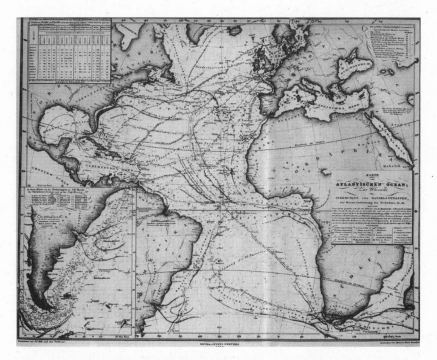

1.1 Heinrich Berghaus's map of the Atlantic Ocean, as modified for A.K. Johnston's *Physical Atlas* in 1848.

model of astronomical science, in order that some of those eternal laws may be made known by which the climatic changes of the firmament are dependent on the liquid and aerial currents of our planet.

But despite his exhortation to apply the physics of Newton and Laplace to the oceans and the atmosphere, Humboldt had no illusions that a quantitative description and explanation of the phenomena of the oceans was at hand, nor was this the opinion of Berghaus and Johnston in 1848. In spite of the certainty of arrows on paper, and in spite of the obvious wisdom of the creator in disposing ocean currents as they were experienced, it was difficult to suggest the causes that lay behind the reality of the cartographer's arrows:

No part of the ocean is in a state of rest – its whole surface is perpetually agitated – one of the wise provisions of the great Author of the universe for preserving its purity. Not only do these currents prevent stagnation in the sea, but by dispersing heat and moisture they temper the climate over a vast portion of the globe. The existence of ocean currents has been long known; yet it is only since the later improvements in navigation, especially by the introduction of the chronometer and lunar observations, about the year 1770, that they have been treated as a regular and extensive system ... Notwithstanding the many attempts to account for the formation of these great systems of aqueous circulation, the subject is still involved in considerable obscurity. The prevalent opinion is ... that they are produced by the pressure of the trade-winds and the consequent accumulation of water near the Equator. Among other causes to which they have been attributed by recent authors, there are, in addition to the trade-winds, the changeableness of atmospheric pressure, the temperature of the air, the difference of temperature or saltness of the sea, the periodical melting of the Polar ice, the unequal amount of evaporation at different latitudes, and the local forms of the continents ... Doubtless many of these causes may combine to produce the result, but the laws which govern the motion of fluid in large bodies, when moving through each other in currents of different temperatures, are not yet sufficiently understood to warrant a satisfactory explanation of these wonderful phenomena.

Under these circumstances, the confusing complexity of the sea, a practical approach was necessary:

Future investigations will, in all probability, develop a system of marine circulation in perfect harmony with the other ascertained laws of the universe, but, in the meantime, we shall content ourselves with describing only what is certainly known.[4]

And what was 'certainly known' in 1848, although so little, was a synthesis of knowledge gained through centuries.

Explorations and Their Outcomes

To twenty-first-century marine geologists and geophysicists, the oceans are an aqueous barrier lying between them and their real objects of study, the sediments and lithosphere of the Earth. Similarly, to early traders carrying goods around the Mediterranean or in and out of the Baltic Sea, or to those beginning the slave trade along the coast of West Africa in the fifteenth century, the oceans represented a medium to be traversed and surmounted, frequently dangerous obstacles to commercial riches. Speculations about the causes of ocean currents were few until the eighteenth century, although by the early sixteenth century knowledge of the extent of the global oceans and of the major current systems near the equator was widespread among seafarers and cartographers.

In a series of books, J.H. Parry has investigated and chronicled exploration of and trade over the oceans from the fifteenth through the eighteenth centuries.[5] He argues that exploration of the new worlds of Africa, Asia, and the Americas required sturdy, manoeuvrable ships, satisfactory techniques of navigation (including charts), stable and increasing financial backing, and encouragement by governments.[6] The capture of Constantinople by the Ottoman Turks in 1453 cut off European land trade to the East or made it prohibitively dangerous and expensive, but the appearance of Claudius Ptolemy's *Geography* in sumptuously produced volumes beginning in the 1470s stimulated the European imagination – and European greed – and suggested other routes to the East.[7] Even without the stimuli of necessity and the imagination, it seems unlikely that the seafarers and commercial adventurers of Iberia and the Italian peninsula (and later Britain and Holland) would have been delayed long in exploring Africa, the East Indies, the New World, and eventually the Pacific. The prospects of major payoffs were simply too great.

In 1419 Portuguese navigators had ventured only as far as Madeira. Less than eighty years later, they had rounded the Cape of Good Hope and crossed the Indian Ocean to Calicut. Although the story of a great school of navigation in southwestern Portugal built by Prince Henry 'The Navigator' is a myth,[8] there is no doubt that map-making and the art of navigation flourished during the expansion of trade by Portugal and its European neighbours throughout the fifteenth century. These navigators had to learn – perforce – about coasts, winds, and currents for the most practical of reasons: survival and profit. But there was an unexpected outcome. Magellan's voyage for the Spanish Crown across the Pacific (and that of his surviving crew eventually around the world) between 1519 and 1522, intended to oppose Portuguese influence in the Spice Islands (Moluccas), had the unintended consequence of demonstrating that all the navigable oceans were connected – 'all the seas of the world are one,' as J.H. Parry has said. Seven years after the survivors of Magellan's expedition straggled back to Sevilla, Diogo Ribeiro's world chart of 1529 showed the extent, the interconnectedness, and some of the details of an oceanic world that provided continuous access to all the lands of the Earth.[9] Unlike its predecessors, Ribeiro's is a world map, and a recognizably modern chart of the oceans; that is, it is congruent with twentieth-century conventions about the form and dimensions of the Earth. Later world maps, such as that prepared by Abraham Ortelius in his *Theatrum Orbis Terrarum* of 1570 (which included the first printed map of the Pacific), added detail and style, but not significant qualitative differences.[10] But what is missing? The major oceans are featureless cartographic wastes in these and later maps of the sixteenth century, with only a few exceptions. One was the Gulf Stream (chanced upon by the Spaniard Ponce de León in 1513 and thereafter an express route westward by homebound Spanish ships), which was indicated on a world map of 1529 and repeated or copied later in the century.[11] Regular depiction of surface ocean currents did not become standard practice until the middle of the next century.

Even though maps and charts of the era do not tell us much about sixteenth-century knowledge of the oceans, a written work from England, *A Booke Called the Treasure for Traveillers* (1578), by the Gravesend innkeeper and man of property William Bourne (1535?–1582) is more forthcoming.[12] Writing for the edification of artisans, Bourne summarized a wide range of observations about the sea and its coasts in a way much akin to later physical geography. For example, he noted the force of the sea and its link with coastal erosion, accounted for the tides by the pull of the moon (which itself was carried by the Aristotelian *primum mobile* – see the next section of this chapter), discussed the salt of the sea, and gave an account of ocean surface currents, which in general flowed from east to west and were tidal in origin. But westerly directed currents could be blocked by continents, whereupon they were deflected to the north and south; for example, the Atlantic Current originating near the Cape of Good Hope, some of which flowed along the equator and was deflected into the Gulf of Mexico, from which it issued as an eastward-directed current through the Florida Strait.[13] Bourne's was only one expression, although a very valuable one for historians, of ideas about oceanic circulation with origins in the sixteenth century and earlier that were developed and extended into a global scheme of surface ocean currents during the seventeenth century.

Currents and Continuity

Though they may not have been depicted, detailed knowledge of currents along a few important trade routes goes back to at least the ninth century, long before the Atlantic and the Pacific were traversed. No later than AD 846, Arab traders and navigators were aware of the seasonally reversing equatorial currents of the Indian Ocean, and especially of the monsoon winds. Making the connection was by no means straightforward, for as recent commentators have noted, the set of an ocean current is masked by strong winds.[14]

It was the Southwest Monsoon winds that allowed ready access to India from East Africa during the summer, and their winter counterparts, the Northeast Monsoons, rather than currents, that made the return voyage fast and reliable.[15] The winds, and their seasonal changes, had been known for centuries when, about AD 846, the Arab geographer Ibn Khurradādbbih (Ibn Khordazbeh in some transliterations) recorded practical knowledge of currents in the northern Indian Ocean. As he wrote, 'In one period the Sea flows during the summer months to the northeast for six months ... In the other period, the Sea flows during the winter months to the southwest for six months.'[16] A century later, in 947, virtually the same information was included by the the historian and geographer al-Mas'ûdi in his book *Meadows of Gold and Mines of Gems*. Apparently the seasonal reversal was known to some navigators (how many is not clear), but very little that was new appeared in this tenth-century encyclopedic work.[17]

What was common knowledge in the Arab world of the northern Indian Ocean began to enter a global picture only in the late fifteenth century. It was of immediate importance to the Portuguese on their way to India after Vasco da Gama's fleet struggled against the south-and-westward-setting Agulhas Current upon rounding the Cape of Good Hope in November 1497, and then in the hands of a knowledgeable Arab or Indian navigator was guided to India in only twenty days using the Southwest Monsoon for a rapid and comfortable passage.[18] Later Portuguese voyagers kept far to the east to avoid frustrating and possibly fatal delays.

During a voyage to the New World in 1498, Columbus encountered a strong westward-setting current along the northern coast of South America.[19] Nearly two decades later, this and more accumulating knowledge was made available to reading audiences by the Italian/Spanish prelate Apostolic Prothonotary and historian of exploration Pietro Martire d'Anghiera (Peter Martyr) (1427–1526) in *The Decades of the New Worlde or West Indies*, parts of which were first published in Latin in 1516, then in English in 1555.[20] In what he refers to as a 'philosophical digression,' Pietro Martire enquired how the constant westward flow of ocean waters could be maintained, and whether the water was deflected back to the east by land or continued on to the west through some passage in the

New World. Citing Sebastian Cabot's observation that the current direction off the northeast coast of North America was westward (the Labrador Current?), Martire concluded that there must be a passage allowing continuity of flow, although there was opinion to the contrary.[21]

Unknown to Pietro Martire in 1516, the Spanish navigator Ponce de León had chanced upon the Florida Current–Gulf Stream system in 1513, clearly a northeasterly current issuing from the Gulf of Mexico. Within a few years, the Spanish were using the clockwise circulation of the North Equatorial Current and North Atlantic to go to and from the New World (Columbus had inadvertently done the same on his first voyage) but without recognizing any return flow of the sea to the east at high latitudes.[22]

The soundly based opinion that the oceans circulated from east to west at low latitudes was reinforced by Barnarde de la Torre's discovery in 1542 of the North Equatorial Current of the Pacific Ocean. But the Gulf Stream was not the only apparent anomaly, for evidence began to accumulate throughout the late sixteenth century of eastwardly directed currents, especially at higher latitudes. Contrary to the opinion of Humphrey Gilbert, Martin Frobisher, on his second voyage to the northern New World in 1577, found evidence of a current from the west and southward drift of ice along the northeast coast of North America.[23] Such observations were frequently not believed; this was the fate of Edward Fenton's observation in 1582 of a strong eastward current (part of the Equatorial Countercurrent–Guinea Current system) in the Gulf of Guinea, West Africa.[24]

If mariners speculated about grand systems of oceanic circulation and their causes during the sixteenth and early seventeenth centuries, there is little evidence of it. But scholars began to be less circumspect. Among these was the German geographer Bernhard Varen (Varenius) (1622–1650?), whose *Geographia Generalis* of 1650 remained in use, in various versions and translations, well into the eighteenth century. In it, Varen described the westward flow of the oceans, driven by the winds, but included accounts of what are now called boundary currents – the Peru Current, the Florida Current, the Mozambique Current, and the Guinea Current.[25] Varen's contemporary, the Dutch/English cleric Isaac Voss (Vossius) (1618–1689), a classical scholar and one of the best-read men of his time, attempted to explain the tides in his remarkable book *De Motu Marium et Ventorum Liber* (1663), which was published in English in 1677 as *A Treatise Concerning the Motion of the Seas and Winds*. This work gives us vivid insight into informed speculation about the phenomena and the mechanisms of the oceans as viewed at the peak of the Scientific Revolution.[26] To Voss, the circulation of the oceans was a single continuous system (a concept termed 'continuity' or 'conservation of volume' in modern oceanographic jargon)[27] in which westward flows are deflected by the land masses, resulting in eastward return motions rejoining the westward currents. As he says,

> The first and chief Motion of the Sea and Winds, is that which between the Tropicks constantly and perpetually follows the Sun; and were it not for the impediment of Land, would with a continued circulation surround the whole Globe of the World.[28]

This flow is continuous for good reasons:

> There may be a most certain reason given, why this motion of the Ocean goes in a course contrary to the first motion thereof. For when the Seas in the Torrid Zone do run perpetually from East to West, and never go back, that this Current may always continue, it is necessary that one of two things should happen, either that these Tracts of Land from whence the Waters begin to depart, should be left wholly bare, or that some succeeding Flood should fill up this vacuity. Now seeing that the former is false, it follows that the other must be true, and that the diminution of waters is made up by fresh Floods which on either side flow thither.[29]

The result, as Voss summarizes it, is 'that all the Waters of the Ocean turn around in a circle, and return to the same point from whence they departed.'[30] According to Voss, then, all the oceans will show similar patterns of gyral circulation, clockwise in the Northern Hemisphere, counterclockwise in the Southern, constantly recirculating. Although he did not say so, with the publication of *De Motu Marium* all the observations of oceanic circulation accumulating since the ninth century could now be fitted into a single comprehensive description.

The insights of Voss, a Protestant in Newtonian England, were certainly not shared by his Jesuit contemporary, the German encyclopedist and polymath, long resident in Rome, Athanasius Kircher (1602–1680), whose book *Mundus Subterraneus* (1665) brought together many old ideas, frequently of Neoplatonic origin, about the surface and the interior of the Earth.[31] But Kircher's 'geographic-hydrographic chart of the motion of ocean currents' (fig. 1.2 is part of this) has the distinction of showing for the first time a comprehensive scheme of surface water

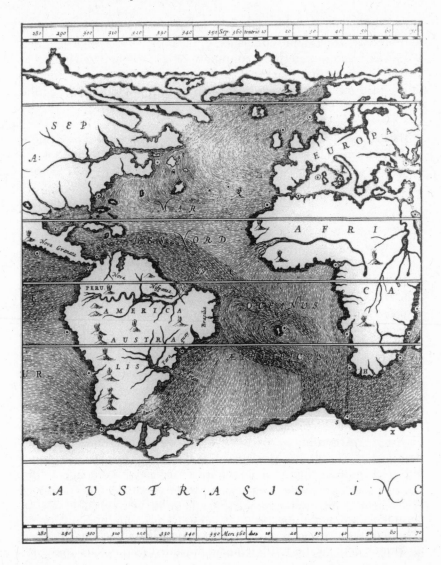

1.2 The circulation of the Atlantic Ocean as depicted by Athanasius Kircher in his book *Mundus Subterraneus* (1665). This appears to be the first depiction of ocean circulation in a world map.

movement. Not without paradoxes – it showed crossing currents in the Indian Ocean – Kircher's map combines myths, such as those of legendary whirlpools, with attempts at a naturalistic graphical representation of currents. Behind it lay a complex of ideas about the driving forces of those motions.

The Causes of Currents and Winds

Practical navigators of the fifteenth and sixteenth centuries discovered, or were soon made aware of, the equatorial current systems that speeded their passages to the west across the Atlantic and Pacific, but we have little or no record of their speculations about the causes of those motions. There was, however, a long tradition of explanation going back at least to the ninth century, when the winds of the Indian Ocean and its currents were first described. These were partly practical – the winds cause water movement – and partly (and frequently inextricably) based on Aristotelian ideas passed on since Antiquity. Only the Moslem world had direct access to the classical authors, including much of Aristotle, until the translations of Arabic works into Latin in the twelfth century. But after that, and especially after the wide spread of printed works in Europe in the fifteenth century, the Aristotelian corpus was readily available for explanation and to be questioned.[32] The basis of much physical speculation about the oceans was Aristotle's *Meteorologica*, and to some extent the curious work *Problemata*, possibly Aristotelian, but likely a compilation by later Peripatetics.[33]

The basis of the Aristotelian scheme is the motion of the sphere of the heavens from east to west around the Earth, following the *primum mobile* (prime mover), which is the outermost of a series of concentric spheres making up the universe. Unmoving at the centre of the universe is the Earth, which responds to the motion of the heavens and which is composed of four simple elements, Earth, Air, Fire, and Water, each in a natural place or attempting to return to it.[34] Thus, superimposed on the motion of the heavens is a series of lesser motions or actions involving the interactions of the four elements. Included in these are the water cycle and the winds.[35]

According to Aristotle's scheme in *Meteorologica* (fig. 1.3), the Sun evaporates water from the sea, leaving the salt behind. The evaporated water condenses as rain, feeding the rivers and soaking into the sponge-like land. Rivers then return water to the sea, maintaining its level, and, although medieval European thought accepted the idea of reservoirs

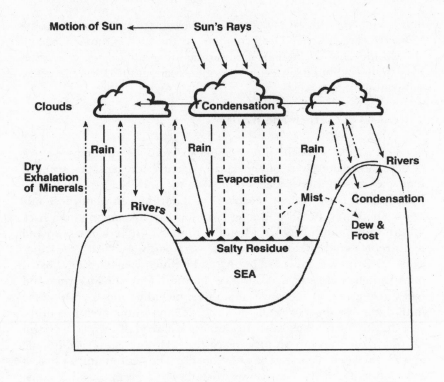

1.3 The hydrological cycle according to Aristotle, based on his *Meteorologica.* Water evaporating from the sea forms mist at low altitudes or is condensed in clouds to rain at higher altitudes. Rain falling on land forms streams and rivers, which are also supplied by rainwater percolating through the Earth and new water condensed from air within the Earth. Rivers are fresh: the salt of the sea arises from a 'dry exhalation' of the Earth raised by the Sun which mixes with the rain falling on the sea. As evaporation from the sea proceeds, the mineral matter remains there. Details of the processes within the Earth are shown on the left, as is the motion of the Sun, as a result of the westward motion of the *primum mobile.*

within the Earth (responsible for rivers and springs, possibly also the constant volume of the oceans), Aristotle did not, regarding the land as merely a sort of filter. The salt of the sea originated in a mixture of rain and air, the air containing what Aristotle termed a 'dry exhalation' raised by the Sun from the land which carried minerals to mix with rainwater, itself a 'wet exhalation' from the sea.[36] To Aristotle, the winds were not air in motion, as Hippocrates had suggested, but were also an exhalation of the Earth in response to the Sun's heat. They were caught up in the motion of the heavens as they rose from the Earth to their natural place, resulting in horizontal motions.[37]

During the seventeenth century, with the development of the concept of air pressure (closely associated with the development of air pumps and barometers), Aristotle's qualitative scheme of atmospheric circulation dropped out of sight, but not without having been expressed in European Renaissance thought. In *Meteorologica* Aristotle had suggested that

> the earth is at rest, and the moisture about it is evaporated by the sun's rays and the other heat from above and rises upwards: but when the heat which caused it to rise leaves it, some being dispersed into the upper region, some being quenched by rising so high into the air above the earth, the vapour cools and condenses again as a result of the loss of heat and the height and turns from air into water: and having become water falls again onto earth.[38]

Eighteen hundred years later, in 1563, the Englishman William Fulke modified Aristotle's scheme slightly to account for the formation of dew, which, like rain,

> ... is drawn up by the sun in the day time, which because it is not carried up into the middle region of the air, abiding in the lower region, by cold of the night is condensed into water, and falleth down in very small drops.[39]

But he does not follow Aristotle slavishly, for Fulke accounts for the salt of the sea differently, claiming that it comes from the earth:

> The saltness of the sea, according to Aristotle's mind, is caused by the sun, that draweth from it all thin and sweet vapours to make rain, leaving the rest as the settling or bottom, which is salt. But men of our time, peradventure, more truly, do not take this for the only and sufficient cause to make so great a quantity of water salt, but say, that, the sea by God's wisdom is

gathered into such valleys of the earth, as were otherwise barren and un-
fruitful, such earths are salt, the sea water then mixed with that earth ...

Far more influential than Aristotle's water cycle was his physics of motion, based upon the action of the *primum mobile*. In 1498, Columbus
expressed his debt to Aristotelian physics succinctly: 'I hold it for certain
that the waters of the sea move from east to west with the sky,' and nearly
a century later Humphrey Gilbert stated that 'the sea runneth by nature
circularly, from the East to the West, following the diurnal motion of the
Primum Mobile.'[40] But it was the Trade Winds, rather than the less well
known currents, that could best be explained by the motion of the heavens, or, as Copernican views prevailed, by the motion of the Earth from
west to east into a sluggish ocean of air or a relatively inert mass of water.
As Isaac Voss summarized common beliefs in the late sixteenth century,
'The first and chief Motion of the Sea and Winds, is that which between
the Tropicks constantly and perpetually follows the Sun; and were it not
for the impediment of Land, would with a continued circulation sur-
round the whole Globe of the World.'[41]

Bernhard Varen, he of the great *Geographia Generalis* (1650), was convinced that the general westward circulation of the oceans was caused by
the Trade Winds, which were generated by the Sun in its (Aristotelian)
motion around the Earth. As the Sun moved from east to west, its heat
expanded the air beneath it, resulting in an inflow of air from the east
in compensation. Confused though he was by the relation between tides
and currents, Varen's hesitant steps toward the idea of a wind-driven oceanic circulation were the first of their kind,[42] suggesting not just that the
equatorial currents were caused by the Trade Winds but that the puzzling
meridional currents (like the northward-directed Peru Current) were
caused by local winds.[43] But how that happened was clearly not settled,
for his equally knowledgeable contemporary Voss placed the action at
the sea surface, which he believed would be expanded and raised by the
Sun's heat, causing a downhill flow to the west, an explanation he applied
to the atmosphere as well, explaining the Trade Winds.[44] Voss's claim that
'the first and chief Motion of the Seas and Winds, is that which between
the Tropicks constantly and perpetually follows the Sun' is already familiar. It is followed by an explanation:

For though the Sun by his heat attract and separate the more subtle parts
of the Water, yet he does not depress or lessen the superfice [surface] and
height of the Sea, but rather dilates and raises it. Wheresoever the Sun

is perpendicular, there is the greatest swelling of the Ocean ... Therefore when the Seas to which the Sun is, or lately before was perpendicular, are raised higher than the surface of the Ocean which reacheth to the West, and hath not as yet felt the beams of this Luminary, it comes to pass necessarily, that the Floods should rowl from a higher surface to a lower. And this is indeed the sole reason which forceth the Ocean Westward. The same thing is to be understood of the winds; for what the Sea suffers, the superincumbent air feels the same.[45]

It was 'the superincumbent air' that forced itself upon practical men of the sea, although by the early seventeenth century no longer in an Aristotelian framework. To natural philosophers, the barometer indicated that an ocean of air surrounded the Earth, and the air pump, used increasingly as an experimental tool during the 1660s, along with the barometer, showed that the air had unusual properties, especially its 'spring' (now called pressure). During this time, as Harold Burstyn has written, '... air and water lost their qualities, so that they became mere bodies subject to the geometrical science of motion.' In this transition, no one played a more important part than the English astronomer and mathematician Edmond Halley (1656?–1743), colleague of the better-studied Robert Boyle, Robert Hooke, and Isaac Newton among the Fellowship of the early Royal Society of London.[46]

Quite aptly, Burstyn has called Halley 'the founder of the modern study of the general circulation of the atmosphere.'[47] And although Halley is best known as an observational and mathematical astronomer – as attested by the comet named for him – Halley's contributions to cartography, geophysics, and meteorology were great.[48] He was responsible for worldwide study and charting of magnetic declination, and for a systematic study and depiction of co-tidal lines around the southern English coast. Halley's sponsorship of Newton's *Principia* before the Royal Society in 1686 (he also paid for its publication) as editor of the *Philosophical Transactions* is well known, but it is less well known that at the same time Halley became interested in the general circulation of the atmosphere and especially the explanation of the Trade Winds.[49] The stimulus was a posthumous manuscript by the French natural philosopher Edme Mariotte (ca. 1620–1684), the *Traité du mouvement des eaux et des autres corps fluides* (1686), sent to London by a correspondent in Paris in April 1686.

Underlying a complex, indeed confused, argument in Mariotte's *Traité*, as Burstyn has shown, is the idea that the main cause of the winds is the rotation of the Earth, as Galileo had suggested. But there are secondary causes, including the distance of the moon and 'the Vicissitudes of the Rarefactions of the Air by the Heat of the Sun, and of its Condensation when the Sun ceases to warm it.'[50] Similar explanations had been used before to explain the Trade Winds, but Mariotte goes on to try to explain the origin of the – by now – very well known westerlies at high latitudes and the return flow to the tropics, involving a meridional flow of air to compensate for air attracted by the moon in its movement from perigee to apogee. He claimed that

> ... in going from the Earth, it [the moon] ought to carry along with it the Air which is next to it; and that, the Air which is below, even to the Land that is under the Torrid Zone: and for this Reason the Air which is near the Poles on each side, must flow thither to preserve the Equilibrium of its Spring.[51]

About a month after receiving it, in May 1686, Halley reported to the Royal Society on Mariotte's *Traité*, which seems to have set him thinking about a consistent, mechanically based system of atmospheric circulation, for at that meeting he also presented a paper titled 'An Historical Account of the Trade Winds and Monsoons ...'[52] Drawing on his experience in the tropics (he had voyaged to St Helena in 1676 to do a survey of geomagnetism and lived there for a year making astronomical observations) and on mariners' accounts, Halley described and depicted the winds of the major oceans, excluding the central Pacific, on which there was little information. Then he approached the causal mechanisms, first by rejecting the Galilean explanation (that the winds were caused by the rotation of the Earth into a stationary mass of air),[53] which failed because of exceptions, namely 'the constant Calms in the *Atlantick* Sea near the *Equator*, the Westerly Winds near the Coast of *Guiny*, and the *Periodical* Westerly *Monsoons* under the Equator in the Indian Seas'[54] What could replace it? Halley tells us that

> it remains ... to substitute some other cause, capable of producing a like constant effect, not liable to the same Objections, but agreable [*sic*] to the known properties of the Elements of Air and Water, and the laws of the Motion of fluid Bodies.[55]

His explanation of that 'other cause' is at first a well-known one. It is that the Sun rarefies the air in the equatorial regions below it, and to maintain equilibrium air rushes in from the east as the Sun moves westward. But Halley knew well, as had generations of navigators before him, that the Trade Winds blew from the northeast and the southeast, not directly from the east, as the rarefaction idea suggested. The reason, according to Halley, was that as well as air rushing in from the east, cold dense air from north and south of the equator moved in to maintain equilibrium; the resultant compound flows were the Northeast and Southeast Trade Winds.[56]

But these were only elements of a continuous system, a meridional cell of circulation, described by Halley:

> But as the cool and dense Air [from north and south] by reason of its greater Gravity, presses upon the hot and rarified [at the equator], 'tis demonstrative that this latter must ascend in a continuous stream as fast as it Rarifies, and that being ascended, it must disperse itself to preserve the *Aequilibrium*, that is, by a contrary Current, the upper Air must move *from* those parts where the greatest Heat is: So by a kind of Circulation, the North-East Trade Wind below, will be attended with a South Westerly above, and the South Easterly with a North West Wind above ...[57]

Thus the Trades were part of a continuous system of circulation involving return winds high above the Earth. Then, in a brilliant inference, based on the continuity of atmospheric circulation and the dynamics of air in motion, he accounted for the reversing winds – the monsoons – of the Indian Ocean: the Northeast Monsoon is the result of winter cooling over the Asian land mass; the Southwest Monsoon replaces it when the Asian land mass heats in summer and intense upward convection begins, drawing in air from the southwest.[58]

Not every element of Halley's scheme of atmospheric circulation met with approval; critics suggested that the winds created by the Sun's heat along the equator ought to be from the *west*, not the east, since the densest air lay to the west, not along the path already followed by the Sun.[59]

Meridional cells of atmospheric circulation (and their implications for wind direction) were not the only global-scale phenomena suggested by Halley; in a little-known paper, he proposed details of the water cycle. The basic scheme was Aristotelian: the Sun evaporates water from the oceans; condensation as dew or rain on land or over the sea returns water to the sea, maintaining the constant volume ('Aequilibrium') of the oceans. But the details were appropriate to the mechanical philosophy, for water condenses when the air rises and cools, and, in addition, as

Halley suggests, there must be a quantitative relationship between the extent of mountains and the amount of water condensed from air rising over them:

> ... it may pass for a rule, that the Magnitude of a River, or the Quantity of Water it Evacuates, is proportionable to the length and height of the Ridges from whence its Fountains arise.[60]

In this Halley has not quite left Aristotle behind, for, in attempting to account for variations in precipitation, and particularly for the occurrence of rain rather than dew, or the reverse, he suggests that in the water vapour–laden atmosphere ''tis possible and not improbable, that some sort of Saline or Angular Particles of Terrestrial Vapour being immixt with the Aqueous, which I take to be Bubbles, may cut or break their Skins or Coats, and so contribute to their more speedy Condensation into Rain.'

According to Margaret Deacon, 'Halley's direct contributions to the study of the sea were disappointingly few.'[61] This is a fair comment, and yet his hypothesis of meridional atmospheric circulation provided a descriptive physical framework for the driving forces of surface ocean currents until the origins of dynamic meteorology and oceanography in the nineteenth century, just as his emphasis on the sufficiency of evaporation and precipitation to maintain the water balance of the oceans provided a firm framework for later work on the global water cycle.

Among those apparently influenced by Halley's chart of the winds, if not by his more theoretical works, was the enquiring English privateer William Dampier (1652–1715), whose *Discourse of Winds, Breezes, Storms, Tides and Currents* was published in 1699 as a supplement to his widely popular account of exploration and piracy, *A New Voyage around the World* (1697).[62] Devoid of speculation, it was a highly intelligent mariner's summary of the wind systems useful in navigation, distinguishing tides from ocean currents, and taking it for granted that the currents were caused by the winds.[63] This simple, practical approach, combined with Halley's scheme of atmospheric circulation, almost without exception, dominated the next hundred years.

Ships, Voyages, and Instruments, 1700–1845

It may well be true for the terrestrial world, as Roy Porter has suggested, that in the early eighteenth century cartography cloaked profound ignorance:

By 1700 neither crushing weight of new evidence nor any urgent reorienta-
tion of interest and experience among naturalists had compelled a radical
reformation in environmental visions. One feature of this was the undi-
minished ignorance and pseudodoxy about so many facets of the globe.
The interiors of most continents were still dark to Europeans. The 'North
West Passage' and the 'Terra australis nondum incognita' were confidently
engraved onto maps – vindicating Swift's cynicism:

So geographers in Afric maps
With savage pictures fill their gaps.[64]

But, with the exception of the Arctic Ocean, parts of the Pacific, and the
seas surrounding Antarctica (that continent was not sighted until 1820),
this was not true of the oceans, which had been traversed for two centuries. As the main highways of the Western mariners and traders, their
currents and the winds associated with them were at least adequately
known and increasingly well charted. The ocean *was* better known than
the land in 1700, but mechanisms of circulation (except for the general
link with the winds) were hard to come by, and indeed not much sought.

We might expect the eighteenth century, the century of Newtonianism
and of rational enlightenment, to have been a turning point in study and
understanding of the oceans. But as Margaret Deacon's analysis shows
us, scientific interest in the oceans was sporadic and undirected during
this period, despite a flurry of activity in the last quarter of the century.[65]
Electricity and magnetism, observational astronomy, celestial mechanics, and systematic biology grew or were revolutionized, but the oceans
remained outside the mainstream of natural philosophy. Nowhere is this
more evident than in Georges Cuvier's report in 1810 on the progress
of natural science, prepared for the Emperor Napoleon, in which he
devotes only a few sentences to the temperature of the sea in the context
of the temperature of the Earth.[66]

Despite this, the eighteenth century did provide some of the means
for more sophisticated study of the oceans than was possible in Halley's
time. The growth of 'the quantifying spirit,'[67] partly an application of
eighteenth-century Newtonianism, partly *sui generis* out of statecraft, refined national administration, and commerce, provided new means of
measurement and a new attitude toward global phenomena, especially
late in the century. Changes in instruments alone, as J.L. Heilbron has
shown, were impressive: astronomical and surveying instruments, barometers, thermometers, clocks, and electrical instruments were all greatly
improved in precision, often by as much as two orders of magnitude.[68]

And at least one redesign of an old instrument, the thermometer, came to play an important part in knowledge of the oceans, making it possible to determine systematically for the first time the temperature of deep oceanic waters.

James Six (1731–1793), an early-retired member of a silk-weaving family in Canterbury, southeast England, was elected a Fellow of the Royal Society in 1792 for the invention of a self-registering thermometer, capable of recording the maximum and minimum temperatures reached during a period of time.[69] Widely interested in natural philosophy – he investigated astronomy, electricity, and meteorology – Six designed his thermometer for use in studying the weather, but also produced a more rugged version for use in the sea, one that apparently never was used. Six's thermometer (fig. 1.4) was quickly noted, widely copied, but not used in the sea until the Russian A.J. Krusenstern's voyage from Europe to the Far East in 1803.[70] Thereafter, maximum-minimum thermometers of similar design ('Six thermometers') were used in nearly every investigation of the open sea throughout the nineteenth century. The implications of these uses will be discussed in the next chapter.

If the eighteenth century was the century of 'trade and dominion,' as J.H. Parry has termed it,[71] succeeding 150 years of exploration, piracy, and exploitation, so too was it the century of change in ships themselves and their uses. The ship's wheel replaced the tiller just after 1700, sails and rigging became more sophisticated and standardized, hulls were frequently sheathed with copper to improve their durability, and the devastating problem of disease afloat (notably scurvy) came under control.[72] Navigation became easier, especially the determination of longitude, an age-old problem. The use of nautical almanacs, the improvement of the lunar distance method, and the development (slowly) of the marine chronometer provided the ultimate solutions to problems of position-finding at sea by the end of the eighteenth century,[73] and perhaps provided an essential element in determining the direction of ocean circulation.[74]

Comparing voyages at the beginning and near the end of the eighteenth century is instructive. William Dampier, as captain of *Roebuck*, on his voyages to New Holland (western Australia) between 1698 and 1701, then on *St George* to the South Pacific from 1703 to 1706/7, and Woodes Rogers (with Dampier aboard) on *Duke* to the eastern Pacific from 1708 to 1710, were explicitly and formally privateers, bent on raiding Spanish shipping.[75] Edmond Halley's command of the little *Paramore* from 1699 to 1700 was an exception, being planned and executed to determine

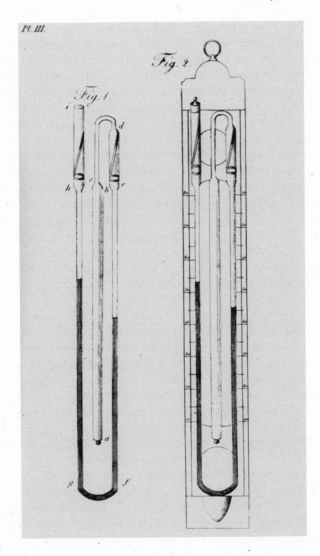

1.4 A maximum-minimum thermometer designed by James Six. Thermometers of this design were used widely throughout the nineteenth century until replaced by reversing thermometers in and after the 1890s. From James Six 1794, p. 59, in Austin and McConnell 1980. With permission of Jill Austin and Anita McConnell.

declination of compass at sea, as a step, it was hoped, in solving the determination of longitude.[76] Four decades later, George Anson also intended to raid Spanish shipping, so that his ambitious and catastrophic circumnavigation from 1740 to 1744 – during the War of Jenkins' Ear – was no exception to the pattern.[77] By contrast, beginning at the latest with the first voyage of James Cook to the South Pacific, 1768–71, in the last half of the century virtually every major expedition of many European nations routinely carried out scientific measurements of some kind. The difference was not just the (temporarily) more peaceful years of the late eighteenth century; a new attitude had evolved quickly, in which, as Richard Sorrenson has claimed, the ship itself became a scientific instrument, inhabited increasingly regularly by natural philosophers with specified duties.[78] Halley, as magnetician and astronomer on *Paramore*, struggled alone with scientific measurements, not to speak of mutiny. Nearly seventy years later, the astronomers and naturalists of Cook's voyages (whose work is described shortly) were sent out with at least the ostensible support of the ships' officers and the authority of scientific societies at home.

If there is a striking contrast between the uses of ships on the oceans at the beginning and end of the eighteenth century, equally there is a dramatic difference in the *kind* of science that could be carried out at sea. Nowhere is this clearer than in the work of a lone Bolognese in the first decade of the eighteenth century and that of the astronomers and naturalists of Cook's voyages six decades later. Luigi Ferdinando Marsigli (1658–1730), virtually a picaresque hero, who spent most of his life as a professional soldier-engineer, was born in Bologna and educated at the University of Bologna in mathematics and natural history by Gemiano Montanari (1632–1687) and Marcello Malpighi (1628–1694).[79] Early in his career, in 1679–80, Marsigli accompanied a Papal diplomatic delegation to Constantinople, where he studied the circulation of the Bosphorus (see discussion in the next chapter).[80] After joining the Hapsburg army of Emperor Leopold I, Marsigli fought the Turks in the War of the Spanish Succession and mapped the boundary of Austria with the Ottoman Empire. Throughout, he observed and collected in the natural and human world around him, amassing collections of insects, plants, minerals, archeological objects, and books which became the basis of his museum and research institute in Bologna.[81]

After surrendering his post to the French in 1704, Marsigli was court-martialled (he narrowly escaped execution) and went into exile in Provence, where he made a base in the little village of Cassis, near Marseille, from 1706 to 1707. There he carried out a noteworthy study of the sea along the northern Mediterranean coast, ranging from hydrography and geology to the chemistry and biology of the coastal waters. Some of his results were presented to the French Académie des Sciences (and later to the Royal Society of London), but were only published fully many years later, with the help of the Dutch natural philosopher Hermann Boerhaave (1668–1738), as the monograph *Histoire physique de la mer* (1725), the first comprehensive study of any area of the oceans.[82]

A study of the Mediterranean coast and the sea from Marseille to the Spanish border, Marsigli's *Histoire* incorporated his observations of the seabed and its sediments, surface and subsurface temperatures, specific gravity of the waters (Marsigli had carried a hydrostatic balance with him to Constantinople in 1679), and the organisms he could observe near shore or collect in deeper water using a simple dredge. Hampered by the difficulty of working from chartered fishing boats, and harassed by pirates, Marsigli obsessively gathered as much information as time, the circumstances, and his budget allowed. He had a questing intellect – and a style of work that gives real meaning to the word *virtuoso* – in the culture of the Italian Galilean tradition. John Stoye characterizes Marsigli as having 'a span of interests so wide that he was not just a soldier, not just a dabbler in the sciences or antiquities, but a personage quickly attracting notice wherever he went.'[83] Certainly he would be noticed even if *Histoire physique de la mer* had been all he published, for it stands alone as an example of how the sea was viewed and studied in the early eighteenth century. The approach was encyclopedic, almost without intellectual boundaries, and idiosyncratic even in its time.[84]

A little more than a century after Marsigli, sea-going travellers, especially naval officers, could govern their scientific work at sea according to the *Admiralty Manual of Scientific Enquiry*, first published in 1849.[85] Nothing like it existed in 1768 when James Cook began his great voyages of charting and discovery; rather, it was Cook's voyages that provided early models of how scientific work could be carried out at sea and what its aims were to be, resulting eventually in the *Admiralty Manual*. This was particularly true of the second voyage, 1772–5, sent out to search for Antarctica, in which Cook and his crew in *Resolution* circumnavigated the Southern Ocean at high latitudes. Aboard *Resolution* were the astronomer William Wales (1734?–1798) and the naturalists Johann Reinhold Forster (1729–1798) and Georg Forster (1759–1794); the astronomer

William Bayly (1737–1810) was aboard *Adventure* (which did not complete the southern circumnavigation) under Tobias Furneaux. Each shipped out with a specific set of instructions prepared by the Board of Longitude and the Royal Society.[86]

Cook's second voyage, in addition to searching for a great southern continent, had the important goal of testing a marine chronometer based on John Harrison's design. So it is not surprising that the instructions to the astronomers emphasized 'daily, and more frequent, determinations of latitude and longitude using a watch [the marine chronometer], azimuth compass, Hadley's sextants, and the *Nautical Almanac*,' and, in addition, that they should 'wind and make comparison of the several watches, and with various solar altitudes, the distance of the moon from the sun, and with the fixed stars.'[87] They were also to make observations of magnetic declination and dip, record the temperature of the sea and the air (they used a standard thermometer in a valved case, referred to widely as Hales' bucket),[88] record the positions of places on coastlines, and perform a variety of other tasks. Details of the work on the voyage, analysed by Morton Rubin, show that they also kept extensive weather records, including barometric pressure, temperature, and wind, observed the aurora, tried to measure atmospheric electricity, and tabulated currents and tides. In short, this was a program of observations not markedly different in *principle* from that undertaken by Marsigli seventy years before, but with the benefit of more and improved instruments (among which ships may be included), the advice of scientifically inclined colleagues (a scientific/technical establishment), and the belief that certain kinds of studies, especially of practical astronomy, the weather, and magnetism, were more important than others. However, despite these differences from Marsigli's work and working style, there is no evidence that studies of the ocean were of any greater importance to Wales and Bayly than they had been to Marsigli, who studied everything. To Wales and Bayly, who recorded surface temperatures at many locations and deeper ones from about 180 to 300 metres at four locations near the Antarctic Convergence and in the Southern Ocean, observations of the ocean were clearly secondary to much else. To them and others, the ocean and its currents were minor components – or, more accurately, peripheral components – in a scientific program that was oriented to the atmosphere and practical problems of navigation. Mechanisms of oceanic circulation, if they were considered at all, were of little greater interest in the 1780s than they had been in 1699 when Edmond Halley set out for the Southern Hemisphere in *Paramore*.

Winds, Currents, and the Internal Forces of the Earth

We come back to winds, for until the middle of the nineteenth century they were of far greater interest than the oceans to natural philosophers of all stripes. The fourth decade of the eighteenth century saw the resolution, although tentative and incomplete, of a problem that had baffled Edmond Halley, namely the direction of the Trade Winds from the northeast and southeast rather from due north and south. The explanation of this phenomenon proved to have general geophysical significance, although this did not become obvious until the mid-nineteenth century.

Halley had given an explanation of the direction of the Trade Winds in terms of the combined motion of the Sun and of equatorward-flowing air along the Earth's surface, but under criticism had withheld judgment and left the problem behind. It was approached again in 1735 by George Hadley (1685–1768), a London barrister who, according to his biographer, 'appears to have been more occupied with mechanical and physical studies than with the law,' and who became the keeper of the Royal Society's meteorological records.[89] In a paper of just over four pages, 'Concerning the Cause of the General Trade Winds,' published in the Society's *Philosophical Transactions*, Hadley began by stating that

> ... the Causes of the General Trade Winds have not been fully explained by any of those who have wrote on that Subject, for want of more particularly and distinctly considering the Share the diurnal Motion of the Earth has in the Production of them ...[90]

He agreed that the Sun was the prime cause of atmospheric circulation, heating air most strongly near the equator and causing it to rise. Cooler air then moved in from all directions, rather than from a preferred direction, as Halley, and before him Voss, had suggested. Hadley, a good Newtonian, next pointed out that any westward motion of the atmosphere following the Sun would be countered by the rotation of the Earth until atmosphere and Earth were rotating at the same speed, and thus, 'for this reason it seems necessary to show how these Phaenomena of the Trade-Winds may be caused without the Production of any real general Motion of the Air westwards.' He continued:

> This will readily be done by taking in the Consideration of the diurnal Motion of the Earth: For, let us suppose the Air in every Part to keep an equal Pace with the Earth in its Diurnal Motion; in which case there be no

relative Motion of the Surface of the Earth and Air, and consequently no Wind ...

But the Sun's heat causes air to rise at the equator, and as a result,

> ... let the air be drawn down thither from the N. and S. Parts. The Parallels are each of them bigger than the other, as they approach to the Equator, and the Equator is bigger than the Tropicks ... and the Surface of the Earth at the Equator moves so much faster than the Surface of the Earth with its Air at the Tropicks. From which it follows, that the Air, as it moves from the Tropicks toward the Equator, having a less Velocity than the Parts of the Earth it arrives at, will have a relative Motion contrary to that of the diurnal Motion of the Earth in those Parts, which being combined with the Motion, towards the Equator, a N.E. Wind will be produced on this Side of the Equator, and a S.E. on the other.[91]

Thus Hadley attributes the direction of the Trade Winds to the difference in velocity between equatorward-directed surface winds and the increasingly higher eastward velocity of the rotating Earth as the moving air approaches the equator. And there is a corollary: surface air leaving the high latitudes must be replaced by a return flow above (completing the meridional circulation cell suggested by Halley in 1686). It, too, will be affected by the Earth's rotation and consequently will blow from a westerly direction, accounting for the mid-latitude belts of westerly winds.[92] Only five years after the publication of Hadley's paper on the Trade Winds, and probably quite independently, similar reasoning was applied to the oceans by the Edinburgh mathematician and Scottish Newtonian, Colin Maclaurin (1698–1746). The occasion was a prize essay competition on the tides, mounted by the Paris Académie des Sciences. A brief section of Maclaurin's essay included the proposition that meridionally directed ocean currents will be deflected for exactly the same reason recognized by Hadley, the difference in the velocity of the Earth as water moved north or south.[93] These insights into geophysical motions on the Earth's moving surface, profound but incomplete,[94] were little known, or unknown, for at least the next century, even to physically trained natural philosophers, but especially to the practical navigators and hydrographers whose knowledge became incorporated into increasingly complex charts of the oceans.

In 1832, two years after the death of James Rennell (1742–1830) (fig. 1.5), his great monograph *Investigation of the Currents of the Atlantic Ocean*,

1.5 James Rennell (1742–1830), whose *Currents of the Atlantic Ocean* (published posthumously in 1832) gave powerful support to the idea of wind-driven ocean circulation. From Markham 1895, frontispiece.

and of Those Which Prevail between the Indian Ocean and the Atlantic was published in London, the product of decades of study of ships' logbooks in his London home. Late twentieth-century writers have said of this work that 'the great breadth of Rennell's study of the entire Atlantic Ocean, truly unprecedented, must accordingly be remembered as the era's seminal piece of work – as it was an invaluable model for subsequent researches into the surface circulation of the World Ocean.'[95] But the reality of Rennell's book is more complex, and its significance lies as much in being a stimulus to a very different view of the oceans and the Earth in the thought of Alexander von Humboldt, who used Rennell's ideas to promote his own distinctive world view.

Rennell learned surveying at sea with the Royal Navy between 1756 (the peak of the Seven Years' War) and 1763, first in France, then in the East Indies.[96] In India he left the R.N. to join the East India Company, where he spent the next thirteen years surveying the rivers of Bengal in preparation for the Company's expansion into the interior, working from 1767 as surveyor-general of Bengal.[97] After serious injuries in a skirmish in 1776, Rennell resigned his position, returning to England in 1777/78 via the Cape of Good Hope, where he spent some time making current measurements in the Agulhas Current. The result was a chart showing the course of the current around the southern tip of Africa, published in 1778.[98]

During his long career in England as the doyen of English geographers, Rennell continued work on ocean currents, first on the local scale, then more broadly, using ships' logbooks from a wide circle of acquaintances and eventually from Royal Navy surveyors.[99] It is clear that his aim was more than documentation of ocean currents; it was reform of the whole system of navigation, moving it from purely celestial navigation and dead-reckoning to chronometric navigation combined with accurate information on ocean currents. In 1793, making the case for a current that swept northwestward out of the Bay of Biscay toward the Isles of Scilly, Rennell wrote:

> I have reason to suppose that our own chart of soundings is very bad; and indeed how can it be otherwise, considering the imperfect state of the art of marine surveying, at the time when it was made? A set of timekeepers will effect more, in the course of a summer, in the hands of a skilful practitioner, than all the science of Dr Halley [mathematical astronomy] during a long life. For who could place a single cast of soundings in the open sea without the aid of a timekeeper?[100]

Rennell's ideal was the determination of ships' positions using precisely determined coordinates before and after drift by currents. It was such information that he compiled, during years of study, into the manuscript of *The Currents of the Atlantic Ocean*, a project that seems to have arisen out of a series of editions of sailing directions prepared for the Admiralty.[101] The result was a synoptic treatment of ocean currents in the Atlantic underlaid by an important and influential base of theory, namely the primacy of the winds in causing ocean currents.

Rennell's conception of the Atlantic circulation began with the Agulhas Current and the previously little known South Atlantic. Part of the Agulhas Current flowed northward (much did not – Rennell recognized what is now called the Agulhas Retroflection, toward the east), merging eventually with the equatorial current system. The complex equatorial current then split at the South America coast, part passing northwestward into the Caribbean Sea and the Gulf of Mexico, the rest turning southward as the Brazil Current perhaps as far as the southern tip of South America. Eastward currents prevailed in the subtropical Atlantic, and perhaps, too, much farther south, allowing the Atlantic circulation to be continuous and giving it continuity (it was likely) with waters of the other oceans.[102] To the north, Rennell's 'Main Equatorial Current' was the source of water issuing from the Gulf of Mexico through the Florida Strait as the Gulf Stream, which left the North American coast toward the Azores. Although data were sparse for the higher latitudes of the North Atlantic (those regions were far north of the usual trade routes), there Rennell recognized at least the currents known today as the Labrador Current and the East Greenland Current.

Rennell's patient *tour de force* of data gathering, cartographic depiction, and induction was recognized rapidly as a masterwork. As Peterson, Stramma, and Kortum commented recently,

> RENNELL's great contribution was to advance significantly the descriptions previously given by VOSSIUS (1663) and DE BRAHM (1772)[103] about the large-scale interconnections between the surface currents. This work proved to be a landmark since in it a wealth of new information was synthesized for the first time. It was immediately hailed by his contemporaries and served as a foundation upon which subsequent developments would be built for decades to come.[104]

But it was more than a mere foundation, for Rennell underpinned his charts of circulation with a body of theory based upon the primacy of the wind in causing ocean currents, mainly through differences of water level. In 1793, in his discussion of circulation out of the Bay of Biscay (the 'Rennell Current'),[105] he said:

> ... as water pent up, in a situation from which it *cannot escape*, acquires a higher level, so in a place where it *can escape*, the same operation produces a current: and this current will extend to a greater or lesser distance, according to the force with which it is set in motion; or, in other words, according to the height at which it is kept by the wind.[106]

Such was the case in the Gulf of Mexico, where the broad, wind-driven drift (a 'drift current' in Rennell's terminology) of the Equatorial Current was concentrated, much like the concentration of a tidal bore in a narrowing estuary, 'for the volume of water cannot *retreat* when pressed laterally; because beside the effect of its own *momentum*, it is impelled forwards by the accession of fresh supplies, constantly brought by the same power which originated the movement.'[107] To this was added the constant effect of the northeast Trade Winds, combined with the concentrating effect of the continental coasts, in raising the water level of the Caribbean Sea and, downstream of it, the Gulf of Mexico. As Rennell summarized his argument, 'Can it be supposed otherwise than that the wind alone is the agent which produces the Gulf-Stream, in the first instance?'[108] Then, from areas of high water level, under appropriate circumstances such as confinement by coastlines, the water flowed out as 'stream currents,' the most dramatic example of which was the Gulf Stream in and seaward of the Florida Strait. On a worldwide scale, all surface ocean currents could be explained by these physical causes, which were drawn directly from Rennell's early work on Indian rivers and his observations of the effect of the wind on canals.[109]

Alexander von Humboldt and the Internal Forces of the Earth

Work on the oceans was not new to Alexander von Humboldt (1769–1859) when he visited Rennell in London in 1827 or 1828 and carried away with him at least some of Rennell's information on currents.[110] Humboldt's interest in the sea may have been aroused by an early trip to the Belgian coast and to the island of Helgoland in 1790 with his friend Georg Forster.[111] But whatever its origin, it became the object of Humboldt's distinctive approach to the natural world, which he characterized as 'the earnest endeavour to comprehend the phenomena of physical objects in their general connection, and to represent nature as one great whole, moved and animated by internal forces.'[112] Humboldt took pains to distance himself from *Naturphilosophie*, and although a superficial examination of his writings suggests links with German Romanticism, this is primarily style and not substance. His emphasis on the material basis of life and on quantitive study of the Earth[113] provided an example – often termed 'Humboldtian science' – that was highly influential in the development of quantitative science throughout the nineteenth century.[114]

The thermometer and the barometer were standard instruments to Humboldt from the beginning of his extensive travels in the 1790s; in addition, he carried magnetic and surveying instruments, electrometers, chronometers, a pendulum, absorptiometers, a cyanometer (for measuring the intensity of the colour of the sky and sea), telescopes, hygrometers, microscopes, and all the paraphernalia of the zoological and botanical collector, along with much else when conditions permitted.[115] With these, and in his writings, he attempted, according to Roy Porter, 'a precisely measured "physique du monde," focussed on the areal associations of physical and organic phenomena, seeking "unity in diversity."'[116] Instruments and the quantitative representation of the results that they yielded resulted in 'knowledge of the chain of connection, by which all natural forces are linked together, and made mutually dependent upon each other,' according to Humboldt himself.[117]

All of Humboldt's scientific travels involved some study of the oceans or of local seas.[118] Beginning his epic voyage to the Americas in 1799 (it ended in 1804), he noted the direction of currents between Spain and the Canary Islands, the Atlantic Equatorial Current, and later (in September 1802) measured the temperature of the cool waters off Peru, speculating about their origin as a current from the south.[119] On his way from Cuba to the United States in the spring of 1804, he noted the temperature of the Gulf Stream and estimated its velocity from the ship's drift.[120] Later, during his service as a chamberlain to the King of Prussia, Friedrich Wilhelm III, Humboldt visited the Caspian Sea in October 1829 during an extensive tour of central Asia and Siberia, and in 1834 he voyaged from Swinemünde to Königsberg (now Kaliningrad) along the southern Baltic coast, noting the variation of sea temperature characteristic of the region.[121] His early observations, from the equatorial Atlantic, the Peru Current, and the Gulf Stream, were mentioned frequently in Humboldt's writings, especially *Ansichten der Natur* (dating from 1808), his contribution to Heinrich Berghaus's *Allgemeine Länder- und Volkerkunde* (1837), and in his great work *Cosmos* (1845). Much less well known is a manuscript on the sea titled 'Über Meeresströmungen in allgemeinen und über die kalte peruanische Strömung der Sudsee im Gegensatz zu dem warmen Golf- oder Florida-Strome,' still unpublished, on which he worked for many years, and which may have originated in a series of lectures in Berlin in 1827/28, if not earlier.[122] It includes ideas on the theory of ocean currents, information on the major cold and warm currents, and much discussion of the North Atlantic circulation, especially the Gulf Stream,[123] based on a wide variety of sources, including Rennell's *Currents of the Atlantic Ocean*.

Despite the importance of the manuscript 'Über Meeresströmungen,' most of Humboldt's ideas about ocean circulation had reached final form when his *Relation historique* of the voyage to the New World was published in 1814.[124] Even *Cosmos*, thirty years later, adds little to the earlier account, which began with his notes on the set of the ship to the east when it was north of the Canary Islands (in the Azores Current–Portugal Current system) and the westward-flowing Equatorial Current.[125] Humboldt accepted the idea that the equatorial currents were driven by the Trade Winds (assisted at times by the tides) and mentioned with approval Halley's conception of a meridional cell of atmospheric circulation resulting in those winds. At its western end, the Equatorial Current entered the Caribbean, and eventually the Gulf of Mexico, where the coastlines acted like a dike, directing the water into a clockwise circulation which was narrowed like a jet at the Florida Strait, issuing as the Gulf Stream and turning north. In the *Relation historique*, Humboldt concentrated on the eventual westward flow of the Gulf Stream toward the Azores, then onward toward the Strait of Gibraltar and the Canary Islands, mentioning briefly another branch toward Europe assisted by westerly winds. Later, in *Cosmos*, he explained the effects of the northern branch in warming the European climate.[126] Even in 1814, long before he had seen Rennell's data, Humboldt regarded these currents as part of a 'perpetual eddy' in the North Atlantic which could be studied quantitatively – its time of circulation determined, for example – and suggested a program of study of the Gulf Stream to afford a synthesis: 'La découverte d'une groupe d'îles inhabitées offre moins d'intérêt que la connaissance des lois qui enchaînent un grand nombre de faits isolés.'[127] And where could one find a greater store of isolated facts or fewer unifying laws than in studying the oceans?

Of all speculations on the causes of ocean currents presented during his time, Humboldt's are the most broadly based and the best substantiated by the physics of the day. Motive forces varied from place to place, but according to Humboldt they could include the 'external force' of the winds, differences in temperature and salinity, the melting of ice and snow, latitudinal differences in evaporation, and possibly changes of atmospheric pressure from place to place. He suggested that constant winds should eventually cause water motion into the great depths of the ocean; this could be tested by studies of the Gulf Stream's deep waters. And the Earth's rotation – at least potentially – could cause deflection of north-south currents, although Humboldt doubted that it really happened because there was such a variety of current directions and because, as he claimed, the slow motion of water (he seems to imply inertia) compared to air should mean that as it changed latitude it readily took on the angular velocity appropriate to each latitude and would not be deflected.[128] The canvas had broadened remarkably as a result of his wide-ranging intellect.

Without doubt, the most influential and contentious part of Humboldt's account of the causes of ocean currents was his claim that deep cold water in the tropics, known at least since the 1750s[129] through casual observation and since the first decade of the nineteenth century thanks especially to James Six's thermometer, was the result of convective flow from the polar oceans caused by low temperatures (and not counteracted by increased density caused by salinity):[130]

> The existence of these cold layers in low latitudes demonstrates as a result a deep current which proceeds from the poles towards the equator: it demonstrates also that saline materials, which alter the specific gravity of water, are distributed in the ocean in a way that does not cancel the effect produced by differences of temperature.[131]

In *Cosmos* he explored the idea of 'lower polar currents' further, stating that 'without these submarine currents, the tropical seas at those depths could only have a temperature equal to the local maximum of cold possessed by the falling particles of water at the radiating and cooled surface of the tropical sea,'[132] and adds a footnote on the effect of salinity:

> The submarine current, which brings the cold polar water to the equatorial regions, would follow an exactly opposite course, that is to say, from the equator toward the poles, if the difference in saline contents were alone concerned.[133]

Even by 1845, when the first volume of *Cosmos* was published, Humboldt's support of a deep cold density-driven circulation from the polar regions, bolstered by new evidence, was little noted. The winds had it as the sole driving forces of oceanic circulation – as they did until more than fifty years after Humboldt's last writings on the subject. A Humboldtian explanation of oceanic circulation, representing nature 'as one great whole, moved and animated by internal forces,' awaited treatment by those with a totally different philosophy of nature involving detailed study of its components rather than a global synthesis.

From Humboldt to Centres of Calculation

With Humboldt, and especially with *Cosmos*, my discussion returns to the starting point of this chapter, Heinrich Berghaus, Alexander Keith Johnston, and the *Physical Atlas of Natural Phenomena*, for they and it incorporated maps by Berghaus that were intended to illustrate the five volumes of *Cosmos*, just as Humboldt had included maps with all his previous monographs. But somehow the publication of the two works was separated,[134] affecting the presentation and impact of Humboldtian ideas on ocean circulation. Although Humboldt summarized previous ideas, added to them his own hypotheses, and presented them in a clear and vivid style (the *Relation historique* is especially noteworthy), the Humboldtian approach was not the one that prevailed in the study of the oceans after his time. Albert Defant noted this in a few brief words in 1960: 'Eine physische Weltbeschreibung in Humboldt'schen Sinne heute nicht mehr zu leisten ist.'[135] After Humboldt's time, increasingly, the physical processes of the oceans were studied separately from the rest of the global system, in contrast to what Humboldt had sought. Why this is so requires a brief examination here.

A distinctive feature of late eighteenth-century science was the increasing temporalization of nature, as well expressed by Roy Porter.

This revolution [during the eighteenth century] is part of an emerging integrating vision of the environment: the growth of a *historical* conception of the earth, and hence of a genetic explanation of its phenomena. Temporalizing nature promised to settle many problems. Before its demise, the Great Chain of Being was temporalized. Palaeontology, the new science of fossils, gave life a history, and thereby rationalized extinction. Evolutionary theories ... suggested a succession of life-forms. Natural history was becoming the history of nature. And, integrating the different strands, the earth itself assumed a defined past, through the rise of a historical geology to interpret the strata archives.[136]

As the following chapters show, giving them a temporal axis was not the kind of development followed by the physics of the oceans, even in the hands of a great synthesizer like Humboldt. Humboldt broke from the eighteenth century – his own century of birth – to treat the oceans and much else in the global system ahistorically and atemporally right from the start. This is just what his successors, the physicists and geophysicists of the late nineteenth century – the century of his death – did too. But they went in a new direction, one that Humboldt would have found totally uncongenial (even though it arose from his methods), by using the mathematical approach to the physical behaviour of fluids to *analyse* and *attempt to predict* rather than unify. Only forty years after Humboldt's death in 1859, the result was a specialized, quantitative science of the physics of the oceans rather than a Humboldtian synthesis. This physics of the ocean was detailed, mathematically circumscribed, and professionally self-contained in a way that was not foreshadowed in his own remarkable life at a time of transition in the sciences.

Theodore Porter gives us an evocative (and provocative) account of quantification that applies directly to what follows:

> To quantify a quality is not merely to solve an intellectual problem. It is to create what Latour calls a center of calculation, surrounded by a network of allies ... The quantification of qualities is as much an administrative accomplishment as an intellectual one ... There is strength in numbers, and anyone who proposes to wield them effectively must ask not only about their validity but also about how the world might be changed by adopting new forms of quantification.[137]

Humboldt's attempt to unify knowledge of the oceans and to integrate it into a unified, global earth science did not succeed for just these reasons. The world was changed by the narrowing effect of quantification, manifested in a new kind of physics that built its boundaries out of numbers, not out of the limits of the globe itself, and that developed in communities of scientists attempting to solve problems peculiar to their local circumstances.

2

Groping through the Darkness: The Problem of Deep Ocean Circulation

... within the last decade the advance of all knowledge of all matters bearing upon the physical geography of the sea has been confusingly rapid – so much so, that at this moment the accumulation of new material has far outstripped the power of combining and digesting and methodising it. This difficulty is greatly increased by the extreme complexity of the questions ... which have arisen.

Charles Wyville Thomson 1878, p. 452

The Problem of Ocean Circulation

Even though the direct influence of Alexander von Humboldt faded after the middle of the nineteenth century, his influence lived on in the application of quantitative methods when scientific travellers reached distant regions of the world. Within the laboratory too, independently of Humboldt, physical science took on new confidence as it found mathematical representations of electricity, magnetism, and heat, and refined physical constants for many phenomena. But this kind of quantification did not extend to the oceans until the late decades of the century. This anomaly defies easy explanation, but at least one reason can be found in well-established ways of thought about the ocean involving persuasive common-sense ideas of its mechanisms.

The American hydrographer and superintendent of the U.S. Naval Observatory, Matthew Fontaine Maury (1806–1873), who was well steeped in natural theology, regarded the oceans, and especially currents, as worthy of knowledge for many practical reasons:

In the beautiful system of cosmical arrangements and terrestrial adaptations by which we are surrounded, they perform active and important parts; they not only dispense heat, and moisture, and temper climates, but they prevent stagnation in the sea; and by their active circulation, transport food and sustenance for its inhabitants from one region to another, and people all parts of it with life and animation. Yet, on this interesting subject former observations have thrown only just light enough to make visible the darkness through which we are groping.[1]

Maury's teleology was regarded as something of an anomaly in his time, if not positively harmful to the developing science of the young American Republic,[2] but his insouciant excursions into physical problems of the sea without any background in physical science were typical of many other attempts to comprehend and explain ocean circulation throughout the last decades of the eighteenth century and at least the first seven decades of the nineteenth. Physical scientists resolutely stayed at their benches and ignored the sea, while a variety of others, among them biologists, geologists, geographers, and ordinary well-educated citizens, did not hesitate to pronounce on the causes of ocean currents and involve themselves in protracted arguments which were often merely commonsense statements of the apparently obvious.

There were some beliefs in common among the few who concerned themselves with the physics of the ocean. Certainly the most widespread was the view that the surface currents of the oceans were driven by the winds, just as Rennell had indicated. F.W. Beechey, in his chapter on hydrography in the widely used *Admiralty Manual of Scientific Enquiry* of 1851, stated categorically that 'monsoons, and zones of trade and variable winds ... and other disturbances of the atmosphere ... are the principal causes of the many currents that sweep over the face of the earth.' Like Maury, he saw purpose in the arrangement, for, as he put it, '... as both [wind and current] perform an important part in the economy of nature, an additional interest attaches to correct knowledge of them.'[3] Coincidently perhaps, a few months later the Manchester alderman Thomas Hopkins exhorted his audience at a meeting of the local Literary and Philosophical Society to reject the (Galilean) notion that the rotation of the Earth caused ocean circulation and to accept that the wind alone could be responsible for surface and deep currents:

... when there is ample space for the water to move forward, the wind readily produces a current, and it is evident from the nature of the force that is in action, that the current will, in deep water, extend to depths proportioned to the length of time that the wind has acted on the water which is in motion.[4]

A decade later, the eminent natural philosopher Sir John Herschel, summarizing physical geography for the *Encyclopaedia Britannica*, took up the refrain: 'every wind that sweeps the ocean drives along before it the surface water,' both water and air being affected by the mutual friction of the interaction.[5] The geographer A.G. Findlay, reviewing ocean circulation before the Royal Geographical Society in 1853, had accepted the prevailing ideas of his age, including the wind-driven surface circulation,[6] but he saw undoubted evidence of widespread deep currents which could not be accounted for so readily: '... *as yet*, we know little or nothing of what is going on at great depths below the surface ...'[7]

Beyond that, darkness reigned. The deep water of the oceans was well known to be very cold, but how cold and why? In particular, how could deep temperature measurements, which accumulated in an unprecedented way in the early decades of the century, be interpreted and incorporated into a global scheme of ocean circulation? They provided the ammunition for the first and last great non-quantitative debate about the causes of ocean currents.

The Currents of the Deep Oceans

Opinions varied widely about the stillness of the abyss, a region so remote from the realm of the winds. But there were reports from seamen as early as the seventeenth century of subsurface currents in the heavily travelled Øresund (The Sound) between Denmark and Sweden,[8] and at the same time suspicion that the well-known inflow into the Mediterranean Sea through the Strait of Gibraltar might be compensated by some kind of outward flow below the surface, allowing the Mediterranean to remain in equilibrium. For two centuries, between the 1660s and the late decades of the nineteenth century, the Strait of Gibraltar piqued the curiosity of engineers, geographers, and natural philosophers, leading to sporadic but intense debate about the possibility of deep currents.[9]

During the British occupation of Tangier during the late seventeenth century, engineers involved in the construction of its harbour turned their attention to currents in the Strait. Among them, Henry Sheeres (?–1710) was convinced, based on simple experiments, that evaporation from the Mediterranean was sufficient to require an inflow of Atlantic water (the well-known surface current into the Strait) but that there was no deep outflow into the Atlantic.[10] As evaporation proceeded from the Mediterranean, water flowed downhill as a shallow surface layer from the higher level of the Atlantic to compensate for the water lost to the atmosphere. His colleague Richard Bolland thought otherwise, suggesting a return flow of excess water from the Mediterranean as a deep westward current, and he suggested a test (never carried out) by lowering a weight or a drogue[11] below the surface to determine the direction of flow. The argument, at least as far as the late seventeenth century was concerned, seems to have been decided in favour of the evaporation idea because of the influence of Edmond Halley. Marshalling evidence for his theory of the water cycle (mentioned in chapter 1), Halley in 1687 published an empirically based estimate that all the rivers flowing into the Mediterranean would only replace about one-third of the water lost to evaporation; there was no need of, or cause for, a return undercurrent.[12]

A generation later, the German engineer J.S. von Waitz (1698–1776), highly skilled in salt mine operations and the commercial extraction of salt from brines for his employer, the Landgrave of Hesse-Cassel, made new estimates of evaporation from the Mediterranean and of the input of river water, concluding that inputs well exceeded evaporation. The result, based on his knowledge of the density of fresh water, was that a double flow could be expected:

> The lighter water from the Atlantic runs in, becomes saltier and heavier through evaporation, sinks to the bottom and there, by reason of its increased weight, pushes aside the lighter water already standing outside and so finds a natural outlet.[13]

Although he had never seen the Mediterranean, or indeed perhaps any ocean, Waitz was not loath to generalize his scheme of density-driven circulation to the major oceans, suggesting that fresher water from high latitudes would flow along the surface toward the equator, where evaporation would increase its density; dense water would then sink and flow along the bottom toward higher latitudes (a scheme not unlike that suggested by Maury a century later).

In his remarkable application of his salt-works skills to ocean circulation, Waitz was aware of the work of a predecessor, the Bolognese polymath Luigi Ferdinando Marsigli (see chapter 1). During 1679–80, accompanying a diplomatic delegation to Constantinople, Marsigli had become interested in the water circulation of the Bosphorus, linking the Black Sea to the Sea of Marmara and eventually the Mediterranean. There, the local fishermen and mariners knew of a subsurface current flowing into the Black Sea, below the outflow of water at the surface. In a series of experiments with weighted lines and floats, Marsigli demonstrated the double flow, and with hydrometers found that the surface water was relatively fresh, while that running counter to it below was more dense (and thus more saline). He concluded that the explanation of the double current structure was the displacement of lighter Black Sea water flowing toward the Mediterranean by heavier Mediterranean water flowing into the Black Sea, showing in an ingenious experiment with a partitioned tank that just such a flow could be produced when salt water was allowed to run below fresh.[14]

Marsigli's notable work on the Bosphorus circulation was known in England by 1684, and certainly to Waitz in the middle of the next century, but it fell into obscurity and had no influence on debates about the nature of ocean circulation that took place in the middle and late decades of the nineteenth century.[15] Doubts persisted that deep currents were significant, if they existed at all, as evidenced in the opinions of the most knowledgeable Mediterranean hydrographer of his time, T.A.B. Spratt (1811–1888) of the Royal Navy. By 1857, early in a thirty-year career of surveying the Mediterranean, Spratt had concluded (contrary to Maury) that he could find no evidence for undercurrents. This was true notably in the Bosphorus and Dardanelles, where Spratt concluded that the only phenomenon acting was a rapid surface current running over static water below.[16] If Mediterranean water lay below that from the Black Sea, it could be attributed to occasional incursions caused by the wind. Even under the assault of new information on the Mediterranean circulation during the late 1860s and 1870s, including the Strait of Gibraltar, the Dardanelles, and the Bosphorus, Spratt clung resolutely to his conclusion that deep currents were the result of mistaken interpretation rather than facts of nature.[17] The remoteness of even relatively shallow subsurface water, the difficulty of making any kind of direct measurements of current flow, the ambiguity of indirect indications like temperature and salinity, and the weight of past opinions, all played roles in making the conservative solution – that ocean currents were caused by the wind and occurred nearly entirely at the surface – easy to accept.

And yet eminent authorities persisted in exhuming or reinventing the idea of deep currents driven by differences in temperature. Foremost among these was certainly Humboldt, who, as I discussed in chapter 1, envisioned the sea as complex, driven by the wind and also by differences in temperature and salinity.[18] In all likelihood, Humboldt's belief that cold would result in the sinking and flow of deep oceanic waters came originally from the multi-talented natural philosopher *cum* court advisor and politician, Benjamin Thompson, Count Rumford (1753–1814).[19] Thompson, who had a deep-seated interest in practical and theoretical aspects of heat, suggested in 1798 that seawater, unlike fresh water, would continue to contract and increase in density below the freezing point of fresh water and would then sink, with implications for ocean circulation. Cold water

> descends to the bottom of the sea, cannot be warmed *where it descends,* [and] as its specific gravity is greater than that of water at the same depth in warmer latitudes, it will immediately begin to spread on the bottom of the sea, and to flow towards the the equator, and this must necessarily produce a current at the surface in an opposite direction ...[20]

This idea, through the mediation of Humboldt, lingered on the margins of scientific debate about the oceans, but its relevance became of greater and greater concern as evidence of very cold temperatures in the oceanic abysses began to accumulate during the early decades of the nineteenth century. The question of what temperatures were found in deep water and how they were distributed became a critical one to a small band of hydrographers and natural philosophers in the 1860s.

Deep-Water Enigmas – Temperature and Density

After the explorations of James Cook in the late eighteenth century, it became commonplace for temperature observations to be made at sea, sometimes directed by the ship's officers but increasingly frequently by a ship's naturalist or physician. They took with them the general opinion that temperature in the sea decreased with depth, although to an unknown degree, just as Captain Henry Ellis had reported in 1751 off West Africa, Constantine Phipps in the Arctic in 1773, and as, earlier, Marsigli and Donati had shown in the Mediterranean. Indeed, the two astronomers with Cook, William Wales and William Bayly, accompanying *Resolution* and *Adventure* from 1772 to 1775, had made a few determinations of subsurface temperatures in the tropics and the Southern Ocean showing that, in general, cool water lay below the surface to the greatest depths they sampled, nearly 300 metres.[21] But there was little to go on for several decades; for example, when François Péron observed temperatures (the deepest at 654 metres) during a French voyage under Nicolas Baudin from 1800 to 1804 there were only seventeen previous subsurface observations for comparison.[22]

James Six's maximum-minimum thermometer (see chapter 1) made its debut on a voyage of exploration in the hands of the physicist J.C. Hörner (1774–1834) on Krusenstern's Russian expedition around the world from 1803 to 1806, on which it supplemented the retrieval of water samples using a valved bucket.[23] But for at least a few decades, some investigators continued to use the kind of device that Péron employed, a standard thermometer well wrapped in insulating material and enclosed in a metal or wooden case. The advantage of Péron's device was its ability to survive rough handling, but it had to be left at depth for lengthy periods to allow the thermometer to reach the ambient temperature. The unprotected Six thermometer, although fragile, allowed much more rapid observations and more of them. Not surprisingly it became the instrument of choice, although not without creating confusion about deep ocean temperatures that took several decades to sort out.

The problem with the Six thermometer is well illustrated by the results of the first voyage of the French corvette *Astrolabe* from 1826 to 1829 under Jules Dumont d'Urville (1790–1842) to survey the Australasian area.[24] Favourably inclined to scientific investigation (he had done botanical investigations under Duperrey on *Coquille* from 1802 to 1804), and determined to get it right, Dumont consulted the great academician François Arago about the correct methods of measuring temperature at sea and had one of his officers trained under Arago.[25] Somehow the question of whether or not the high pressures in the great depths might affect the thermometers never came up, and the officers of *Astrolabe* recorded the results of 66 measurements below 15 metres to a maximum depth of about 1,900 metres. The results were given to Arago, but when he had done nothing with them after two years, Dumont prepared the scientific results of the expedition himself.[26] Taking the temperature results at face value, and adding to them the results of other expeditions, he found a striking pattern of deep water temperatures: below about 200 metres, the oceans were full of water at a constant temperature of about 4°C. At high latitudes, somewhere between 40° and 50° north and south, at least at some seasons, this cold deep water extended to the surface, establishing a zone later called the 'circle of mean temperature.'[27]

Dumont's great generalization, which affected ideas about the deepsea environment and oceanic circulation for thirty years, was at least partly based upon a misconception about the density of water near and at its freezing point. Rumford, in the 1790s, recognized that the well-known property of fresh water that its density was greatest above its freezing point (at 3.98°C) did not apply to seawater, which continued to increase in density until it froze.[28] This fact was well established experimentally by the Swiss-English chemist Alexander Marcet (1770–1822) in 1819 and by the German physicist G.A. Erman (1806–1877) nine years later.[29] There is evidence that Dumont knew of Erman's work but chose to ignore it.[30] Nonetheless, the idea that seawater behaved like fresh water and thus would sink to the bottom when its temperature was 4°C was widely held. It was supported by the widespread use of Six thermometers that were not protected against the effect of pressure,[31] including those used by Charles Wilkes (1798–1877) on the U.S. Exploring Expedition of 1838–42 and by James Clark Ross (1800–1862) during the British exploration of the Southern Ocean in *Erebus* and *Terror* from 1839 to 1843.[32]

Information on deep-sea temperatures that took into account the effect of pressure on thermometers had been available when Dumont proposed the 4°C theory. It came from the careful work of the Russian physicist Emil von Lenz (1804–1865),[33] who as a student had accompanied Otto von Kotzebue on his second circumnavigating voyage for the Russian Crown between 1823 and 1826.[34] Well prepared by his teacher in St Petersburg, the physicist G.F. Parrot (1767–1852), Lenz worked at sea, not with Six thermometers, which fortuitously had proven unreliable on the earlier Kotzebue voyage, but with thermometers well protected within self-closing containers like Hales' buckets. These he carefully calibrated experimentally so that inaccuracies due to the warming of the water sample during the recovery of the bucket could be calculated. The result was a series of extremely accurate deep temperature measurements, showing that water well below 4°C occurred in the depths.[35] A decade later, the French navigator Abel du Petit Thouars (1793–1864) in the frigate *Vénus*, accompanied by the very skilful physicist Urban Dortet de Tessan (1804–1879), made fifty-nine deep temperature determinations using Six thermometers specially protected in cases against the effects of pressure, and carefully worked out corrections when these failed. Not all their attempts were successful, but the results showed temperatures as low as 1.4°C at very great depths (3,740 metres) and others at lesser depths still well below 4°C.[36]

Lenz, by then a distinguished researcher on electricity and magnetism in St Petersburg, returned to the subject of deep-sea temperatures in 1845, drawing on his own work and other data that had accumulated since the 1830s.[37] He made it clear that any temperature data from thermometers unprotected against pressure would have to be corrected or discarded. When reliable determinations alone were considered, it was clear that the 4°C theory could not be correct and that it was an artifact of the effect of deep-ocean pressures on unprotected thermometers. Instead, there was evidence that much colder polar water flowed toward the equator from the high latitudes of both hemispheres,[38] where it rose toward the surface (accounting for the shallowing of cold water at the equator noted at least since the time of Péron) and began a return journey toward the poles. Thus the global circulation of the oceans took the form of two giant vortices meeting at the equator and taking in both deep cold water and shallower and much warmer water moving in the opposite direction (fig. 2.1).[39]

Lenz's careful assessment of the problems of deep temperature measurement, notably the effect of pressure on thermometers, along with his model of global oceanic circulation, resulted from the best physical science of its time applied by a master experimentalist. But it lay hidden from view for thirty years until new kinds of marine investigations brought renewed criticism of the 4°C theory and the idea of global oceanic circulation reappeared at the heart of two opposing theories.

Shattering Misconceptions: W.B. Carpenter's 'Magnificent Generalization'

The greatest nineteenth-century controversy about the nature and causes of ocean circulation began, not with physics or physical geography, but with biology, specifically the search for living fossils in deep water off the British Isles. John Vaughn Thompson, an Irish army surgeon of Cork, discovered the tiny pentacrinoid juvenile stage of the crinoid, *Antedon* in 1823, but the implications of his discovery, the possibility of relating the stalked and unstalked crinoids, were not taken up in Britain until the 1860s when the zoologist Charles Wyville Thomson (1830–1882) and the physiologist W.B. Carpenter (1813–1885) (fig. 2.2) began to work together on the morphology and embryology of echinoderms.[40]

In 1864, Thomson and Carpenter were aroused to ambitious explorations beyond their sheltered coastal waters by the discovery of the stalked crinoid *Rhizocrinus lofotensis*, representing a group known mainly from the fossil record, and other apparent living fossils in deep Norwegian waters by G.O. Sars (1837–1937), a fisheries inspector for the Norwegian government and son of the distinguished Norwegian zoologist Michael Sars (1805–1869).[41] After a visit by Thomson to the Sarses, father and

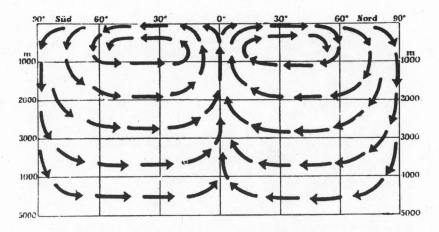

2.1 The scheme of abyssal circulation suggested by Emil von Lenz in 1848, indicating that cold deep water sank at high northern and southern latitudes, moved toward the equator, and rose there as it warmed. From Wüst 1968, p. 112. With permission of the Musée océanographique de Monaco.

son, in Oslo in 1866, Thomson and Carpenter began to look for ways to explore deep water beyond the depths plumbed by even their most enthusiastic amateur colleagues.[42] Their main need was a ship that was capable of working effectively and relatively safely on the open ocean in the closest deep water accessible from British ports. Carpenter, who had close connections to the Gladstone government through his position as registrar of the University of London and as a vice-president of the Royal Society, began to campaign for the use of Royal Navy ship and its crew.

In May 1868, Thomson wrote to Carpenter describing the Sarses' discoveries, linking them to the widespread belief, originating mainly from the Manx/Scots zoologist Edward Forbes (1815–1854), that the depths of the oceans below a few hundred metres were azoic.[43] Mainly though, it would be possible, as Thomson wrote, to 'give us an opportunity of testing our determination of the zoological position of some fossil types by an examination ... of their recent relatives.'[44] By August 1868 the two were ready to go to sea in the R.N.'s old paddle-wheel surveying vessel *Lightning*. Carpenter wrote to his dredging friend, the clergyman-naturalist A.M. Norman (1831–1918), that 'our object is to make out *everything that we can* as to the distribution of life at great depths in these northern seas; and we shall carefully preserve *everything* that gives the least promise of organic mixture.' In preparation, Carpenter wrote, 'we

2.2 W.B. Carpenter (1813–1885), whose preoccupation with density-driven circulation of the oceans lay behind the *Porcupine* and *Challenger* expeditions and a lengthy, bitter debate with James Croll. From J.E. Carpenter 1888, frontispiece.

are going to fill ourselves as full of coal this afternoon as our old tub (the *first steam vessel* in H.M. Navy) will hold; and then shall proceed pretty nearly due north, towards the Faroe Banks, dredging in different depths as we come to them. Having a donkey-engine, we hope to work to 500 or 600 fathoms. We shall then put into the Faroes; and shall go as much farther N.W. as our our stock of coal, and the power of replenishment at Thorshavn, may render desirable.'[45] This choice of dredging grounds and the direction led quickly to the resolution of Forbes's hypothesis and to a startling discovery about deep-water temperatures.

Both Thomson and Carpenter had gone to sea on *Lightning* firm in the belief that they would find deep-water temperatures in the open sea of 4°C, the presumed temperature of maximum density of seawater. Within days of leaving their Scottish port, Oban, despite the balkiness of the vessel and unusually bad weather, they had found animals as deep as they dredged, 1,189 metres, and the truly surprising result that deep-water temperatures were 0–0.5°C northeast of a line between the Faroes and the Shetland Islands and much warmer, 4.5–8.5°C, to the southwest in the open North Atlantic. Not only did they disprove Forbes's azoic hypothesis, but they had to abandon their misapprehension about deep-sea temperatures and the uniformity of the deep-sea environment.[46] A year later, in a more suitable vessel, the R.N. surveying ship *Porcupine*, they carried out a much more comprehensive Royal Society–supported series of cruises to the northwest of Scotland and into truly deep water southwest of Ireland.[47] Their aims were ambitious, including, in addition to deep dredging and temperature measurements, determinations of gases, salinity, the organic matter in seawater, and the extinction of light in the water column. The ship's equipment was designed or recommended by a committee of the Royal Society, which had among its members the instrument maker C.W. Siemens (1823–1883) and the physicists John Tyndall (1820–1893) and Charles Wheatstone (1802–1875). *Lightning* in 1868 had on board unprotected thermometers, but on *Porcupine* were the first Miller-Casella protected thermometers (the result of collaboration of the University of London chemist W.A. Miller [1817–1870] and the instrument-making firm Casella).[48]

Thomson and Carpenter, with the help of the Welsh conchologist John Gwyn Jeffreys (1809–1885), kept *Porcupine* at sea from 18 May to 15 September 1870. During the second leg, under Thomson, they dredged southwest of Ireland at the greatest depth to that date, 4,465 metres, recovering from that vast depth a variety of invertebrate animals and ringing the final death-knell of Forbes's azoic hypothesis.[49] And, on the final leg, under Carpenter, the ship was directed north to the boundary between cold and warmer deep waters between the Shetland and the Faroe Islands, with the aim of investigating 'the Physical and Biological conditions of the two Submarine Provinces included in that area, which are characterized by a strongly marked contrast in Climate ... and to trace this climatic disparity to its source'[50] The results confirmed the previous summer's findings that to the northeast there was a large area of cold deep water, which increasingly appeared to Carpenter to be the result of a cold Arctic current carrying water far to the south into the deep interior of the Atlantic. By the end of *Porcupine*'s cruise in 1869, Carpenter had embarked upon the development of his 'magnificent generalization' that the cold temperatures were part of a large-scale general oceanic circulation.[51]

The Implications of a 'General Oceanic Circulation'

Carpenter's hypothesis, simple but bold, had developed very shortly after the return of *Lightning*. Referring to the ideas of Humboldt and an influential textbook by the physicist Heinrich Buff, in his account of the first summer's work Carpenter postulated 'deep currents bringing cold water from Polar Regions to replace the warmer water that is continually flowing as (notably) in the Gulf-stream, from the Equatorial towards the Polar Regions, as well as to make good the immense loss which is constantly taking place by evaporation from the surface of Tropical seas.'[52] Both hemispheres were involved, for temperature determinations from the depths of the northern Indian Ocean showed temperatures as low as 0.8°C at about 3,300 metres, a finding only explicable by the northward movement of deep water from the Antarctic.[53]

A few months later, in April 1869, Carpenter presented a more ambitious scheme to an audience at the Royal Institution:

> ... the water that is cooled in the Polar seas must sink and displace the water that is warmer than itself, pushing it away towards the equator; so that in the *deepest parts* of the ocean there will be a progressive movement in the *equatorial* direction, whilst, conversely, the warm water of the Tropical sea, being the lighter, will spread itself north and south over the *surface* of the ocean, and will thus move towards the *polar regions*, losing its heat as it approaches them, until it is there so much reduced in temperature as to sink to the bottom, and thus return to its source.[54]

This flow of deep water toward the equator and its eventual return toward the poles he regarded as 'just as much a physical necessity as that interchange of *air* which has so large a part in the production of winds,' that is, as the well-known Hadley Cell of circulation in the atmosphere, responsible for the Trade Winds and the upper-level Anti-Trades that gave them continuity. The implications were wide, for such a circulation of warm surface water to the north would help to moderate climate, while the existence of deep cold water could show the geologist that fossils of cold-water marine animals might indicate they had lived in the deep sea rather than in a shallow-water environment that was colder than at present; changes in depth and temperature could be confused with changes in climate and time. Moreover, the alternations of deep-sea climates caused by changes in the positions of cold and warm currents, if not too fast, might allow animals to adapt and to form new species.[55]

But it was the physics of oceanic circulation, not its biological implications, that captured Carpenter's imagination. The moderate climate of northern Europe could be explained by the existence of what he called 'a vast body of *warm* water' hundreds of metres deep, of which the Gulf Stream was only a westward intensification. Cold deep water originating in the polar regions returned toward the equator from north and south in currents directed by submarine ridges, such as the one that accounted for the remarkable temperature contrast in the region between the Shetland and Faeroe Islands. The general oceanic circulation was a global phenomenon, driven by heat and cold, into which a host of local phenomena could be fitted.[56] Among these was the still problematic question of the circulation in the Strait of Gibraltar.

In the summer of 1870, *Porcupine* was made available again to Thomson, Carpenter, and Jeffreys, this time, under the supervision of a Royal Society committee to investigate the deep-water fauna of the Mediterranean (Jeffreys's interest) and the exchange of water between the Atlantic and the Mediterranean. Thomson was ill, leaving the ship to Jeffreys and Carpenter for three months. Carpenter, who met *Porcupine* at Gibraltar in mid-August, wasted little time in having temperature and density measurements made in the Strait of Gibraltar, in the adjacent Atlantic, and in the Mediterranean Basin as far east as Malta.[57]

Indirect evidence – the temperature and density structure of the water columns in the narrowest part of the Strait and in the adjacent Atlantic and Mediterranean – suggested that while there was an eastward-directed inflow at the surface (well known since the seventeenth century),

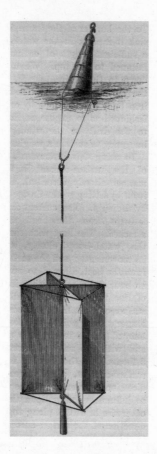

2.3 The current drag (it would now be called a drogue) used on HMS *Challenger*, similar to that used from HMSS *Porcupine* and *Shearwater* in 1870 and 1871 when the outflowing undercurrent in the Strait of Gibraltar was demonstrated. From Thomson 1877, p. 363.

water of relatively low temperature and unusually high density was flowing out of the Mediterranean centred at a depth of about 450 metres. When a small boat was put over, it was carried eastward in the surface current until it lowered a current drag (fig. 2.3) into the region between 180 and 450 metres, whereupon its eastward motion was slowed and eventually stopped, indicating dramatically the outflow of water.

Later in the summer, just before *Porcupine* returned to England, a similar attempt had an even more dramatic outcome. As Carpenter described the event and his conclusions,

> The 'current drag' was then lowered to a depth of 250 fathoms [approx. 450 metres]; and in a short time the boat was seen to be carried along by it in a direction (W.N.W.) *almost exactly opposite* to that of the middle *in*-current of the Strait ... Thus, then, our previous deductions were now justified by a *conclusive proof* that there was at this time a return current in the mid-channel of this *narrowest* part of the Strait from the Mediterranean towards the Atlantic, flowing beneath the constant surface-stream from the Atlantic into the Mediterranean; and ... that a strong presumption may be fairly raised for the *constant* existence of such a return current, though its force and amount are liable to variation.[58]

Indeed, the existence of that variation – along with further evidence of the inflow – was provided by yet another cruise to the Mediterranean, this time on HMSS *Shearwater* from August to October of 1871. Although variation in both the surface and deep currents was evident, Carpenter detected Mediterranean water flowing along the bottom over the sill in the Strait into the Atlantic and once again observed a small boat with a current drag being carried into the Atlantic against the surface inflow by the Mediterranean outflow. He satisfied himself too that the deep water of the Mediterranean Basin was isolated from the deep polar water so characteristic of the Atlantic, taking up a temperature governed by the climate of the region, not of distant polar areas.[59]

The Mediterranean Sea was unusual because its surface waters were so greatly affected by evaporation, resulting in an inflow of Atlantic water to make up the loss, but the Mediterranean took its place usefully in Carpenter's developing model of ocean circulation based upon global differences in the temperature of the sea and thus of its density. In the conclusion to his report on the cruise of *Porcupine* to the Mediterranean in 1870, he set out the principles that could account for the overall circulation of all oceans, beginning with the effect that density differences, no matter how produced, could have upon the creation of currents.[60] In the case of the Mediterranean, evaporation decreased the level of that sea, causing an inflow from the Atlantic, and a deep outflow of the high density water produced by evaporation in response. In more general terms, surface currents resulted from differences in level of the sea surface, and in response to the flow of water from the base of columns of deep, dense water, whether they were produced by evaporation or by cooling. As he summarized the causes of circulation,

> That if there be at the same time a *difference of level* and an *excess of density* on the side of the shorter column, there will be a tendency to the restoration of the *level* by a *surface*-flow *from the higher to the lower*, and a tendency to the restoration of the *equilibrium* by an *under*-flow in the opposite direction *from the heavier to the lighter* column.

And as a result,

> ... a *vertical circulation* will be kept up by any continuous agency which alters at the same time both the *level* and the *density* of the two bodies of water.[61]

In the case of the Mediterranean and Red Seas, it was the sun that caused the excess of density of one column, resulting in an inflow. On the other hand, the Baltic Sea and the Black Sea continually gained water from rivers and rain, resulting in outflow of light waters at the surface and an inflow of Atlantic or Mediterranean waters as deep currents. Thus a reverse flow could be predicted in each case on the basis of exactly the same physical principle, the flow of water from a higher (and lighter) column of water to a lower (and heavier) one.

The link between the physical forces causing complex horizontal and vertical circulation in these regional seas and the major oceans was another factor (besides evaporation and precipitation) causing increase in density, namely polar cooling:

> The agency of Polar Cold will be exerted, not merely in reducing the bulk of the water exposed to it, and thereby at the same time *lowering its level* and *increasing its density*, but also in imparting a *downward movement* to each new surface-stratum as its temperature is reduced whereby a continual in-draught will be occasioned from the warmer surface-stratum around.[62]

As equatorial water moved northward, according to Carpenter's hypothesis, it would be replaced by slowly ascending and warming polar water from below, completing the cycle of circulation.[63] Behind all lay 'a *vera causa* for a General Oceanic circulation' based upon what he described as 'the immense motor power of Polar Cold.'[64]

In the autumn of 1868, while Carpenter's model of a general oceanic circulation was still in its embryonic stages, he challenged those who doubted the existence of polar and equatorial currents to provide a plausible alternative:

> It may be said that the asserted existence of these Currents is a mere hypothesis, until an actual movement of water in opposite directions has been substantiated. But ... the existence of such deep currents is a necessary consequence of the difference of surface-temperature between Equatorial and Polar waters; and those who raise the objection are consequently bound to offer some other conceivable hypothesis on which the facts ... can be accounted for.[65]

Just such an objector existed, ready to present and tenaciously defend an alternative hypothesis of ocean circulation, not for its own sake, but as a crucial part of a geophysical explanation of how the glaciations began.

Enter James Croll

One reader who paid rapt attention to Carpenter, Jeffreys, and Thomson's report on the cruise of *Porcupine* when it reached Edinburgh in December 1868 was James Croll (1821–1890), a newly appointed clerk of the Geological Survey of Scotland under Sir Archibald Geikie (fig. 2.4). Self-educated in philosophy, mathematics, physics, and geology, Croll had already had a varied career as millwright, carpenter, tea-shop proprietor, hotel-keeper, insurance salesman, and college caretaker when he went to Edinburgh in 1867.[66] Early in 1864, with the library of the Andersonian College in Glasgow available for his use, he became interested in the cause of the glaciations and happened upon Joseph Alphonse Adhémar's book *Révolutions de la mer*, in which an astronomical theory of the onset of glaciations was put forward. According to Adhémar, the explanation lay in the ellipticity of the Earth's orbit and in variations in its axis of rotation (precession of the axis of rotation), which together resulted in the Southern Hemisphere being cooler than the Northern and in periodic onsets of glaciation in the north every 22,000 years (the period of precession). In effect, the length of winter in each hemisphere was governed by variations in the Earth's orbit, resulting in glaciations when winter was longest in each hemisphere.[67] Stimulated by Adhémar's approach, Croll refined the astronomical approach by incorporating calculations of the shape of the Earth's orbit and of the precession of its orbit by the astronomer Urbain Leverrier into predictions of the timing

2.4 James Croll (1821–1890), an exponent of wind-driven ocean circulation, W.B. Carpenter's implacable foe in a bitter dispute over the causes of deep-ocean circulation. From Irons 1896.

of the major glacial epochs recurring at intervals of no less than 100,000 years.[68] His first publication on the subject, in the *Philosophical Magazine* in 1864, resulted in Croll's appointment to the Geological Survey, and his book of 1875, *Climate and Time*, much developing these themes, was his definitive statement of the link between astronomical events and the onset of glaciations.

Croll's approach was based on the premise that 'the true cosmical cause must be sought for in the relations of the sun to the earth,'[69] especially the situation when the Earth's orbit showed the greatest eccentricity and the precession of the orbit resulted in winter in one hemisphere or the other being when the Earth was at the greatest distance from the Sun (Earth at aphelion). Under these conditions, even though the Earth as a whole received as much total radiation as during other parts of the ellipticity and precession cycles, the positive feedback of increased length of the cold season, increased back-reflection from the growing area of snow, and the formation of mist over the cold area (blocking the sun) would contribute to the growth of continental glaciers.[70] The oceans, too, were affected in an intimate way both in response to the increase in glacial ice and in increasing the effect of the orbital and axial variations of the Earth, for as ice increased in the northern Hemisphere the temperature gradient between the polar regions would steepen, causing increases in the intensity of the northeast Trade Winds so that they blew across the equator. This would result in decreased flow of the Gulf Stream to the north and deflection of its parent current, the South Equatorial Current, into the Southern Hemisphere, adding to the cooling effect in the north that had been initiated by astronomical causes.[71]

Between February and October 1870, Croll published three papers on ocean currents to bolster his astronomically based theory and establishing, to his satisfaction, the primacy of wind-driven ocean currents, especially the Gulf Stream, in controlling global climate and even making the Earth habitable. The oceans carried enormous amounts of heat north (he estimated that 20 per cent of the heat of the North Atlantic Ocean was carried by the Gulf Stream), which ameliorated the climates of high latitudes, notably Europe and the Arctic, and reduced the temperature of the Southern Hemisphere.[72] But this was more than a quantitative exercise in physical geography. He captioned the beginning of his second paper on ocean circulation 'Deflection of Ocean-Currents the Chief Cause of Secular Changes of Climate,' and more firmly nailed his colours to the mast:

This is no mere picture of the imagination, no mere hypothesis devised to meet a difficult case; for if what has already been stated [about the significance of ocean currents in carrying heat] be not completely erroneous, all this follows as a necessary consequence from physical principles.[73]

Working from his confident application of mathematics to the heat capacity and heat transport of the oceans, Croll supported his own scheme of wind-driven circulation by taking apart the views most opposed to his own. Among these were the near-incoherent and certainly inconsistent suggestions about the causes of ocean circulation put forward by Matthew Fontaine Maury beginning in 1855 in various editions of his *Physical Geography of the Sea*. These centred on differences in specific gravity caused by evaporation (raising the salinity of the water) which brought about motion – as he said, '... it is a difference that disturbs equilibrium, and currents are the consequence.'[74] Although Maury recognized that there was a general movement of ocean waters from the equator toward the poles at the surface and from high latitudes toward the equator in the depths, his descriptions of the causes were contradictory and inconsistent, claiming in one place that equatorial waters were *less* dense than elsewhere and thus flowed away like oil over water and in another that they were *more* dense (due to evaporation) and thereby flowed away. Not surprisingly, Croll's bafflement is obvious; he complained, referring in general to Maury's views, '... it is somewhat difficult to discover what they really are,' and later, referring to the contrasting specific gravity explanations, 'Lieut. Maury's *two causes neutralize each other*. Here we have two theories put forth regarding the cause of ocean-currents, the one in direct opposition to the other.'[75]

With Carpenter's reports on the cruises of *Lightning* and *Porcupine* before him, along with an edition of Maury's book, it is evident that Croll saw an another threat from a similar quarter, but based upon much more tangible evidence. He responded in a few pages (much less than for Maury, saving his ammunition for later), contesting Carpenter's claim that the great mass of warm water detected in the northeast Atlantic on the two cruises was due, not to the Gulf Stream, but to a general northward movement of water from equatorial and tropical latitudes, an exact mirror image of which occurred in the Southern Hemisphere. Reiterating his claim that the Gulf Stream was the main transporter of heat into high latitudes, gaining much of it from the Stream's origins in the South Equatorial Current – thus from the Southern Hemisphere – Croll presented calculations that the equatorial oceans did not contain enough heat to account for the warmth of the North Atlantic at high latitudes.[76] And he made clear a view of the causes of ocean currents that was to colour the debate until it ended in stalemate several years later. As Croll expressed his interpretation of what he called the 'specific gravity' or 'gravity' argument,

> ... if the intertropical waters of the ocean are expanded by heat, and the waters around the poles contracted by cold, the surface of the ocean will stand at a higher level at the equator than at the poles. Equilibrium thus being disturbed, the water at the equator will tend to flow towards the poles as a surface-current, and the water at the poles towards the equator as an undercurrent.

His opinion on what he saw as Carpenter's need for a higher sea level at the equator than at the poles was forthright: 'This, at first sight, looks well, especially to those who take but a superficial view of the matter.'[77] But a deeper examination showed that the explanation of density-driven currents that Croll attributed to Maury and Carpenter was physically impossible. Using tables of the relation between the volume of seawater and its temperature and some assumptions about the little-known decrease of temperature with depth, Croll calculated that the height of columns of seawater at the equator and at a high polar latitude differed by no more than 18 feet, resulting in a slope along a meridian of 6,200 miles of only 1 in 1,820,000.[78] This in itself was of little significance, but he coupled the calculation of slope with the experimental results of the French engineer P.-L.-G. Du Buat (1734–1809), whose posthumous *Principes d'hydraulique* of 1816 presented experimental evidence using flow channels that at slopes less than 1 in 1,000,000 all motion ceased because of the viscosity of the water.[79] As Croll concluded, ' ... the inclination afforded by the difference of temperature between the sea in equatorial and polar regions does not exceed the half of this, and consequently it can have absolutely no effect whatever in producing currents ...'[80]

Thereafter Croll's approach to the problem of ocean currents centred on the principles sketched in his early publications supporting wind-driven circulation and the primacy of the Gulf Stream in the cause of the glaciations. First, for differences of specific gravity (from whatever cause) to cause currents there must be a slope of the sea surface down which water could run. But the slope was insufficient – if it existed. Second, the Gulf Stream was quite adequate to provide the heat reaching the high latitudes of the Atlantic. If, as Carpenter suggested, there were a general poleward motion of equatorial water, temperatures at high latitudes should be much higher than they actually were. Finally, Croll argued that the Gulf Stream originated primarily in the Southern Hemisphere, depleting it of warm water. Any symmetrical general ocean circulation centred on the equator (as suggested by Lenz and Carpenter) would equalize the temperature between the two hemispheres. This was manifestly not the case.

Here were principles for a clash of hypotheses about ocean circulation with very important implications for Croll. He was quite forthright in his explanation of the problem: '... if the theory [of density-driven circulation] is correct, it militates strongly against the physical theory of secular changes of climate ...,' that is, against his theory of glaciations.[81] Croll's theory and the theory of general oceanic circulation being developed by Carpenter made contradictory predictions, for Croll predicted decreased transport of warm water away from the Northern Hemisphere as a glaciation developed. But Carpenter's theory would entail increased sinking of cold polar water compensated by increased transport of warm water to the north, cancelling the effect needed for the onset of glaciation.[82] With so much at stake, Croll prepared to fight for the survival of his brainchild.

The Carpenter-Croll Controversy

For five years, Croll doggedly and relentlessly opposed all Carpenter's evidence for a general ocean circulation based upon latitudinal differences in temperature. After the first exchanges, the debate takes on a mind-numbing sameness, but this forgotten controversy is not without importance, for, at the very least, as Margaret Deacon has pointed out, without the Carpenter-Croll controversy there almost certainly would have been no *Challenger* expedition.[83] The debate is useful, too, in illuminating the many largely non-mathematical approaches to ocean circulation during the mid-nineteenth century. It allows one to address, although not solve, the problem of why, at least in Britain, the relatively well developed practical sciences of hydraulics and theoretical hydrodynamics played such small parts in theories of ocean circulation until nearly the turn of the century.

The viewpoint against which Croll fought was expressed by an anonymous author writing in *Nature* in June 1871, who referred to

... that great Oceanic Circulation, which, while it eludes all ordinary means of direct observation, seems to produce a far more important effect, both on terrestrial climate and on the distribution of marine fauna, than that of the entire aggregate of surface-currents which are more patent to sight.[84]

The reference, although not explicit, was to Carpenter's 'General Oceanic Circulation' of cold deep water and warm water at the surface. Croll, perceiving the threat to his glacial hypothesis, consolidated an attack on Carpenter's hypothesis in a series of lengthy papers in the *Philosophical Magazine*, along with shorter barbs in *Nature*. These culminated in several chapters in his book on the glacial hypothesis, *Climate and Time*, published in 1875.[85] At the outset, it was clear, as he said, that

> ... more than nine tenths of all the error and uncertainty which prevail, both in regard to the cause of ocean currents and to their effect on climate, is due, not ... to the intrinsic difficulties of the subject, but rather to the defective methods which have hitherto been employed in its investigation – that is, in not treating the subject according to the rigid methods adopted in other departments of physics. What I most particularly allude to is the disregard paid to the modern method of determining the amount of effects in *absolute measure*.[86]

Straightforward computation using common-sense physical principles could clear away the misapprehensions arising from misguided hypothetical schemes such as Carpenter's general oceanic circulation. Here lay the particular gritty strength of Croll's approach, and the roots of its ultimate failure to dislodge exponents of Carpenter's scheme. As Croll clearly recognized, there was no established body of theory that compelled common assent; in his own words, 'at present the question cannot be decided by a reference to authorities.'[87] He was on his own, within the bounds set by the presuppositions of his theory of the glaciations,[88] confident of his ability to use his good Scots practicality to defeat the enemy.

Two themes dominate Croll's arguments between 1871 and 1875: the ability of the Gulf Stream unaided to account for the large volume of warm water in the central and northeast Atlantic; and the absolute necessity of a sloping sea surface if the return part of Carpenter's general circulation from the tropics to high latitudes were to be possible. Without a full heat budget of the ocean, available to no one in their time, it was impossible to decide between Croll's claim, based on simple calculations, that the Gulf Stream was quite capable of carrying the requisite volume of water north, and in the course of doing so, ameliorating the climate of western Europe,[89] and Carpenter's reply that the effects of the Gulf Stream were lost far from European shores and that the great volume of warm water in the north had to be due to a general northward movement of surface water across a wide front from the tropics to high latitudes.[90]

Despite his occasional reservations that convective water flow could occur without a sloping sea surface,[91] Carpenter was eventually trapped into defending his model of circulation on the basis that sufficient slope existed to allow a return flow from the tropics to high latitudes. Croll's attack was incessant: for a density-driven surface flow to occur there must be sufficient slope for the effect of gravity to overcome the friction of water molecules upon themselves. His initial calculations (mentioned earlier), based upon the thermal expansion of water and the laboratory work of Du Buat, indicated that the slope from the equator to high northern latitudes of the Atlantic was too little to allow any appreciable movement, if it existed at all.[92] Moreover, according to Croll, even if water did move from equator to pole, it would be at equilibrium when it arrived and thus would have no tendency to sink and begin a return deep current.[93] Croll returned to variants of this argument frequently, asserting the necessity of a slope sufficiently steep to overcome friction if any return flow to high latitudes were to occur, a situation he referred to as 'an obvious principle of mechanics.'[94]

As he repeatedly threw scorn on Carpenter's qualitative model of circulation, by 1874 Croll began to develop a more complex scheme of wind-driven circulation to account not just for the simple and obvious relationships, such as that between the Trade Winds and the equatorial currents, but also for more difficult cases such as ocean currents running against the wind direction and for the deep-ocean currents themselves. Croll summarized his 'Wind Theory of Oceanic Circulation'[95] in a few words:

> ... the surface-currents of the ocean are not separate and independent of one another, but form one grand system of circulation, and ... the impelling cause keeping up this system of circulation is not the *trade-winds* alone, as is generally supposed, but the *prevailing winds of the entire globe considered also as one grand system.*[96]

Although there was general worldwide agreement between wind direction and the direction of ocean currents, there were circumstances that could lead to opposing winds and currents; for example, the Humboldt Current along the west coast of South America, which ran against the direction of the Trades, impelled by winds in another area.[97] Deep counter-currents could readily be explained, too, as the response of water boxed in by land taking 'the *path of least resistance'* and flowing below and in an opposite direction to surface currents. Such was Croll's explanation of at least part of the deep return flow from the north in the Atlantic and for the subsurface ouflow in the Strait of Gibraltar, which, Carpenter's demonstration notwithstanding, was simply a reflux of surface Gulf Stream water finding its way out of the Mediterranean, a *cul-de-sac*.[98] He explained how winds blowing equatorial water into high latitudes

> ... must tend to destroy static equilibrium by making the equatorial too light and the temperate and polar columns too heavy ... The effect must be to produce a constant ascent of the equatorial column and an *inflow* of cold water below equal to the *outflow* above. In short, the wind must produce a system of circulation precisely the same as that supposed to take place by difference of temperature.[99]

Thus the winds alone, when considered in their global extent, were quite sufficient to account for all the phenomena of ocean circulation.[100]

It had been Carpenter's hope, in a rather undefined way, that the results of the scientific circumnavigation by HMS *Challenger*, beginning late in 1872, would provide new and definitive information confirming the general oceanic circulation.[101] But although the hydrographic results of the great expedition, as they returned to England from the ship's port stops, made it clear that all the major ocean basins were full of deep cold water, just as Carpenter's scheme predicted, it was Croll who benefited most from *Challenger*'s temperature, salinity, and specific gravity determinations. *Challenger*'s 'crucial test'[102] of a vertical oceanic circulation was highjacked by Croll and turned against Carpenter, the architect of the idea and of the expedition.

Carpenter, early in his arguments with Croll, had stated that the omnipresence of deep cold water all over the world, and particularly in the Pacific (which had no connection with deep water in the north, but a broad connection to the south polar regions), would be good inductive evidence of the 'General Oceanic Circulation.' He provided the evidence available from *Challenger* and the U.S. cable-survey vessel *Tuscarora* in a lecture to the Royal Geographical Society in June 1875, showing that temperatures in the deep Pacific were consistent with a far southern origin, and that isolated basins in the Indonesian region were cut off by their sills from the deep flow from the south.[103] A year earlier, lecturing to the same group, he carefully depicted the thermal stratification of the Atlantic, showing the vast extent of shallow warm water (far too great to be attributable to the Gulf Stream), reduction of surface temperature and salinity at the equator (evidence for the return of cold deep water to the surface), and the presence everywhere of cold water at the bottom. On this basis, he described the general oceanic circulation as 'an established Doctrine of Terrestrial Physics.'[104] Probably not to Carpenter's surprise, Croll turned both kinds of evidence against him.

The wide extent of deep cold water was of the least interest to Croll. He pointed out that it could equally well be explained by his scheme of global wind-driven circulation and refluxes of deep water from the high latitudes. But the Atlantic temperatures provided a surprise and, as he saw the evidence, a total contradiction of the 'General Oceanic Circulation.'[105] Using the data upon which a meridional temperature section of the Atlantic depicted by Carpenter had been based, Croll calculated that the surface of the North Atlantic was higher at mid latitudes than at the equator, making it impossible for there to be a continuous downslope from the equator toward the north. There was, in fact, a rise of 3½ feet from the equator to 38°N, giving the North Atlantic a domed shape along its axis to the north.[106] Similarly, the North Pacific sloped downward toward the equator from 52°N, implying anomalous surface flow from north to south across the warm equatorial region.[107] The implications were clear:

> ... it is mechanically impossible that, as far as the North Atlantic is concerned, there can be any such general movement as Dr. Carpenter believes. Gravitation can no more cause the surface-water of the Atlantic to flow towards the Arctic regions than it can compel the waters of the Gulf of Mexico up the Mississippi into the Missouri. The impossibility is equally great in both cases.[108]

Finally, goaded by a response from Carpenter indicating that Croll's calculations were irrelevant to the real physical basis of the 'General Oceanic Circulation,' namely, the constant absence of any kind of static equilibrium between equatorial and polar water columns,[109] Croll answered in exasperation, and perhaps exultation:

> The point at issue is now simply this: *Does it follow, or does it not, from the*

temperature-soundings given in Dr. Carpenter's own section, that the North Atlantic at 38° is above the level of the equator? If he or anyone else will prove that it does not, I shall at once abandon the crucial test argument and acknowledge my mistake; but if they fail to do this, I submit that they ought at least in all fairness to admit that in so far as the North Atlantic is concerned, the gravitation theory is untenable.[110]

Firmly committed to a scheme of fluid flow based upon the simple mechanical analogy of particles rolling down hill, and despite his success in keeping Carpenter off balance, Croll had reached the limit of his ability to counter Carpenter's 'magnificent generalization.' After 1875, preccupied solely by his theory of glaciations, dogged by ill-health, and with renewed interests in philosophy and theology, he wrote little more about the problem of ocean circulation.[111]

W.B. Carpenter under Siege

James Croll's attack on Carpenter's 'General Oceanic Circulation' was always defensive, dictated by his need to preserve the system of oceanic and atmospheric currents that would support his theory of the glaciations. The debate was spirited, even bitter at times, and Carpenter's irritation at what he regarded as Croll's obduracy grew with time, resulting in one self-important but anonymous commentator accusing Carpenter of an *ad hominem* attack rather than a decorous exchange of views.[112] Carpenter indeed had the irritating habit of bolstering his views by referring to the eminent physicists who would give credence to his ideas, including Sir John Herschel, Sir William Thomson, G.G. Stokes, Sir George Airy, and Henrik Mohn, rather than by engaging Croll's criticisms directly. As time went on, it is evident that considerable personal antipathy had been aroused and that the two thoroughly disliked each other.[113] Underlying all is the possibility that Croll's lower social status, as well at his attack-dog like tendencies in scientific debate, gave a particularly bitter tone to the controversy. In a gentle rebuke to both, John Young Buchanan, *Challenger's* chemist, writing in 1874 during the voyage, suggested that more would be gained by careful attention to the detailed physics of the atmosphere and of seawater in specific locations than by grand generalizations.[114]

But Carpenter's personal failings should not prevent the close examination of what he considered to be evidence for the causes of ocean circulation. Although Carpenter remained wedded to the overall scheme of a 'General Oceanic Circulation' until the end of his life, his ideas evolved into a form considerably more complex than realized by Croll or by récent commentators on the Carpenter-Croll controversy. Provoked and refined by Croll's onslaught, Carpenter's ideas, especially after 1872, showed increasing physical insight, although always in a purely qualitative way. Part of the heat of the debate between Carpenter and Croll lay in their differing approaches to physical reality: Carpenter repeated frequently that his model of a 'General Oceanic Circulation' was a provisional hypothesis, subject to verification, ideally, though not realistically, by direct measurement:

> ... I claim for the doctrine of the General Oceanic Circulation no higher a character than that of a 'good working hypothesis,' consistent with our present knowledge of facts, and therefore *provisionally* adopted for the purpose of stimulating and directing further inquiry.[115]

Croll scorned hypotheses, preferring, as he said in response, to deal with reality:

> I am unable to agree with Dr. Carpenter on this latter point. It seems to me that there is no necessity for adopting any hypothetical mode of circulation to account for the facts, as they can quite well be accounted for by means of that mode of circulation which does *actually exist.*[116]

Carpenter's subtlety was no match in argument for Croll's self-confident pragmatism, but it led eventually to a more sophisticated view of ocean circulation than that of his Scots opponent.

Carpenter's view of the course and causation of ocean currents, as he summarized it anonymously in 1872, involved the efficacy of a 'General Oceanic *vertical* circulation' with the 'action of Polar Cold as its *primum mobile.*'[117] Despite the evidence marshalled by the German geographer August Petermann (and accepted by Croll) that the Gulf Stream could be traced by its temperature signature even into the Arctic Ocean north of Europe, Carpenter was sceptical: how could the effect of a limited oceanic current, even one as significant as the Gulf Stream, match the effect of a broad, ocean-wide northward movement of ocean water?[118] And having established the reality of a great general circulation of the major oceans, especially the Atlantic, the model could be applied, indeed tested, against the circulation of peripheral seas, such as the Mediterranean, the Baltic, and the Black Sea, which were cut off from the effects of the

Gulf Stream. Everywhere the evidence showed the occurrence of vertical circulation and of return currents brought about by differences in temperature (and thus of density), although direct demonstration had only been possible in the entrance to the Mediterranean.[119] In his most complete exposition of the theory of the general oceanic circulation, based on the cruise of HMSS *Shearwater* in 1871, Carpenter emphasized six supporting points:

1. the omnipresence in the Atlantic of warm upper and cold lower water layers separated by a sharp discontinuity;
2. by contrast, the absence of significant horizontal vertical stratification in the Mediterranean, Red, and other isolated seas, due to the absence of sinking due to cooling of surface water;[120]
3. the prevalence of cool water, between 47°F and 52°F, in the equatorial regions of the Atlantic, explicable by the return of cold water toward the surface, a phenomenon not seen in the Mediterranean, where the deepest water was warmer than at intermediate depths of the tropical Atlantic;
4. the ubiquity of cold deep water over the bottom of the Atlantic, in the Indian Ocean (north and south of the equator) and the China Sea, and below the Gulf Stream from Newfoundland to the Florida Strait, indicating a polar origin of deep water;
5. the presence of very cold water, 29.5°F to 32°F, in the 'Lightning Channel' (now known as the Faroe-Shetland Channel), directly indicating movement of polar water into the mid-latitude Atlantic from the north;
6. warmer water, above the temperature expected at that latitude, just to the south of the polar water in the 'Lightning Channel,' indicating a general movement of warm water toward the northeast, as part of the compensating return flow of the 'General Oceanic Circulation.'[121]

But if polar cold was the *primum mobile* of the 'General Oceanic Circulation,' was it necessary to have a sloping sea surface to allow a return flow toward the poles? Croll insisted on it, but Carpenter vacillated, at one time claiming that only the disturbance of equilibrium brought about by intense cooling was necessary, at others that disturbance of equilibrium would result in a drop of sea level at the cold end of the circulation and a rise near the equator (see earlier discussion). Part of the solution lay in the viscosity of water, which if were very low, as Carpenter presumed, would be no impediment to the flow of water without any perceptible slope in response to cooling and warming.[122]

But if that argument was insoluble, Carpenter was encouraged by the realization that in envisioning the big picture of oceanic vertical circulation, he had a distinguished predecessor. For several years, the eminent geologist Joseph Prestwich (1812–1896) had been compiling and evaluating all the available temperature records from the deep sea in an attempt to see how sediments, organisms, and temperature were related in the present as a key to environmental relations in the geological past. On 18 June 1874 he presented a summary of his work to the Royal Society, including the mention of the long-forgotten observations and conclusions of Emil von Lenz, based upon his temperature observations during Kotzebue's circumnavigation of 1823–6.[123] Carpenter, who had learned of the results a little earlier, was captivated:

> As I have never claimed any originality in regard to the doctrine of oceanic circulation, which I have advocated solely as an important scientific truth, it has afforded me nothing but the most unalloyed satisfaction to find that the doctrine, which appeared to me ... the 'common sense of the matter,' was put forward nearly thirty years ago by one of the most eminent physicists of his day, as a necessary deduction from the facts of observation.[124]

In a thorough review of his position before the Royal Geographical Society the same month, incorporating the results becoming available from all over the world as a result of the *Challenger* expedition, Carpenter reiterated the significance of Lenz's symmetrical model of vertical ocean circulation and went further, to set out clearly a relation he had begun to accept three years earlier, the different roles, but equal realities, of *both* a vertical temperature-driven circulation and a horizontal wind-driven one. Early on, in 1871, he regarded the two systems as completely distinct –

> ... it is fully recognized by myself, that the *current* movements of *surface*-water are, for the most part, produced by the agency of winds; but these movements, I contend, all belong to a *horizontal circulation, which tends to complete itself* ... [125]

– a viewpoint that he maintained throughout the peak of his controversy with Croll during 1874–5. But a middle position was beginning to appear in Carpenter's arguments, even under the relentless stress of Croll's position, for as Carpenter said in October 1875,

... it is scarcely fair in Mr. Croll, therefore, to continue speaking of the 'wind theory' and the 'gravitation theory' of Ocean Circulation as if they were antagonistic, instead of being not only compatible, but mutually complementary – the wind-circulation being *horizontal*, and the thermal circulation *vertical*.[126]

Several years later, when the controversy was only a bitter memory, Carpenter made it clear that his view of ocean circulation, greatly affected by the *Challenger* results, and no doubt too by the heavy pressure from Croll, had taken on a new form. Describing the significance of the *Challenger* results to general readers in 1880, he linked the wind-driven circulation of the Gulf Stream, the large pool of western North Atlantic warm water revealed by *Challenger*'s temperature sections (the Sargasso Sea), and his 'General Oceanic Circulation' into a single system:

As the superheating of the upper stratum of the mid-Atlantic [he was referring to the Sargasso Sea] is dependent on the influx of Gulf Stream and other water exceptionally warmed in the Equatorial Water, the *thermal effect* of its north-east flow is mainly dependent upon the Gulf Stream and its adjuncts, while its *movement* is kept up by the Polar indraught [by which he meant the portion of the surface 'General Oceanic Circulation' that entered the Arctic and warmed the high latitudes of the eastern North Atlantic]. Thus neither the General Oceanic Circulation nor the Gulf Stream could alone produce the result which is due to their conjoint action. The Gulf Stream, without the Polar indraught, would remain in the mid-Atlantic; and the Polar indraught, without Gulf Stream water to feed it, would be almost as destitute of thermal power as it is in the South Atlantic.[127]

Croll would scarcely have been pleased to see his Gulf Stream pressed into the service of even a greatly modified 'General Oceanic Circulation' – although he never commented in print on Carpenter's new model – but to Carpenter the expansion of his model was the endpoint of a decade's struggle and a concession in a small but important way to Croll's influence.

In other respects, too, Carpenter's views of ocean circulation were transformed by the kind of detail that *Challenger*'s results offered. Rather than refining the generalities of the 'General Oceanic Circulation' on a broad geographic scale, by 1875 he saw the necessity of physical investigations in specific places, especially at high latitudes where deep water should form. Responding to Croll's criticism that in the Antarctic (where Carpenter had suggested much of the 'glacial' deep water of the south and equatorial Atlantic must sink), a temperature inversion made it impossible for sinking to occur, Carpenter countered that

> ... the descent of the cooled surface stratum cannot take place in the polar summer ... but that it takes place wherever and whenever the surface-cold is sufficient to check surface-liquefaction, and to cool down water of ordinary salinity to a temperature below that of the subadjacent stratum ... [128]

Winter deep water formation in the Antarctic had to remain an unexamined problem during Carpenter's lifetime, but north polar exploration and competition brought him the opportunity to look for evidence of convective deep water formation in the Arctic. The British North Polar Expedition of 1875 under George S. Nares (the first captain of *Challenger*, 1872–4) attempted an approach to the pole via West Greenland, Baffin Bay, and Smith Sound.[129] Carpenter's son Herbert accompanied the supply ship *Valorous* when it sailed to meet the expedition's vessels in West Greenland during the summer of 1875, taking temperature measurements, several below 1,000 fathoms, as far north as Disko Island. The results showed the Atlantic water cooling as it moved north at the surface into Davis Strait, a thick, apparently immobile 'intermediate layer,' and deep cold water that appeared to have cooled and sunk somewhere to the north, perhaps in Baffin Bay.[130] Here lay one source of the cold abyssal water that drove the 'General Oceanic Circulation,' admittedly limited in extent, adding to the relatively southward flow in the 'Lightning Channel.'[131]

Here Carpenter's active investigation of the deep-sea circulation and its relation to the 'General Oceanic Circulation' ended. The argument with Croll had burned itself out, and both participants went on to other things.[132] In Carpenter's case, he seems to have recognized, despite the originality of his new ideas about circulation, that the mere assembling of new temperature data from the *Challenger* expedition or any of its successors could not add significantly to knowledge of ocean circulation without new techniques, techniques that neither he nor any of his scientific colleagues could or would bring to bear on the problems presented by the oceans.

Attacking the Oceans

From a twenty-first-century viewpoint, it is difficult not to admire Carpenter and Croll for the courage involved in their non-mathematical

approach to the oceans (save for Croll's resolute determination that simple arithmetic alone would be sufficient to give total physical insight).
But, of course, this twenty-first-century viewpoint is wrong-headed – in
mid-nineteenth-century Britain there was no way to know *a priori* how
ocean circulation could be explained or if the explanations needed
to go beyond the physics demonstration bench and the calculation of
simple heat budgets. A more profound but difficult question is why the
very sophisticated mathematical physics of nineteenth-century England
was never applied to the kinds of problems that Carpenter and Croll attacked so bravely.

Part of the answer lies in the backgrounds of Carpenter and Croll
themselves. Croll should not be considered only a dogged (and doctri-
naire) Scots controversialist. His theory of the ice ages was considered
with the greatest seriousness by geologists, notably James Geikie, and
astronomers, among them the eminent Robert Ball.[133] In the same year,
1876, he became both a Fellow of the Royal Society and an LL.D. of the
University of St Andrews. Even today his name is linked with that of the
Serbian geophysicist Milutin Milankovic (1879–1958) in the theory of
astronomically driven climatic cycles.[134] Croll was widely read – he had
educated himself in philosophy, theology, physics, and evolution – and
he kept up with at least some of the European literature, although with
the help of friends as translators.[135] He was convinced that the rational
mind, supported by the acceptance only of phenomena that could be
unequivocally demonstrated, would lead to certain truth. In fact, Croll's
scientific quest was a moral one, as he revealed in a letter to Osmond
Fisher in August 1871. Commenting on his dislike of the British Association (one of the favourite venues of W.B. Carpenter), he explained why
he stayed away:

> My chief reason ... was that I dislike all such public displays. The truth is,
> I have very little sympathy with the leading idea of the British Association,
> viz., that science is the all-important thing. I don't believe anything of the
> kind. There are more noble and ennobling studies than science. You can
> hardly expect one who has devoted twenty years of the best part of his life
> to the study of mental, moral, and metaphysical philosophy to have much
> sympathy with the narrow-mindedness of the British Association.[136]

Or, for that matter, how could he sympathize with the attitude and scientific philosophy of a paragon of the establishment like W.B. Carpenter?
But Croll seriously underestimated, and certainly never understood,

the scientific philosophy of his opponent. Apparently a creature of the establishment, Carpenter had studied medicine in London and Edinburgh, working as a zoologist and only briefly as a physician. His text, *Principles of Comparative Physiology*, published in 1839, became a classic and was repeatedly revised. First as Fullerian professor in the Royal Institution, then as professor in University College, London, and as registrar of the University of London from 1865 to 1879, Carpenter kept up work on invertebrate nervous systems and increasingly on the relation between mind and body in animal physiology. His approach was reductionist but strongly conditioned by a Unitarian faith that became stronger in mid-life and that was characterized by his belief that vital activities, although completely physical, were an expression of God's will. He was among the first in Britain to apply the principles of conservation of energy to animal physiology, seeking a way to understand life through the unification of vital and physical processes.[137] As one commentator has expressed Carpenter's system of beliefs, 'clearly, the inorganic and organic worlds together constituted a closed system of power or force' in which there was a continuous exchange of matter and energy between living and non-living components of the world within the bounds set by available energy, an approach characterized as a 'natural theology of force.'[138]

The last complete expression of Carpenter's philosophy of nature was the lecture 'Man the Interpreter of Nature,' his presidential address to the British Association in Brighton in 1872.[139] In it, at the beginning of his most bitter debates with Croll, he expressed his belief that science 'is *a representation framed by the Mind itself* out of the materials supplied by the impressions which external objects make upon the Senses; so that to each Man of Science, *Nature is what he individually believes her to be,*'[140] thereby emphasizing the difficulty of interpreting observations of natural phenomena and the possibility of error. It was common sense that eventually dictated scientific truth where the evidence was ambiguous:

> In a large number of scientific cases ... our Scientific interpretations are clearly matters of *judgment*; and this is eminently a *personal act*, the value of its results depending in each case upon the qualifications of the *individual* for arriving at a correct decision. The surest of such judgments are those dictated by what we term 'Common Sense,' as to matters on which there seems no room for difference of opinion ...'[141]

And although he emphasized the possibility of human fallibility in interpreting nature, Carpenter retained his belief in force as 'one of those elementary Forms of Thought with which we can no more dispense, than we can with the notion of Space or Succession.'[142] But universal, though provisional, truths might not hint at the surprises hidden in nature, among which he cited the unusual properties of water as it was cooled and the change of its properties when salt was added, the basis of the then discredited 4°C theory.

It is certainly too facile to see an immediate link between these aspects of Carpenter's philosophy of nature and his theory of a 'General Oceanic Circulation' – and yet the hints are there. His short account of the 'General Oceanic Circulation' to the British Association meeting in Edinburgh in 1871 was titled 'On the Thermo-Dynamics of the General Oceanic Circulation.' The account is a routine one of cooling and convection, but it is clear that Carpenter saw the problem in a framework that was larger than merely the application of heat and cold, as a manifestation of a great law of nature.[143] This was true, too, of his long insistence that the *primum mobile* (his term) of oceanic circulation was heat (more accurately *polar cold*), a manifestation of force, rather than the epiphenomenon of the winds, as Croll would have it. Carpenter's philosophical subtlety and his commitment to a physiology and natural philosophy of energy made it inevitable that he and Croll would find themselves on quite different planes of discourse.

And yet the two opponents were united in a curious way: in their ignorance of mathematical fluid dynamics and attempts to apply them to another but similar system of circulation, the atmosphere. And why was there so little help from physicists in attempting to resolve the very real physical problems that lay behind the Carpenter-Croll controversy? Some clues can be found in Croll's correspondence, for in 1875, involved in the 'crucial test' exchange with Carpenter over the height of the North Atlantic sea surface,[144] he submitted his results to a physicist, G.C. Foster, asking if there could be a northward surface flow up the slope of the sea surface.[145] Foster, although generally favourable to Croll, referred to the difficulty of inferring circulation from limited data, and pointed out that the really significant information needed was the distribution of *pressure* throughout the water column, not merely the calculation of height, a clue that Croll did not heed or did not understand.[146] Another of Croll's authorities, equally knowledgeable mathematically, the geophysicist Osmond Fisher (1817–1914), while sitting on the fence in the dispute, also pointed out, although indirectly, that solutions should be sought within the water column but disclaimed special knowledge of fluid motion.[147] And the knowledgeable Dutch hydrographer Marin Jansen (1817–1914) of the Royal Netherlands Navy, approached with the same problem, pointed out that the Gulf Stream, too, would have difficulty running uphill, and that he found inadequacies in all theories of ocean circulation.[148]

Both Carpenter and Croll ran into the same problem. Eminent authorities, notably superstar physicists, were willing to pay polite attention to questions about the physics of the ocean but were not really interested in those problems. Their attention was elsewhere: in mathematical models of the ultimate constituents of matter, in formulating laws uniting electricity and magnetism, in the applicability of fluid mechanics to 'perfect' fluids, such as the ether or to organ pipes – nearly everything except the ocean.[149] And although there was a long history of attention to hydraulics, much of it mathematical, no one, at least in Britain, seems to have had the slightest interest in applying a significant body of theoretical fluid mechanics to the dynamics of the ocean.[150] This is clear in one of the most influential texts of its generation – one contemporaneous with the declining years of the Carpenter-Croll controversy – Horace Lamb's *Treatise on the Mathematical Theory of the Motion of Fluids* (1879).[151] Based on a series of lectures that Lamb had begun to deliver to Cambridge mathematicians in 1874, it was resolutely theoretical. There is no mention anywhere of any geophysical fluid – the atmosphere or ocean water – for reasons that are made clear in the first sentence of this long-lived classic:

> The following investigations proceed on the assumption that the fluids with which we deal may be treated as practically continuous and homogeneous in structure; *i.e.* we assume that the properties of the smallest portions into which we can conceive them to be divided are the same as those of the substance in bulk.[152]

Although Lamb went on to explain that this greatly simplifying assumption was for convenience only, and that with some adjustment, the behaviour of 'ordinary matter' would obey the same principles, it was clear to physicists, and increasingly to non-physicists like Carpenter, that the distribution of properties in the ocean was far from homogeneous: heat and salt (and thus the distribution of density) varied in complex ways in both horizontal and vertical directions. The problem was not expressed in this way, but it is clear from the response of physicists that not only were there bigger fish to fry – solving the problems of matter and energy

– but that the oceans were too complex to be approached lightly. In their largely untutored – and certainly non-mathematical – innocence, Croll and Carpenter had stumbled into a physical problem well beyond their competences.

Elsewhere, never mentioned by the protagonists or their advisors, physicists had approached the properties of another fluid in motion, the atmosphere, in a way uniting thermodynamics and fluid motion. This lay outside the intellectual horizons of Carpenter and Croll, and apparently also those of the outstanding group of British mathematical physicists that they consulted. It is clear, for example, that Croll never understood the effect of the Earth's rotation, despite his correspondence with the American mathematician William Ferrel (1817–1891), who suggested that Croll's and Carpenter's approaches to ocean circulation could be united.[153] Nor did Carpenter recognize the significance of the work of Henrik Mohn, the Norwegian meteorologist, with whom he corresponded about the influence of the Gulf Stream at high northern latitudes,[154] in applying mathematical approaches to the atmosphere and the ocean. It was in Scandinavia, beginning in the late 1870s, that the problems tackled qualitatively by Carpenter and Croll began to be resolved by a mathematical approach to physics that could be applied to the real world.

3

Boundaries Built with Numbers:
Making the Ocean Mathematical

Since we last talked I have had a chance for the first time to gather my thoughts on the problems of meteorology and hydrography. And this has led to a complete transformation of my view. I could no longer withdraw from the answer to that question: what do I really want [to do]? I found the answer could be only one: I want to solve the problem of predicting the future states of the atmosphere and ocean ...

Vilhelm Bjerknes to Fridtjof Nansen, 4 September 1904,
in Friedman 1989, p. 55

'Combining and Digesting and Methodising'
Information on the Oceans

Charles Wyville Thomson said it well in 1878: the accumulation of evidence about the circulation of the oceans was 'confusingly rapid,' and the problems being faced by the theory builders like Carpenter and Croll were so complex that it was difficult to see where solutions might be found.[1] Thomson, and his co-organizer of the *Challenger* expedition, W.B. Carpenter, believed that the accumulation of data would in itself provide insight into the nature and causes of oceanic circulation, even though they were without any clear view of how the data could be used to provide that insight.

There was little precedent in the experience of either to give them guidance. Implicit in their approach to scientific problems by the 1870s was the Humboldtian view that nature could be revealed and systematized through measurement. Here lay the nub of the problem faced by Carpenter and Croll: how could measurement alone, in the absence of some new world picture produced by abstraction – in modern terms, a new physical model – do more than increase confusion and debate? Carpenter, in his late years, began to suspect that the problem of ocean circulation might be too complex to be dealt with using simple arithmetic. But his time was short, and although he knew of work going on in Scandinavia and had corresponded with the Norwegian meteorologist Henrik Mohn, Carpenter had exhausted his mental and physical resources in the years before his death in 1885. Nonetheless, his instinct about looking to Scandinavia for help in physical science was sound.

Ocean Sciences in Scandinavia

It seems anomalous that the Scandinavian countries played a pre-eminent role in the development of quantitative ocean science. These were small countries in the mid-nineteenth century: Sweden, the largest, with about 4 million inhabitants; both Norway and Denmark with only about 1.8 million. The economy of each centred on agriculture and forestry, although in Norway fisheries bulked large as an occupation, providing food and income to supplement the production of mostly small-holding agriculture. In that country, fisheries (notably cod and herring) provided 45 per cent of commercial exports at least through the end of the 1860s. The fisheries remained important even when the Norwegian economy began to expand with the increase of industry and especially the expansion of that country's international shipping later in the century. In Sweden, even though it began to industrialize before its Nordic neighbours, the fishery, especially for herring, remained important in the economy of many communities around its southern coast.[2]

Scandinavian fisheries fluctuated widely in the nineteenth century, at their low points bringing real hardship, even starvation, during the crop failures of the cold 1870s and 1880s. For example, the Norwegian cod fishery, which had expanded throughout the first five decades of the century, failed in the traditionally rich Vestfjord (Lofoten Islands) in 1860. It then fluctuated, seemingly erratically, for some decades following, and reached a critically low level again during the 1890s. Norwegian herring, which had been productive through the middle of the century, began to fall off in the 1870s and 1880s, coinciding with a series of cold years (and with agricultural failure throughout much of the region). In Sweden, where the Bohuslän herring fishery had failed in 1808, catches began to rise again in 1877 and remained high until nearly the turn of the century, seemingly in concert with the decline of herring in Norwegian waters.[3]

With each crisis came government interest in studying the fisheries and their environment; thus, many of the strands in the development of the marine sciences in Scandinavia are plausibly related to fisheries crises during the nineteenth century. One example is the appointment in 1864 of the zoologist G.O. Sars (1837–1927)[4] to investigate the Lofoten cod fishery, resulting in his discovery that cod eggs are pelagic and thus would be at the mercy of changes in ocean circulation. They culminated with the appointment in 1900 of Johan Hjort as director of fisheries for Norway and the involvement, mainly through Hjort, of Norway in the formation of the International Council for the Exploration of the Sea (ICES) in 1902 to investigate the biological and physical causes of fishery fluctuations. Similarly, in Sweden the return of herring to Bohuslän in 1877 and thereafter resulted in Gustaf Ekman (1852–1930)[5] being asked by his government to investigate, leading by a series of circumstances to his collaboration with the chemist Otto Pettersson (1848–1941)[6] in hydrographic studies of Swedish waters, then the waters around all of Scandinavia and the northern British Isles, and eventually to the deep involvement of Pettersson and Sweden in the organization of ICES. In both Norway and Sweden, the scientists like Sars, Hjort, Gustaf Ekman, and Pettersson who redirected their careers to fisheries and to international oceanography clearly viewed fluctuations of oceanic properties – temperatures, salinities, and currents, and their link with weather – as critical ones in understanding why these resources varied as they did.[7]

But not every foray from pure science into study of the ocean is explicable so easily, as the example of Fredrik Laurents Ekman (1830–1890) indicates.[8] Ekman, a chemist, and from 1870 until his death professor of chemistry in the Stockholm Tekniska högskola (Technological Institute – later Kungliga Tekniska Högskolan), became interested in the coastal waters of southern Sweden in the late 1860s, concentrating first on the temperature and salinity relations of estuaries and on instruments. In a series of papers beginning in 1870, he identified the salt wedge entering estuaries and postulated 'reaction streams' – now referred to as estuarine entrainment – in which the outflowing fresh water mixed with the deeper inflowing salt water resulting in increased inflow (and the renewal of nutrients in the surface waters, as later investigators revealed).[9] In 1875 he broadened the scope of his studies, applying his knowledge of 'reaction streams' to oceanic circulation on a much larger scale. His lengthy and purely qualitative scheme of the causes of oceanic circulation involved current-causing factors that raised sea level, such as warming, evaporation, and precipitation, and those that moved the water by mechanical action, such as wind and river inflows, as well as the effects of currents on the adjacent water. Taking encouragement from the work of Croll and Carpenter as well as the grand schemes of Maury, Ekman nonetheless found critical flaws in the ideas of each. He concentrated instead on his contention that any disturbance of equilibrium would result in both a surface current and in response a deep countercurrent (a 'reaction stream'). All the causative factors interacted to produce the worldwide oceanic circulation.[10]

Ekman's theoretical work, much in a tradition that was familiar to Carpenter and Croll (and although it was mentioned in Otto Krümmel's *Handbuch der Ozeanographie* in 1911),[11] languished in the face of practical challenges, such as his involvement in attempts to survey synoptically the hydrography of Scandinavian waters. These began in 1876 with sections of the Kattegat and Skagerrak by Gustaf Ekman and August Cronander, followed the next summer by a survey of the Baltic using two Swedish ships directed by F.L. Ekman himself.[12] Thereafter, the extensive hydrographic work carried out in Scandinavian waters, and increasingly in German and Scottish seas, was directed at least nominally toward fisheries problems. This was a subject in which F.L. Ekman showed no signs of interest. Where, then, did his interest in the sea originate?

It may be significant that until the last decade of the nineteenth century most Scandinavian interest in the 'physics' of the sea – its physical properties and circulation – was centred in Sweden. Sweden, among the Nordic countries, was the first to industrialize,[13] and apparently the first to respond to the need for engineering expertise in industry. Although there were no technical colleges in Norway until 1914, reflecting that country's relatively late industrialization,[14] Sweden was training engineers much earlier. F.L. Ekman was a teacher in the Stockholm Technological Institute beginning in the late 1850s, followed in succeeding decades by his younger relative Gustaf Ekman, as well as the future mover and shaker of European marine sciences, Otto Pettersson. Although the proximate causes of F.L. Ekman's interest in the sea are difficult to discern, it seems very likely that the level of technical education in Sweden and the economic importance of the sea and its resources combined to intrigue wide-ranging intellects in that country even before fisheries crises became important and fashionable. This contention is supported from another quarter: a physical scientist, from a discipline of the utmost practical importance to Norway, meteorology, was the first to investigate the circulation of the sea quantitatively.

Henrik Mohn and the North Ocean

On 19 March 1874, the zoologist G.O. Sars and the meteorologist Henrik Mohn (1835–1916)[15] sent a lengthy memo to the Home Department of the Norwegian government in which they proposed a concerted investigation of the open sea adjacent to Norway, extending to Iceland in the west and beyond Jan Mayen to Spitsbergen in the north. Behind it lay deep-seated concerns in Norway about the variation of herring and cod fisheries and the possibility that resources so vital to its people were linked in some way to the weather, and especially to variations in the Norwegian Sea.[16] Claiming that '... as to the open sea, what we yet know is meagre in the extreme,' Sars and Mohn believed that what they called 'the physical and biological conditions of our native country' would become explicable if the open sea, the breeder of Scandinavian weather and the habitat of commercially important fishes, were better known. Their rhetoric aimed to convince nationalistic politicians of the importance of scientific work on the ocean: 'From the facts as yet determined respecting the sea that laves our shores, we may safely assume, that to its waters is Norway indebted for her existence as a habitable and civilized country.' More specifically, as they suggested, 'the results of such an investigation might be expected to throw light upon the physical conditions determining our climate, and, above all upon the biological characteristics of our migratory fishes.'[17]

Sars and Mohn proposed a series of summer cruises into the Norwegian Sea based upon the work of the British *Lightning* and *Porcupine* from 1868 to 1870 and the *Challenger* expedition of 1872–6, then in progress, encompassing depth determinations, sea temperature, water chemistry, current measurements, meteorology, and the Earth's magnetic field, plus the systematics and distribution of plants and animals. After the approval of their plan in March 1875, they selected a scientific staff and by December had arranged to charter a steamer, *Vøringen* (fig. 3.1), from Bergen for the summer of 1876. The first summer's work, mainly in the south, went well, and in the summers of 1877 and 1878 they worked north to Spitsbergen, completing the work of their Norwegian North-Atlantic Expedition in August 1878.[18]

Henrik Mohn (fig. 3.2), who in 1876 was director of the Norwegian Meteorological Institute and professor of meteorology in Christiania (now Oslo), had begun in 1867 to collect sea and air temperature measurements made for him by sealing captains on their annual trips north. Initially, the data collected on the North-Atlantic Expeditions were just

3.1 The Norwegian steamer *Vøringen*, used by Henrik Mohn and G.O. Sars during the Norwegian North-Atlantic Expeditions of 1876–8. From Wille 1882b.

3.2 Henrik Mohn (1835–1916), whose application of quantitative methods from meteorology to the Norwegian Sea as a result of the Norwegian North-Atlantic Expeditions of 1876–8 provided the first dynamic analysis of circulation in any ocean. From *Folkebladet* No. 3, 15 February 1904, courtesy of Dr Vera Schwach.

an extension and expansion of this program, a part of the routine work of Mohn's Institute, but by about 1880 they had begun to take on new significance in the context of mathematical meteorology.

Mohn's background was nothing if not varied. Notably, it was resolutely quantitative and mathematical. As an undergraduate, he studied astronomy, physics, and meteorology, and for his first advanced degree, mineralogy. His first university position was in astronomy, but in 1866 he was appointed to the new chair in meteorology in Christiania. Thereafter, he worked tirelessly on practical meteorology and the climatology of Norway. As one of Mohn's biographers has written, 'At his death he was regarded as the grand old man of European meteorology.'[19] There was another side to Mohn, the theoretical one, linked closely to the development of cyclone models in European meteorology, and less successfully (according to his successors) to the development for the first time of dynamic oceanography.

When Mohn first ventured fully professionally into meteorology in 1866, that field was being transformed by the application of physics, notably thermodynamics, to the explanation and prediction of atmospheric motions. The first synoptic charts of the weather appeared in the 1860s, and professional societies and research networks began to proliferate at the same time. National weather services, like the one headed by Mohn in Norway, appeared throughout western Europe and in the United States by and during the 1870s.[20] Although there was no unanimity of opinion about the dynamics of the atmosphere, the conception developed of storms in mid-latitudes as convective, rotating systems within which significant horizontal and vertical motions and large releases of heat occurred.

Henrik Mohn was among those maintaining and elaborating this 'thermal theory of cyclones,'[21] involving the ascent of warm air and the condensation of water vapour in cyclonic systems. At first basing his work upon the demonstration by the Scottish meteorologist Alexander Buchan that cyclones had a counter-clockwise circulation in which intense convection and the resulting release of latent heat provided energy,[22] in 1870 Mohn proposed that the cyclonic circulation was asymmetrical and driven by the input of warm air into the leading edge of the cyclone as it moved. This qualitative model of the structure and dynamics of cyclones, promoted in his textbook *Grundzüge der Meteorologie* of 1875 and other publications,[23] formed the basis of a theoretical treatment of cyclonic weather systems a few years later when Mohn collaborated with the mathematician Cato Guldberg (1836–1902)[24] to develop equations of motion for temperate-latitude cyclonic systems on a rotating Earth.

Guldberg and Mohn's great contribution, in the words of Gisela Kutzbach, lay in solving the problem that before 1876 'no attempt had been made to analyze quantitatively, and within a dynamical framework,' namely, 'the relationships between the vertical and horizontal motions.' For the solution of this problem, '.., it was first necessary to describe mathematically the motion of air in terms of the forces acting on it.'[25] Stimulated first by the work of the Danish engineer L.A. Colding on the motions of the sea and atmosphere,[26] Guldberg and Mohn became aware in 1874 of the work of the American mathematician and theoretical meteorologist William Ferrel (1817–1891)[27] on the effect of the Earth's rotation and on the mathematical theory of cyclones. Ferrel, who had demonstrated the importance of the Earth's rotation on motions over the Earth's surface in a series of papers beginning in 1856,[28] and who had tried unsuccessfully, if perfunctorily, to show James Croll the error of his ways (see chapter 2), burst into European consciousness late, showing in his paper of 1874 how to relate barometric pressure and wind speed. Encouraged that they were on the right track, Guldberg and Mohn pushed ahead with their analysis, the first part of which was published in 1876 in Christiana as *Études sur les mouvements de l'atmosphère.*[29]

In this publication and its successors, Guldberg and Mohn developed equations of motion for the air in cyclonic systems, involving horizontal input of air and vertical convection at the centre, the velocity of the rising air and the storm's geometry determining the velocity of the incoming horizontal current.[30] The primary factors involved in their analysis were the pressure gradient, the effect of the Earth's rotation, the centrifugal and tangential forces involved in the movement of the air, and the effects of friction. Two equations represented, respectively, the radial and tangential horizontal motions of the of the air in the system.[31] Then, from these equations of motion, they developed a relation soon afterwards named the baric wind law by the Russian/German meteorologist Wladimir Köppen (1846–1940) and now known as the geostrophic equation.[32] From this it was, at least in principle, an easy matter to calculate the horizontal velocity of the wind in a cyclonic system given the distribution of atmospheric pressure and the density of the air.

It seems that climatology, rather than any theoretical approach to the circulation of the sea, had been Mohn's concern at the inception of the Norwegian North-Atlantic Expedition in 1876 – but in the same year, the first of his great collaborations on dynamic meteorology with Guldberg was published, and the two began to work out the implications of their model for a variety of weather systems.[33] Although it is not known when Mohn first began to see that a dynamic analysis of the sea was possible, it is likely that it was not long after 1878, judging by the immense, time-consuming labour that went into his treatment of the physical results of the expedition. He commented later that ocean circulation had been far from his mind when the Norwegian North-Atlantic Expedition began:

> The reader must bear in mind that, when, in 1876, we started on our Expedition, I had no definite idea of being able to advance so far with my studies as ocean circulation. All was then an unknown field of research, and depth, temperature, and the chemical properties of water laid claim to the first place in our labours.[34]

Certainly by 1880 he was, as he said, 'occupied with the study of ... currents.'[35] It took no great stretch of Mohn's scientific imagination to see that theoretical approaches to circulation could overcome the paucity of direct observations of ocean currents, particularly given the success of the Guldberg-Mohn equations in his meteorological work. As he said in introducing the section on ocean circulation of his monograph *The North Ocean: Its Depths, Temperature and Circulation*, in 1887,

> ... some parts of the sea have never been visited either by seafarers in general or by men of science ...
>
> ...
>
> We have, however, another mode of attacking the subject, which, with a probability greatly superior to that afforded by direct wind-observations, leads to the end in view. The meteorological investigations of the last decennial periods give, as their best founded result, the connection existing between the distribution of atmospheric pressure and direction and velocity of the wind. And this law, the baric wind-law, applies – with respect to the strata of air nearest the earth – in its full rigour precisely to the phenomena of the surface of the sea.[36]

In 1885, Mohn's first monograph on ocean currents appeared, applying the methods of mathematical meteorology to the currents of the Norwegian Sea – the North Ocean. This was expanded and extended two years later in his definitive monograph, *The North Ocean*.[37] These provided for the first time a dynamic approach to ocean circulation combining the effect of the wind with the effects of variations in density within the water column so that the circulation of both the surface layers and the deep waters could be calculated.

Based upon the hydrographic information gathered by the Expedition and upon extensive meteorological data from northern Europe and Iceland,[38] Mohn's technique, simple in concept but extremely laborious in execution, was based on the principle that, as he said, 'the surface of the sea is kept in continual motion, alike by the prevailing winds and by the unequal distribution of density.'[39] Though the data on winds across the Norwegian Sea were limited, he was able to work with the pressure distribution available from coastal stations and the Expedition to calculate, using the baric wind law, the mean winds and their velocities across the region. Given the wind field, he then calculated the effect of the wind on the surface water, which was to move it in the same direction as the wind but at lower speed. This yielded a cyclonic circulation similar to that of the wind over the Norwegian Sea (fig. 3.3). Taking into account the Earth's rotation imposed changes in the height of the sea surface relative to a standard reference surface (he used what he called 'the surface of level' – the geoid – that is, a subsurface level of equal gravity). As a result, he calculated that the surface of the Norwegian Sea was concave, about 0.8 metres lower in the middle than around the edges, due to effect of the wind alone.[40]

Mohn's great originality came in the next steps, which involved determining the effects of density differences (derived from temperature and salinity measurements made during the Expedition and on other expeditions) on the height of the water columns throughout the Norwegian Sea. The density patterns also resulted in a concave shape, which, when added to the effects of the wind, gave a total difference between the margins and the centre of the gyre of 1.4 metres.[41] The net surface currents (derived using a version of the baric wind law) resulting from both wind and the density field of the water, were also cyclonic, but, of course, a little stronger than those attributable to the wind alone. Similar calculations could be made for the deep waters, by determining the distribution of pressure at a variety of depths – Mohn used 300, 500, 1,000, and 1,500 fathoms (approximately 560, 930, 1,900, and 2,800 metres) – to calculate, once again using a version of the baric wind law, the direction and velocity of the currents that would result from pressure gradients at great depths.[42]

Mohn did not overestimate the significance of this achievement. He was well aware that because of deficiencies in instruments (especially the thermometers)[43] and the widely scattered stations in a very large ocean,

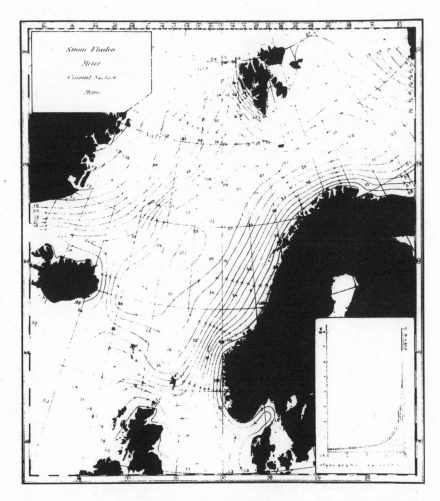

3.3 The currents of the Norwegian Sea as depicted in Henrik Mohn's *The North Ocean* of 1887.

errors could result. Nonetheless, he recognized the significance of what he had done for the future of ocean circulation studies:

> ... I have sought to explain the motion of the water in the North Ocean as produced alike by the normal winds and the differences in the density of the water. The former cause predominates, but the latter too has full

significance. I regard this research as an attempt, not indeed as a mere hypothesis but as a series of ideas carried out by consistent computations, which, proceeding from known forces and their mode of action, have led to results that in many respects and on many heads agree with the actual results given by the observations.[44]

He was hopeful that his lead would be followed by others:

The great deep-sea Expeditions despatched of late years by several of the chief nations of Europe and America, will yield material for investigating the currents in the great oceans of the world, better in many respects than what I have had at my disposal for the study of the North Ocean. I very much wish that some one would undertake a work of the kind, and that my Memoir might serve to guide his researches.[45]

Flawed though it may be in modern terms (see note 42), Mohn's work on the Norwegian Sea was a *tour de force* of conceptualization, analysis, and computation, aptly described as 'the first attempt at treating all known forces together in a single picture of ocean circulation.'[46] But it was a solitary structure of physical thought about the oceans until the end of the century. This was due in part to Mohn's inability to follow it up because of other work, and possibly due to his discouragement with the possibilities of dynamic methods in meteorology.[47] It was not solely due to neglect of his accomplishment by others.

The first person to exploit the possibilities of Mohn's method was Otto Krümmel (1854–1912), professor of geography in the University of Kiel from 1884 to 1911.[48] Krümmel's knowledge of oceanography was synoptic – certainly the most complete of his era – and his advanced texts, beginning in 1887, summarized the latest advances in the field (including Mohn's techniques) for several generations of students in Germany and abroad.[49] By 1910, Krümmel saw the dynamics of ocean circulation, not as a matter of single causes, but of 'current components' involving both wind and density, much as Mohn had done.[50] Rudolf Engelhardt, one of Krümmel's doctoral students, applied Mohn's method to the density layers of the Baltic Sea in 1899. Another, Georg Wegemann, extended Mohn's analysis to the wind-driven surface layer of the northern North Atlantic, a study applied to the deeper waters by Gerhard Castens in 1905. The last analysis of this kind was by Krümmel's student Wilhelm Wissemann in 1906 in a study of the Black Sea. Finally, Mohn's technique re-emerged briefly in Alfred Merz's and Georg Wüst's work on Atlantic deep circulation in 1922 (see chapter 5).[51] But outside Kiel and Scandinavia, where theoretical approaches to the ocean were relatively familiar, Mohn's great work was ignored or not known. Why?

Deficiencies of the method probably had little to do with the problem.[52] As Mohn himself recognized, the measurements of temperature and salinity were problematic, and, as was soon apparent, gave some incorrect results.[53] These were correctable flaws, as were the problems identified by later critics, such as the problem of how to determine where the deep reference surface lay on which the calculations were based. Mohn had accepted unquestioningly the work of the mathematical physicist Karl Zöppritz (1838–1885), which purported to show that the wind would affect the surface waters rapidly but that its effects would be transferred very slowly to great depths (Zöppritz estimated that mean winds would take more than two hundred years to cause deep currents), but this played no critical part in the formulation of his work.[54] Long after Mohn's time, the great hydrodynamicist V.W. Ekman suggested that the 'monumentally self-contained' nature of Mohn's system of calculation had prevented its adoption by others, who were more content to correct small deficiencies using band-aid solutions (he seems to have been referring to Krümmel's students) than to make the technique the *modus operandi* of a school of researchers.[55] As the next sections of this chapter will make plain, this must be incorrect, because Mohn had happened upon the germ of a physical approach that was developed successfully by others, although independently. More plausible is the suggestion by the German physical oceanographer Hermann Thorade, writing in 1935 to commemorate the centenary of Mohn's birth, that in Mohn's time oceanography was more an 'extensive' geographical science than an 'intensive' geophysical one, and that 'exact dynamic ways of thinking did not meet with approval.'[56] It is true that Mohn had no direct scientific successors in ocean studies, and that his work, striking in its originality, fell upon unprepared ground. Prepared ground did exist only a decade and a half later when another Scandinavian initiative in 'intensive' geophysical analysis of the oceans spread rapidly in Europe. Very aptly, Thorade described Mohn's work as 'an abandoned mine: the veins have been exploited and their products have been incorporated into daily use, perhaps worn and recast, so that their presence is no longer recognizable.'[57]

Electricity, Magnetism, and the Ether – Unlikely Parents of Dynamical Oceanography

Action at a distance gave birth to mathematical physical oceanography through the work of the Norwegian physicist and applied mathematician Vilhelm Bjerknes. More accurately, an intense debate about the means by which light, magnetism, and electricity were transmitted, a debate that dominated late-nineteenth-century physics, gave issue to mathematical methods in physical oceanography. The link is far from direct, for had Vilhelm Bjerknes not taken up the cause of an outmoded physics espoused by his father in Sweden, a scientifically peripheral part of northern Europe, physical oceanography's history as a quantitative science might have been quite different. Bjerknes's failures as a classical physicist directed him into the study of the atmosphere and the oceans. With him, he brought a distinctive view of the mechanical nature of reality and a mathematical approach, to his great fortune, that adapted well to real geophysical fluids like air and ocean waters. The development of mathematical physical oceanography, as discussed in the preceding sections, developed in the context of control of fisheries by the atmosphere and the oceans. Other concerns, such as the utility of weather forecasting to agriculture, shipping, and aviation played larger and larger roles after 1898. Vilhelm Bjerknes, his colleagues and students in Norway and Sweden, were the key players. These Scandinavians appropriated both the weather and the sea to their goals (to adapt Robert Marc Friedman's expression).[58] Scientifically, their approach was firmly grounded in an increasingly unfashionable kind of late-nineteenth-century theoretical physics.

By contemporary standards, the nineteenth century was one of undiluted triumph in physics. Chemical properties, light, magnetism, and electricity all came to be understood theoretically and in practice, the theories involving either or both of a deep mechanical or mathematical insight.[59] Naturally the success of some scientific programs brought the decline of less-favoured ones: careers being linked to scientific programs, it is natural that scientists espousing less favoured or unfashionable ones should experience the decline of their viewpoints or even of their careers. Paradoxically, a scientific viewpoint of great significance for the development of mathematical physical oceanography arose from an unfashionable approach to electromagnetism in the late nineteenth century. The Bjerkneses, father and son, found themselves on the less-favoured side in the last great physical debate of the nineteenth century.

The ferment of physics in the nineteenth century, especially in the late decades, has been well explored by historians of science.[60] Subjects of enquiry and styles changed quickly, and scientists with strong personalities promoted new, fashionable approaches ruthlessly.[61] To those left behind, like Russell McCormmach's fictional physicist Viktor Jacob at the time of the First World War, the result was bitterness, confusion, and anxiety.[62] Especially at the end of the nineteenth century, as Vilhelm Bjerknes discovered, it was easy to be trapped hopelessly on one side of an intellectual divide. It was with electricity and magnetism that the greatest problems arose. The underlying puzzle was how forces could act at a distance. Increasingly, pure 'action at a distance' theories of electromagnetism became unfashionable. In their place, various models of the plenitude of space were developed, filling space with more or less mechanical media, particulate or continuous, by which forces such as electricity and magnetism and radiation such as light could be propagated.[63]

Ether theories have a long history, but in the nineteenth century they began to have real importance in physics before the middle of the century with Michael Faraday's work on magnetic induction of electricity. Faraday envisioned electricity and magnetism propagating through closely contiguous particles linked by 'an atmosphere of force'[64] constituting an electromagnetic field. This 'atmosphere of force' was what William Thomson (later ennobled as Lord Kelvin) and James Clerk Maxwell only a few years later developed into a complex mechanical/mathematical field theory.[65] In doing so, they (particularly Maxwell) modified the 'luminiferous ether,' believed to be the medium in which light was transmitted, to a universal medium responsible for the transmission of light, electricity, and magnetism.[66]

William Thomson, throughout his long life in physics, remained a staunch supporter of pictorial models in physics, not as direct representations of physical reality, but as what might now be called working or heuristic hypotheses. Thomson's widely quoted remarks in a lecture in Baltimore in 1884 illustrate his approach:

> I never satisfy myself until I can make a mechanical model of a thing. If I can make a mechanical model I can understand it. As long as I cannot make a mechanical model all the way through I cannot understand; and that is why I cannot get the electromagnetic theory.[67]

'The electromagnetic theory' was that of his younger colleague James Clerk Maxwell, who used mechanical models at first as aids to the creative imagination, but who later abandoned them for abstract mathematical representations of electromagnetism. Maxwell shared a Scottish education and Cambridge mathematics with Thomson,[68] but his treatment of light and electromagnetism, indeed his whole approach to the representation of reality, diverged from Thomson's between 1856 and 1873. The result was the mathematical unification of light, electricity, and magnetism in late-nineteenth-century physics, but also the creation of a deep and uncomfortable rift in the field.

The result was a mathematically abstract representation of electro-magnetism, grounded in the properties of vortices and a continuous ether, that had two profound outcomes: first, it claimed that light and electricity had the same velocity and thus were aspects of the same physi-cal reality; second, it provided the banner of a purely electromagnetic physics which could dispense with mechanical models of the kind that William Thomson, along with many other traditional physicists, includ-ing the Bjerkneses, found both compelling and essential.

Maxwell's achievements provoked both admiration and discomfort, not only in Britain but on the Continent, where different theories of the ether and of electrodynamics had been developed.[69] Action at a dis-tance lived on in Europe, especially in the electrodynamics developed by Hermann von Helmholtz (1821–1894), which was accepted there as a clarification of Maxwell's work.[70] But mechanical models and abstract algebraic formulations of purely electromagnetic phenomena were ex-tremes of a spectrum of conceptual approaches, none easy to accept unequivocally. Arthur Eddington, writing, in an obituary of the British ether theorist Joseph Larmor (1857–1942), of the state of physics at the end of the nineteenth century, captured the situation well:

> The ever-urgent problem of the ultimate relation of matter and electricity and ether, and the search for a unifying conception which would explain how they came to possess their fundamental properties had occupied the greatest minds of the time; and it was hard to see any direction in which new light might be found. The ground had been gone over again and again, and impassable barriers seemed to have been reached. Classical physics was near the end of its tether.[71]

But upon very close inspection, impassable barriers may prove to be per-meable. This was the hope and belief of Carl Anton and Vilhelm Bjerknes during the last two decades of the nineteenth century. How their physics reached its impassable barrier and transcended it in an unexpected way stems in important ways from Vilhelm Bjerknes's link with one of the great post-Maxwellians, Heinrich Hertz (1857–1894).

Hertz is the quintessential hero of Maxwellian physics. Interested in electrodynamics since his student days in Berlin with Helmholtz, Hertz exploited the implication of Maxwell's theory that electromagnetism could be transmitted wave-like through space, showing in 1888 that electromagnetic waves could be propagated from a transmitter to a receiver in his laboratory. It was a superb example of laboratory physics, with the implication that Maxwell's theory was vindicated and that it was likely that a transmitting medium, the ether, truly existed.[72] But Hertz, like his mentor Helmholtz, was dissatisfied with Maxwell's electrodynamics. Helmholtz had proposed a drastic modification of Maxwellian electrodynamic theory in which action at a distance occurred instantaneously, not in an intervening medium at the speed of light, by the transmission of polarizing forces by intervening bodies. After his experimental verification of the transmission of electromagnetic waves, Hertz undertook a recasting of Maxwellian dynamics into axiomatic form, leaving behind the mechanical and untestable remnants that had persisted into Maxwell's late work. The result was an electrodynamics that Ernst Mach approvingly called 'a physics free of mythology."[73]

In 1890, about a year after he had moved from Karlsruhe to a professorship in Bonn, Hertz's interest turned away from experimental physics and field equations toward mechanics, which he proposed to reformulate along axiomatic lines, much as he had done with Maxwellian theory. His book *Die Principien der Mechanik*, published after his death in 1894, was a theoretical work, devoted to kinematics rather than dynamics, from which force and energy had been removed as primary terms. Its only primitive terms were mass, space, and time. Action at a distance was removed by postulating a set of hidden variables linking the observables.[74] Thus force was replaced by the interactions of masses, some visible, others hidden but detectable through their effects. In this way, a rigorous mechanical theory of the ether might be achieved, removing the ambiguities and uncertainties of Maxwellian theory. As Martin Klein has said, 'Hertz's system was to furnish the mechanics of that electromagnetic ether whose existence his own experiments did so much to establish.'[75]

Hertz's *Mechanics* was regarded as profound and elegant, but coming at the time it did, and without concrete examples of how it might be applied, it was set aside by theoretical physicists. Philosophers and historians of science, not physicists, now read Hertz's last work. Its lasting effect, greatly transformed, was on the Norwegian physicist Vilhelm Bjerknes, who had worked in Bonn with Hertz on electrical phenomena, but who brought from his father's work on hydrodynamic analogies with electromagnetic effects an almost obsessive concern with ether

3.4 Vilhelm Bjerknes (1862–1951), whose circulation theorem of 1898 provided a means of calculating motions of both the atmosphere and the oceans. From Friedman 1989, with permission of the Norwegian Academy of Science and Letters.

theories and the mechanical explanation of action at a distance. Vilhelm Bjerknes was encouraged and inspired by Hertz's *Mechanics*.[76] Through him and the effect of Hertz's book, mathematical physical oceanography and predictive meteorology developed from an unfashionable branch of late-nineteenth-century physics.

Vilhelm Bjerknes: Early Career of a Physicist

Vilhelm Bjerknes (fig. 3.4) came to hydrodynamics and mechanics honestly, from his father, Carl Anton Bjerknes (1825–1903), who held chairs of mathematics in Christiania from 1866 until 1903. As a youth, C.A. Bjerknes had read Leonhard Euler's *Letters to a German Princess* (1770),[77] which he remembered when he heard the mathematics lectures of P.G.L. Dirichlet (1805–1859) in Göttingen in 1856. Euler had argued that action at a distance of the kind found in Newtonian physics was to be rejected: seeming action at a distance had to be due to an intervening medium. Dirichlet suggested that, on theoretical grounds, bodies in a perfect (viscosity-free) fluid should move without restraint, eliminating the problem that the ether would hinder their movement. If the ether were such a perfect fluid, actions at a distance would be due to its effects in transmitting forces. Using moving bodies in fluids, C.A. Bjerknes worked for most of his life to demonstrate hydrodynamical analogies to actions in the ether, attempting to provide physical evidence for its reality.[78] His son, Vilhelm Frimann Koren Bjerknes (1862–1951),[79] became involved in his father's work while he was still a student in Christiania, and he was present when his father's hydrodynamic laboratory models received special recognition during the Paris International Electric Exhibition of 1881. When Hertz's work on the vindication of Maxwell's theory was published in 1888, the Bjerkneses could see the importance of their work in providing further support of an ether-based electrodynamics.

But C.A. Bjerknes was slow in publishing his results, and eventually this task fell to his son. In 1889, Vilhelm Bjerknes, after graduating in Christiania, went to Paris, where, still a student, he heard Poincaré's lectures on physics. In 1890 he moved on to Bonn, to join the newly installed Hertz in his Physical Institute. With Hertz, Bjerknes began to work on electrical resonance, a study that occupied him until 1895.[80] Vilhelm Bjerknes received a doctorate from Christiania for resonance studies after his return to Norway in 1892. The following year, he was appointed lecturer in mechanics in the Stockholm högskola (later Stockholm University); in 1895 he became professor of mechanics and mathematical physics in the högskola. It was at this time that the burden of completing and publishing his father's work fell on him. Imbued with mechanics since his youth and encouraged by the publication of Hertz's *Mechanics* in 1894, Vilhelm Bjerknes could see the urgency of providing mechanical models of the ether at a time when theoretical physics was in rapid transition from mechanical representations of reality to purely electromagnetic ones. Working furiously, he arranged to have his lectures on hydrodynamic forces, prepared for the högskola, published in Leipzig just before the old man died.[81]

A problem arising in Vilhelm Bjerknes's lectures extending his father's work was that his theoretical results contradicted the theorems of Helmholtz and William Thomson describing the actions of vortices in perfect, incompressible fluids. Vortices, rather than being conserved, were actually produced when portions of the fluid with different densities from the bulk of the fluid were set in motion. This result came about when the restrictions imposed by Thomson were removed, allowing the incompressible fluid to become compressible while at the same time allowing its density to be governed by factors in addition to pressure. In modern terms, the idealized situation conceived of by Thomson dealt with *barotropic* flow, but Bjerknes found himself dealing with *baroclinic* flow.[82] Coincidentally, the latter was characteristic at times of real geophysical fluids such as seawater and the atmosphere.

Early in his career, in 1897, imbued with ideas of an ether-based physics, Bjerknes devised a circulation theorem applicable to hydrodynamics (he hoped it could be developed and applied to electrodynamics) in which the acceleration of a fluid resulted from the unequal distribution of pressure and density. A year later, in a lecture delivered in Stockholm on 11 March 1898, he extended the analysis to the atmosphere and the ocean, making it clear that although originally he had developed the circulation theorem to provide a mathematical basis for his father's electrodynamic theories, it applied equally well to real fluids with 'material heterogeneities' such as the atmosphere and oceans. [83] Only two months later, on May 11, he took the practical applications farther, developing in detail a 'fundamental theorem of hydrodynamics' and emphasizing particularly in the concluding sections of the printed version its application to the atmosphere and ocean and its suitability for solving by quantitative means 'contentious problems' persisting in the physics of those media.[84]

Circumstances in his academic environment also forced him to consider applications of his theorem to the real world. In this, his colleagues in Stockholm Nils Ekholm (1848–1923) and Otto Pettersson were the main actors. Ekholm, a meteorologist, was apparently introduced to Bjerknes by Svante Arrhenius, professor of physics in the högskola.[85] Ekholm united an interest in cyclone formation with ballooning. When the Swedish balloon expedition to the North Pole disappeared in 1897, his interest in predicting motions of the atmosphere took on sudden urgency. Bjerknes's circulation theorem might be applicable, especially because Ekholm was able to demonstrate the baroclinic distribution of pressure and density in cyclones. Understanding the vertical distribution of properties in the atmosphere could lead to the prediction of their origin and motions using Bjerknes's theorem.[86]

Otto Petersson was equally quick off the mark. In 1898, as the steps leading to the International Council for the Exploration of the Sea were beginning, Pettersson recognized that Bjerknes's theorem might be applied quantitatively to the circulation of the ocean and its overlying atmosphere, thus to the fluctuations of fish stocks that had concerned him for years.[87] In the following year, at Pettersson's request, Bjerknes talked of his work at the first organizing meeting of ICES, held in Stockholm. A year later, again at Pettersson's urging, he described the importance of his work for ocean circulation and meteorology at another organizational meeting, this time in Göteborg.[88] Simple hydrographic sections, those already being employed by some of the collaborating nations and in prospect for an enlarged international study of the northern European seas, could be used to determine what Bjerknes called the 'nature and causes of oceanic motions.' And although his discussion to the delegates in 1900 was qualitative, devised to persuade them of its utility, he was at work on a refined version of the circulation theorem taking into consideration the effect of the Earth's rotation, factors ignored earlier for the sake of simplicity.[89]

In his paper 'Cirkulation relativ zu der Erde' in 1901, Bjerknes systematically and carefully explored the implications of baroclinicity on a rotating Earth. The intersecting isobars and isopycnals of the baroclinically circulating atmosphere or ocean, which he now for the first time called 'solenoids' (reflecting his work on electricity with Hertz in Bonn, also mathematical usage),[90] would result in vertical accelerations on a fixed Earth. Under the influence of the Earth's rotation (the Coriolis effect), the motion was along the length of the resulting three-dimensional parallelograms, yielding as a resultant a horizontal motion, the magnitude of which was determined by the number of solenoids, that is, the closeness of the packing of the intersecting isobars and isopycnals.[91] This formulation, Bjerknes claimed, could be both diagnostic and prognostic of the circulation[92] of the atmosphere and the ocean.

News of Bjerknes's abilities – or, more accurately, his potential – in solving oceanographic problems was not restricted to Stockholm. Fridtjof Nansen (1861–1930), then professor of oceanography in Christiania, asked for Bjerknes's help in solving two problems that had aroused his interest during the north polar voyage of *Fram* between 1893 and 1896, namely the difficulty in making way where fresh water lay over salt (the 'dead water' problem) and the reason for the ship's drift at an angle to the wind. Both were solved in 1900 and 1901 by Vagn Walfrid Ekman (1874–1954),[93] the son of F.L. Ekman, a student from Uppsala who had been attending Bjerknes's lectures in Stockholm (fig. 3.5).[94] With Ekman's assistance, Nansen included calculations of the water motion near the Norwegian coast at the entrance to the Barents Sea and in the North Polar Basin using Bjerknes's theorem, which he knew was being modified by some of Bjerknes's junior colleagues (as described later). This was the first practical application of the technique.[95] Then in 1901 Bjørn Helland-Hansen, Nansen's protegé from Bergen, arrived in Stockholm to learn the circulation theorem from Bjerknes and his assistant Johan Sandström.

With these events, the groundwork was laid for the full-scale use of Bjerknes's circulation theorem in oceanography. Bjerknes now threw himself nearly exclusively into meteorology; he is best known now as the founder of dynamic meteorology and especially for the achievements of the Bergen School of meteorology, whose work reached a peak just after the First World War but remained a dominant force throughout the twentieth century.[96] Both dynamical meteorology and dynamical oceanography owe their origin to Vilhelm Bjerknes's unfashionable mechanical approach to electrodynamics, with its roots in late-nineteenth-century controversies over physical reality, aimed originally at understanding the characteristics of the ether and thus at providing a rational basis for dynamics without the paradox of action at a distance. The Scandinavian setting in which Bjerknes found himself in the late 1890s was critical to the success of his mechanics, but in an unexpected way and direction, for, without the urgency of solving meteorological and oceanographic problems for his colleagues, Bjerknes would probably have driven himself farther and farther along the dead-end road of ether physics. Once his work was redirected, Bjerknes carried himself along the ultimately fruitful way to dynamical meteorology and oceanography. Especially in oceanography, his students and colleagues, especially Ekman, Helland-Hansen, and Sandström, adapted his work, ensuring that it was accepted and became the basis of a new way of looking at the oceans, using mathematics.

Helland-Hansen and Sandström: The Mathematical Investigation of Ocean Currents

Writing in 1901, H.N. Dickson (1866–1922), who had been involved between 1893 and 1896 in hydrographic surveys in the Faeroe-Shetland Channel for the Fishery Board for Scotland, credited Scandinavians, especially F.L. Ekman and Otto Pettersson, with the development of 'mod-

3.5 V.W. Ekman (1874–1954) in 1901, at the time he first formulated the famous 'Ekman spiral,' caused by the action of the wind on the sea surface. Archive, Musée océanographique de Monaco, courtesy of Jacqueline Carpine-Lancre.

ern methods of research' suitable for linking fisheries and the physical conditions of the sea which controlled them:

> Put shortly, the outstanding feature may be said to be the application of the idea of the synoptic chart – the survey of the part of the ocean under investigation in such a manner that the physical or chemical conditions in its waters are known at successive instants of time, at intervals sufficiently short to allow of the changes being continuously traced.[97]

In 1901, Dickson did not recognize the full significance of having such detailed observations, namely their role in the mathematical calculation of oceanic circulation, but his Norwegian successor on a Scottish cruise to the same area in 1902,[98] Bjørn Helland-Hansen (1877–1957), and his colleague, the Swede Johan Sandström (1874–1947), certainly did. They were responsible for modifying Bjerknes's 'fundamental theorem' for use with oceanographic data, capitalizing on Bjerknes's theorem and the quantity and quality of synoptic information on northern European seas that had been accumulating in Scandinavia for the preceding decade.

Bjørn Helland-Hansen (fig. 3.6) came to oceanography by an indirect route.[99] Born and schooled in Christiania, he began to study medicine there in 1898. But early in 1898, while an assistant studying the aurora on an expedition in Finnmark under Kristian Birkeland, he froze his fingers and had to have some amputated. His prospects as a physician ruined, Helland-Hansen turned to the marine sciences, studying briefly in Copenhagen with the chemist Martin Knudsen before joining the new Norwegian Fisheries Directorate under Johan Hjort as a hydrographer. During the next five years, he was at sea constantly, mainly on the new research vessel *Michael Sars*, collaborating at first with Fridtjof Nansen on extensive surveys of the Norwegian Sea.[100] It was at this time, in 1901, that Nansen arranged to have Helland-Hansen, who was naturally gifted mathematically, study with Bjerknes in Stockholm. There he met Johan Sandström and began the collaboration that turned physical oceanography definitively away from its descriptive origins.

Johan Sandström (fig. 3.7) came to their joint work from a very different direction. Bright but untutored, he was sent in 1898 from his job in a sawmill in northern Sweden to study in the Stockholm högskola, where he came to the attention of Vilhelm Bjerknes because of his mathematical talents.[101] Soon essential to the development of Bjerknes's mathematical meteorology, Sandström was employed full-time by him under a grant from the Carnegie Institution of Washington from 1906 until 1908,

3.6 The Norwegian oceanographer Bjørn Helland-Hansen (1877–1957), who, with Johan Sandström, provided the means of converting Vilhelm Bjerknes's circulation theorem into a practical mathematical tool for calculating the direction and velocity of ocean currents. Helland-Hansen's Geophysical Institute in Bergen became the major centre for the early spread of dynamic oceanography. From Mosby 1958, p. 321.

3.7 Johan Sandström (1874–1947), Swedish oceanographer and meteorologist, who, with the Norwegian Bjørn Helland-Hansen, developed Vilhelm Bjerknes's circulation theorem into a practical mathematical tool for calculating the direction and velocity of ocean currents. Courtesy of Dr Artur Svansson, Göteborg, Sweden.

when he left to join the Swedish Hydrographic-Biologic Commission. He headed the Meteorological Division of the Swedish Meteorological and Hydrological Institute from 1919 to 1939. In 1901, Sandström, like Helland-Hansen, found himself in the enviable position of conscious innovator in a new science.

The best-known result of Sandström and Helland-Hansen's collaboration, the monograph *Über die Berechnung von Meeresströmungen* (On the calculation of ocean currents), published in the reports of the Norwegian Fisheries Directorate in 1903, bears little indication of the reasons that the two came to work together. Best known in the English translation prepared by D'Arcy Wentworth Thompson in 1905,[102] it opens with the statement that 'in a discussion of the laws that govern the movements of the sea, we must continually hark back to the fundamental principles of mechanics. But we have so far a choice of ways, that we may either apply these laws directly in their original and simple form, or transmute them first into new and more developed forms, designed for the end in view,' and cites Mohn's work as an instance of an analysis of oceanic motions involving 'very laborious numerical calculations' that could aptly be simplified. In his correspondence with D'Arcy Thompson in 1903, Helland-Hansen remarks of his work in the Shetland-Faroe Channel:

> ... I have tried to work up the results as extensively as possible in order to lay the fundament [*sic*] for future investigations. It is of course easily done to write a paper in the ordinary manner only using the ordinary methods of interpretation of the observations of salinity and temperature. I have made a large lot of dynamical calculations, and in introduction I have given some general remarks on these calculations. I hope, then, to be able to tell new things of interest, and to show that hydrographers ought to pay more attention to the mathematical analysis than hitherto done in order to find the variation and the development of the currents ...[103]

He was making it clear that practical results would ensue from the use of an appropriately designed analysis of physical data from the oceans – and, with the hubris of a bright young man at the forefront of a breakthrough in research methods, that he had little sympathy with the conservatism of his predecessors.

Über die Berechnung von Meeresströmungen is largely a reprise, explication, and expansion of Bjerknes's paper of 1898 on the application of his fundamental theorem to meteorology and oceanography, with a simplified method of calculating the number of solenoids in the ocean (including tables to simplify calculations) and clearly worked examples based upon data from Norwegian surveys in the open Atlantic. It concludes with an analysis that Sandström came back to later in his career, a calculation of the meridional circulation of the Atlantic between the equator and high northern latitudes based upon the postulated baroclinic distribution of density and pressure between the surface and 1,000 metres. But, as events show, Helland-Hansen was not satisfied with this clear presentation and modification of Bjerknes's approach. Even as his 1903 monograph with Sandström was being translated by D'Arcy Thompson, he was at work simplifying the calculations by concentrating on the cases (most common in the ocean, as he saw it) when currents were steady, not accelerated, and when friction could be ignored.[104]

For simplicity, Helland-Hansen considered two stations with vertical lines of data, joined by two horizontal lines, one at or near the surface, the other at a significant depth below it. Then from the earlier paper, or using new sets of tables, the number of solenoids between the two stations in each depth interval, governing the total circulation, could be calculated easily. The effect of the Earth's rotation could also be simplified by considering only horizontal currents perpendicular to the plane of the two stations. Using the new formulation, it was a routine matter to calculate the difference of current speed between surface and the depths across the plane joining the two stations.[105] Drily, he commented that 'at our present stage, in calculating the oceanic movements, I think that equation (3) [the one described above] will give valuable results, until *Bjerknes's* complete equation may some day be practically used with all its factors.'[106] Within two years, Helland-Hansen's reformulation of Bjerknes's theorem was applied again by his Scottish colleague A.J. Robertson to calculate currents in the Faroe-Shetland Channel in 1904–5,[107] and in succeeding decades by many others in Europe and North America in their own regions. As Joseph Proudman wrote in 1953, 'The application of this formula caused an new epoch in physical oceanography, and it has been widely used ever since.'[108]

For decades following 1905, Helland-Hansen's equation, or its variants, served as a relatively straightforward way of converting data on pressure and density (usually calculated from measurements of temperature and salinity) into estimates of current velocity and direction.[109] Physical oceanography now had a mathematical technique and a mode of practice based upon hydrodynamic theory, originating, admittedly, in a curiously convoluted way from classical physics, which were soon put into use in several settings.

4

Evangelizing in the Wilderness: Dynamic Oceanography Comes to Canada

The Editorial Committee has had before it some of the printed signatures of the report on the investigations conducted by Dr. Johan Hjort, of Bergen, Norway, into 'Age and Growth of the Herring in Canadian Waters.' An examination of the report, as printed, shows that it is of a most abstruse and technical character, and will be of use to only a very limited number of persons in the Dominion. From an economic stand-point the value of the report to the fishing industry of Canada will be nil.

Editorial Committee on Governmental Publications
to Deputy Minister of Naval Service, 20 December 1919.[1]

Science in a Scientific Wilderness

It is a surprise – and seemingly a paradox – to find the first full-scale ap-plication of the mathematical analysis of ocean currents begun by Mohn and Bjerknes (as simplified by Sandström and Helland-Hansen) in Can-ada rather than in the more scientifically sophisticated northern Euro-pean nations. In fact, as succeeding chapters will show, nowhere, initially, was there a community of interest sufficiently trained in mathematics or accustomed to quantitative ways of thought, nor perhaps with a pressing need, to give the new method easy acceptance. That it came to Canada first was something of a historical accident having to do with the needs of the country and the ambitions of two fisheries biologists, E.E. Prince and Johan Hjort. Their interests were expressed in the Canadian Fisheries Expedition of 1915 and its report, published four years later.[2] The first oceanographic station in Canadian history, on Bradelle Bank just west of the Magdalen Islands, on 11 May 1915, patterned on the example set by the quarterly cruises of the ICES nations, was the first step in an attempt to apply European techniques of quantitative physical oceanography in the New World.

Canada at the outbreak of the First World War in 1914 was a small but rapidly growing country in the throes of rapid urbanization and industrialization.[3] To Wilfrid Laurier, the Liberal prime minister from 1896 to 1911, the twentieth century would be Canada's.[4] The modern map of Canada was complete by 1905. The country had expanded in terms of both inhabited territory and population as a result of government-encouraged immigration from Europe during the first decade of the new century (population increased from just over 5 million in 1901 to more than 7.2 million in 1911). Both Montreal (which was the nation's largest city) and Toronto had populations of more than half a million and were expanding fast as immigration began to shift from the agricultural western provinces to the eastern cities, where artisans and factory workers could find employment. But regional imbalances were evident: for example, 60 per cent of Canada's population in 1911 lived in the central provinces of Ontario and Quebec, and only 11 per cent in the resource-rich but depressed Maritime Provinces.[5] More than 900,000 left the East Coast region in the decade 1911–21.[6]

On the Atlantic coast, the fishery was a significant part of a larger problem: an abundance of resources such as forests and fish, but no industrial base. Nonetheless, the figures were not inconsequential: the total value of the Canadian fishery in 1909–10 was $30,000,000, of which $16,000,000 came from the Atlantic coast.[7] A significant increase had taken place in the Atlantic fishery between the 1880s and the beginning of the First World War.[8] Groundfish catches were relatively stable, but for a time good incomes were made in the lobster fishery, the basis of a thriving canning industry. But rapid changes took place just after the turn of the century, as the fishing industry had to adapt to changing market preferences, notably the shift from salt fish to fresh, and the introduction of new technologies, including beam trawling, steam trawlers, and gas engines.[9]

Canadian science did not have much to add to knowledge of the eastern Canadian fishery. Science in Canadian universities was sparse: before the First World War, universities were few and mainly concerned with undergraduate education. Only McGill and Toronto had major graduate programs at the turn of the century; for example, at the University of Toronto there were fewer than four hundred students pursuing graduate degrees, and the first research doctorate was awarded in 1900, nearly a quarter of a century after the development of graduate

schools in the United States.[10] There were no research grants until a
limited program began under the newly established Honorary Advisory
Council for Scientific and Industrial Research (the ancestor of the National Research Council of Canada) in 1916, with nearly all grants being
made in physics, chemistry, and engineering.[11] In 1914, under the Conservative government of Robert Borden, Canada seemed a very poor
candidate for any kind of innovation in science, whether related to its
resources or not.

Background of the Canadian Fisheries Expedition

At Confederation in 1867 the responsibility for fisheries was granted to
the federal government. It took responsibility for administering a lais-
sez-faire industry, which on the East Coast was rapidly divided between
big fish companies originating in Britain and thousands of small-scale
and frequently poverty-stricken individual fishermen. Fishery regulations were increased, and implemented by a fleet of Canadian government fisheries patrol vessels serving as a surrogate navy.[12] The Fisheries
Branch, at first part of the Department of Marine and Fisheries established just after Confederation, sat awkwardly with other agencies concerned with shipping, ports, lighthouses, and other aids to navigation,
and was for a time (1914–20) shunted into the Department of the Naval Service, responsible for the new military service. Science, apart from
some rudimentary listing of resources and a small fisheries museum in
Ottawa, was well outside the mandate, as well as the interests, of most
fisheries officials in Ottawa and their political masters.[13]

But there is more to be said. Since 1842, long before Confederation,
the Geological Survey of Canada had shown that research and practicality could be united (admittedly with politically awkward episodes) in a
way that brought credit to the nation and an international reputation in
geology.[14] Situated outside the main departmental lines of administration, for many years it found a way of satisfying the practical needs of
the mining industry while keeping and adding to a staff that was increasingly devoted to the research ethic of pure science. More than fifty years
later, a similar initiative began in the marine sciences, centred around
the need for a Canadian biological station and the desire within the Department of Marine and Fisheries for a scientific expert to work on the
increasingly complex biological problems of the fisheries.[15] Acting on
advice from a well-known Scottish fisheries and marine biologist, W.C.
McIntosh of the University of St Andrews, the department appointed

the English-born biologist Edward E. Prince (1858–1936) to the newly created position of Dominion commissioner of fisheries.[16]

Within a very short time, Prince began to campaign for the establishment of a biological station in Canada, a subject that had been on the minds of Canadian biologists for nearly a decade.[17] As the most fashionable and widely known big scientific innovations of the late nineteenth century, marine biological stations in Europe had proliferated and were actively used by North American biologists to promote their careers.[18] In North America, Canadians were especially familiar with Johns Hopkins University's Chesapeake Biological Laboratory and the expanding Marine Biological Laboratory at Woods Hole, Massachusetts. A station at home would afford them facilities for research – and would afford Prince the opportunity to add the capacity for fisheries-related research carried out by eager university biologists to the accomplishments of his department. With political finesse and hard work in Ottawa, Prince and his colleagues from the universities were successful in obtaining a financial appropriation for a biological station in 1898, to be under the direction of a (very cumbersomely named) Board of Management of the Marine Biological Station for Canada. Its first accomplishment was the construction and use of a floating biological station, which, beginning in 1899, was towed from one site of interest to another until it was damaged in a storm in 1907.[19] The floating station, having proved the utility of fisheries research, at least to the satisfaction of Prince and the university teachers who spent their summers in pleasant East Coast haunts, was succeeded by two small permanent stations in 1908, one on the East Coast at St Andrews, New Brunswick, the other on the West Coast at Nanaimo, British Columbia. The Board of Management, under new governmental regulations streamlining its administration, became the Biological Board of Canada in 1912, and was administered by Prince until 1921, just three years before his retirement.[20]

In the years just before the First World War, Canadian fisheries caused some unease to officials like J.J. Cowie (1870–1943),[21] who had been brought from Scotland by Prince to improve the handling of the herring catch. In a generally sanguine account of the East Coast fishery to the Canadian Commission of Conservation in 1912, he summarized the lack of progress in the industry:

> In the course of the fifteen years from 1870 to 1885, a steady advance was maintained in the value and importance of the fisheries of the four eastern provinces. The value of all kinds of fish caught by the fishermen of those

provinces during the former year amounted to $6,312,409, while in the latter year the value rose to no less than $14,780,584. In all of the four provinces there were employed in 1870, on board of vessels and boats, 27,385 fishermen, while at the end of the fifteen-year period in 1885, the number had increased to 51,498. In looking at the results for the period from 1885 to 1910, the discovery is made that little or no progress had taken place during those twenty-five years. The aggregate value in 1910 shows an increase of only $834,900 over that in 1885, while the increase in the number of men engaged only amounted to 683 in the course of a quarter of a century.[22]

The reasons lay not in any decline of the stocks, for both he and Prince agreed that the capacity of fish to reproduce far outstripped the ability of fishermen to catch them,[23] but in the primitive conditions of the fishery. Commenting on the catch of cod in the Gulf of St Lawrence, he said that 'there can be no doubt that but for the inferior type of boat used, and the fact that many of the fishermen around the shores of the Gulf cease operations during the very height of the season to attend to the work of the farm, the value of the cod fishery of that portion of our coasts could be enormously increased.'[24] Herring were an even more striking case, for the fishermen were catching only a small fraction of the available stock by restricting their operations to nearshore areas rather than following the fish in their movements to open waters. Moreover, the processing of the catch was haphazard, resulting in marketable products of very low quality compared to those from Scotland and Norway.[25]

E.E. Prince, speaking to the Commission of Conservation three years later, expressed his concerns clearly:

> That Canada should possess one of the most wonderful herring resources in the world in her Atlantic and Pacific coastal waters, but that her herring industries should rank as wholly inferior in value and reputation, has been an anomaly difficult to understand, and still more difficult to explain. Why is it that Scottish and Norwegian herring should have such a high reputation that the herring fisheries of these two countries approach the total value of the whole of Canada's fisheries? According to the answers frequently given, the quality of our Atlantic and Pacific coast herring, in a fresh condition, is very inferior ... It is, however, an erroneous assumption that the fresh Canadian herring are not of the very best quality.[26]

The problem lay in the nature of the fishery and in the preparation of the fish once caught. Canadian fishermen restricted their catch to large herring, as a result of the type of nets they used and keeping their catches to nearshore; they then preserved and barrelled the catch poorly. By changing the fishery to an offshore one based on younger and more desirable fish (called 'fat herring' in Europe) and by producing a better product, 'a vast herring industry' could be developed in Canada. The time was ripe, for, as D. J. Byrne, the president of the Canadian Fisheries Association, pointed out at the same meeting, the war had created a fish shortage in Europe that could be made up very profitably by exporting high-quality Canadian fish to Britain and elsewhere. As Prince wrote to his minister, J.D. Hazen, on 10 December 1914, 'The present time is, unquestionably, most opportune for initiating such a development, as the European war has practically stopped the Great North Sea Fisheries, and next year and for some years to come, most countries of Europe and South America will have no fish, or such reduced supplies that Canada could find an illimitable market for her herring and other fish.'[27]

To Prince, contemplating the unsatisfactory state of Canadian fisheries, particularly the herring fishery, so crucial for food and bait, improvements were unlikely from the resources of the Department of Marine and Fisheries alone. Several years earlier, he had attempted to get high-powered help from Europe without success; in 1914 he renewed his attempt to interest the eminent Norwegian fishery biologist (and director of fisheries for Norway) Johan Hjort in coming to Canada to develop the herring fishery upon a base of new knowledge. As Prince saw the situation, it was '... above all things necessary to discover the schools and determine the migrations of the esteemed "fat" herring and, in practice, to adopt better methods of capture and handling, curing, and packing, so that our Canadian herring industry may rise to the front rank in the herring fisheries of the world.'[28]

In 1914, Johan Hjort (1869–1948) (fig. 4.1) was at a crucial time in his career.[29] Born in Christiania (now Oslo), he had toyed with medicine as a student before going on in his real interest, zoology, in Munich and at the Stazione Zoologica in Naples. He returned to Christiania as lecturer in the university, curator of its zoological museum, and in 1894 successor to G.O. Sars as fisheries investigator. After further study in Germany, Hjort was back in Norway again in 1897 as director of the university's marine station at Drøbak and in 1900 was appointed to a new position directing fisheries research, based intially in Bergen. There, where he arranged the construction of the new research vessel *Michael Sars,* Hjort stimulated the research of an outstanding group of biologists, whose interests turned rapidly to the factors causing variations in the abundances

4.1 An officer (right) and Johan Hjort (left) probably on the deck of the CSS Acadia, summer 1915, during the Canadian Fisheries Expedition. Photograph probably by A.G. Huntsman. From Photograph Archives, St Andrews Biological Station, Fisheries and Oceans Canada. With permission of Department of Fisheries and Oceans.

of marine fish species, notably cod and herring.[30] In 1910, when the Scottish marine biologist John Murray, famed for his role on the *Challenger* expedition thirty-five years earlier, offered to pay for the ship to do open-ocean research throughout the North Atlantic, Hjort accepted and accompanied Murray on the cruise. One outcome, under both their names, was the very influential book *The Depths of the Ocean* (1912), which for three decades was the most authoritative treatment of oceanography in English, and which Hjort regarded as a blueprint of the way comprehensive fisheries research should be conducted.

Hjort had been deeply involved, along with his countryman Nansen, in the negotiations leading to the founding of the International Council for the Exploration of the Sea (ICES) in 1902. This alone, without his other activities, would have given him some international standing, but undoubtedly the greatest boost to his reputation – and in some quarters his notoriety – was his development, based first on the early work in Bergen, of the year-class hypothesis of variations in fish abundance. Summarized in his great monograph 'Fluctuations in the Great Fisheries of Northern Europe,' he presented overwhelming evidence that cod and herring catches varied because of the differential survival of only a few years' recruitment, for example, the 1904 year class of Norwegian herring, which dominated the fishery for many years.[31]

The basis of Hjort's wide-ranging monograph was the ability to age fish using seasonal changes in the marks on their scales representing annual growth. This method, originating with the German biologists Carl Hoffbauer and Johannes Reibisch,[32] was exploited by Hjort's co-workers, notably Hjalmar Broch, Knut Dahl, Désiré Damas, and later Einar Lea, to show that not only herring but also cod and haddock could be aged using scale markings and that all showed evidence of variations in the abundances of fish produced in different years. For Norwegian herring, in particular, the 1904 year class outnumbered all others and was still the dominant group being caught in 1914.[33] What was less clear was an explanation of the success of particular year classes like the 1904 one. But Hjort, like his junior colleagues, was convinced that, as he said, the explanation had to be sought in 'changes in the sea itself,' perhaps in the fluctuations of the Gulf Stream (actually the North Atlantic Current) off Norway, and that this was to be an important goal of Norwegian fishery investigations using their research vessel *Michael Sars*.[34]

Based upon the earliest oceanographic sections made from the new vessel, extending from Sognefjord to near Iceland and back to Lofoten, Helland-Hansen and Nansen concluded that, taking time-lags into consideration, there was evidence for close causal links between variations in the temperature of the North Atlantic Current and the condition of cod at Lofoten, the temperature variations themselves apparently linked to what they called 'cosmic causes,' that is, sunspot cycles.[35] Hjort disagreed, showing that the years producing the largest year classes of fish (notably the 1904 year class of cod) were ones in which the fewest eggs had been produced. Clearly the strength of year classes was not related directly to the spawning of the fish, and that to water temperature, but to some other factor or factors occurring later in the development of the young fish.[36] Reviewing the conclusions of the work leading up to his 1914 monograph, Hjort drew the lines clearly between his position and an earlier orthodoxy in fisheries biology:

> The opinion generally prevalent hitherto was that the renewal of the stock of fish took place, as in the case of the increase of any human population, by means of a more or less constant annual increment in the form of new individuals; the results here arrived at, however, indicate that this renewal, in the case of the species investigated [herring and cod], is of a highly irregular nature. At certain intervals year classes arise which far exceed the average in point of numbers, and during their lifetime, this numerical superiority affects the general character of the stock, both as regards quantity and quality, thus again exerting a decisive influence upon the yield of fisheries in both respects.[37]

But if Helland-Hansen and Nansen were wrong about the causes that resulted in variation in fish stocks, what could those causes be?

Hjort's view, based upon an intimate knowledge of cod and herring abundances in the Northeast Atlantic, was that the size of fish populations was determined sometime after spawning (that is, after the egg stage) but before the young fish were large enough for capture in the fishery. Italicizing his opinion for emphasis, he said,

> *The rich year classes ... appear to make their presence felt when still quite young; in other words, the numerical value of a year class is apparently determined at a very early stage, and continues in approximately the same relation to that of other year classes throughout the life of the individuals.*[38]

But how young was that 'very early stage'? There appeared to be a 'most critical period':

> Nothing is known with certainty as to this; such data as are available, however, appear to indicate *the very earliest larval and young fry stages* as most important.[39]

These young stages would be subject to changes in food and in water currents. Taking evidence from observations in nature and from fish culture, Hjort noted that mortality of young fish was highest at the stage when they had just shifted from dependence on the yolk-sac to feeding. If suitable food were not available, the young fish would not survive. Timing was everything:

> If the time when the eggs of the fish are spawned, and the time of occurrence of ... plant growth both be variable, it is hardly likely that both would always correspond in point of time and manner. It may well be imagined, for instance, that a certain – though possibly brief – lapse of time might occur between the period when the young larvae first require extraneous nourishment, and the period when such nourishment is first available. If so, it is highly probable that an enormous mortality would result. It would then also be easy to understand that even the richest spawning might yield but a poor amount of fish, while poorer spawning, taking place at a time more favourable in respect of the future nourishment of the young larvae, might often produce the richest year classes.[40]

But nothing was known about how larval fish nutrition, and survival in general, might be affected by the current regime in which the young fish found themselves. Admitting ignorance, Hjort pointed out that 'it would be especially desirable to ascertain the extent of such movement, and how far the young fry are able to return, of their own volition, to such localities as offer favourable conditions for their further growth.'[41] This was no easy task, nor was it one that could be tackled adequately by single disciplines. Hjort called for solutions to the problem of year-class variation to be sought by the development of what he called 'a very extensive plan':

> A study of the fluctuations in the populations of the sea, both fish and smaller organisms, and thus of the whole organic life existent in the ocean, is therefore the soundest possible basis for marine research, whether with theoretical or practical ends in view. There is moreover, scarcely any other question which is so well calculated to focus the attention of men engaged upon different branches of science, as this must necessarily be the case where several investigators are at work on board the same vessel.[42]

It was at this time that the invitation came from Canada to investigate the Canadian herring fishery. Buoyed by the interest in his hypotheses and on the alert for opportunities to find more evidence for them, Hjort found the invitation to work on Canadian herring too good to resist.[43]

With Prince and the deputy minister responsible for fisheries, G.J. Desbarats, he began the organization of a major study of eastern Canadian herring populations.[44]

The Canadian Fisheries Expedition of 1915

Hjort credited his interest in coming to Canada to study herring to his observations during the *Michael Sars* expedition with John Murray in 1910, when the ship had stopped briefly in Newfoundland:

> On this cruise, from the Sargasso sea to Newfoundland ... we had occasion ... to make certain observations, the results of which still further convinced me of the great and peculiar interest attaching to such a comparison as that mentioned [between fish stocks and their environments on both sides of the North Atlantic]. Indeed, this last cruise in itself sufficed to show in what unique degree the waters of the coasts of Canada and Newfoundland were suited to the study of those very problems which have ranked foremost in the Scandinavian marine researches of the past generation, to wit, the relation between the distribution and life-cycle of the organisms, on the one hand, and the prevalent physical conditions in the sea on the other.[45]

The strong contrast of environmental conditions along the east coast of Canada, he seems to have been saying, would provide a test of whether the abundance of fish was governed to a greater extent by physical conditions, or by the biological factors that he favoured and had provided evidence for in 'The Great Fisheries of Northern Europe.'

At the invitation of the Biological Board, Hjort came to Canada in October 1914. With characteristic vigour, he was soon visiting coastal communities from Massachusetts to Newfoundland, and arranging to have samples of herring sent to him from the whole area. Working through the late autumn and winter in the University of Toronto, as guest of the zoologist A.G. Huntsman[46] (who was also curator of the Biological Board's St Andrews biological station), he began the analysis of age of the preserved fish, based mainly upon scale and length analysis, that would enable him to look for comparisons with European stocks. His report on the winter's work showed, in a preliminary way, that eastern Canadian herring fell into at least four distinct groups from western Newfoundland to southwestern Nova Scotia, each with a restricted geographical range and distinctive year-class strength. To his delight, the fish from western Newfoundland, most like Norwegian herring in their size and growth, also were dominated by the 1904 year class.[47] With such an encouraging start to his work in Canada, by no later than early November Hjort was working out with Prince and the secretary-treasurer of the Biological Board, A.B. Macallum, plans for a full-scale fisheries expedition centring on the Gulf of St Lawrence (where the greatest variation in herring stocks occurred, as well as the greatest environmental variation) in the summer of 1915.[48]

Hjort probably come to Canada at the outset with the idea of an expedition in mind. But, practically minded as always, and realizing the problems that Canadian fishermen faced in getting a good product to market and in preserving bait, he also intended to sell the Canadian government an inexpensive means of freezing fish. Curiously, the deputy minister of the Department of Fisheries, G.J. Desbarats, who was quite willing to give financial and logistic support to an expedition, balked at the practical proposal aimed at aiding fishermen. The expedition nearly foundered on this rock, and was only saved when Hjort gave up his idea of introducing the freezing method to Canada.[49] He concentrated, perforce, on the organization of the expedition, which was intended, as he said, to 'give important information regarding the spawning areas of the most important fishes, inside and outside the gulf [sic] of St. Lawrence, and determine hydrographical and biological conditions in this great fishing area.'[50]

In March 1915, Hjort proposed to Desbarats that the expedition should involve fishing surveys, to be carried out by the Department's herring drifter No. 33,[51] and hydrographic and biological collections, to be carried out by a small staff on at least two fisheries protections steamers doing three series of transects of the Gulf of St Lawrence and its inflowing water. These would be matched to the breeding periods of the main fish species and to seasonal changes in oceanographic conditions:

> An investigation along these lines would give three cross sections of the outflowing waters of the St. Lawrence river and of the Atlantic water flowing into the Gulf and provide full opportunities for definitely determining the old questions of the connection between the polar water (coming southwards along the coast of New Foundland [sic]) and the Atlantic and the Gulf of St. Lawrence water. The investigation would further give important information regarding the spawning areas of the most important fishes inside and outside the Gulf of St. Lawrence, and determine the hydrographical and biological conditions in this great fishing area.[52]

4.2 Canadian personnel of the Canadian Fisheries Expedition, summer 1915. From left to right, Arthur Willey, A.G. Huntsman, and James W. Mavor, apparently posing in the hold of one of the participating ships. From Photograph Archives, St Andrews Biological Station, Fisheries and Oceans Canada. With permission of Department of Fisheries and Oceans.

The ships would be outfitted with the most modern equipment, typical of that used by the ICES nations in the Northeast Atlantic, ranging from herring drift nets to the latest version of Nansen's water bottle carrying state-of-the-art German reversing thermometers.[53] The counts of plankton and fish eggs, the measurements of herring and cod, as well as the chemical determination of the salinity of the water, crucial to determining its density, would be carried out on land at a base laboratory in the little town of Souris, Prince Edward Island. Hjort insisted that along with the personnel he brought from Norway, his assistant Paul Bjerkan[54] (to do the chemistry), and the experienced fishing captain Thor Iversen to handle the herring surveys, there be Canadian scientists involved in all aspects of the work. Huntsman got leave from his university job for the summer of 1915, and he was joined by Arthur Willey, Strathcona Professor of Zoology at McGill University, and a young fisheries biologist, J.W. Mavor, then at the University of Wisconsin (fig. 4.2).[55]

4.3 The CSS *Acadia* at Souris, Prince Edward Island, during the Canadian Fisheries Expedition, summer 1915. *Acadia* was the newly built flagship of the Canadian Hydrographic Service, diverted from charting for part of the summer to do an oceanographic survey of the Scotian Shelf under Johan Hjort. Photograph probably by A.G. Huntsman. From Photograph Archives, St Andrews Biological Station, Fisheries and Oceans Canada. With permission of Department of Fisheries and Oceans.

In the event, the regular series of cruises that Hjort had hoped for in May, June, and July-August could not be carried out because of problems scheduling the government ships *Princess* and *Acadia* (fig. 4.3). Another problem was ice, which choked the southern Gulf of St Lawrence in early May. Eventually, *Princess* crept out of Charlottetown through the receding ice to make two initial oceanographic stations near the Magdalen Islands (Îles de la Madeleine) on May 11, the first of their kind in Canadian history and the most detailed made to that date in North America.[56] Thereafter, the expedition proceeded without undue problems, beginning with an ambitious survey of the Scotian Shelf by *Acadia* in late May and early June, followed by further cruises by both ships until the middle of August, and a final examination of the hydrography of Cabot Strait, the gateway to the Gulf of St Lawrence by Captain Frederick Anderson of *Acadia* in mid-November. Hjort left for Norway in late August, bearing large amounts of data from more than 160 oceanographic stations (fig. 4.4) and the urgent necessity to prepare it for publication by the Canadian government.

Johan Hjort: Solving 'the Biological-Hydrographical Problem'

Delayed by the involvement of Johan Hjort in fisheries problems involving Norway, Britain, and Germany during the war, the report of the Canadian Fisheries Expedition appeared in print in 1919. Despite the scepticism of some Ottawa bureaucrats that it had any value (see the epigraph to this chapter), it was a substantial volume bearing a great deal of scientific information more or less related to Canadian fisheries.[57] First, it was clear, just as Hjort had suspected, that a variety of herring stocks existed in Canadian waters, and that all of them showed the same kind of year-class variations documented in Europe. Einar Lea's detailed examination of the herring collections resulted in a long chapter that was virtually a textbook of scale-analysis and a detailed account of the rate of growth and other biological characteristics of the stocks. Another Norwegian, Alf Dannevig, charted the distribution of fish eggs and larvae. Huntsman and Willey contributed information on the distribution and abundance of the zooplankton (particularly chaetognaths and copepods) of the Gulf of St Lawrence and Scotian Shelf, and H.H. Gran, a distinguished biological oceanographer from the University of Oslo, examined the phytoplankton and protozoan collections, showing that the phytoplankton flora was depauperate compared to Europe, was made up mainly of northern species, and bloomed later than in northern European waters.[58] Hjort's assistant Bjerkan, who had done many of the chemical analyses in Souris, summarized the hydrographic data, tabulated temperature, salinity, and density, and developed a three-dimensional scheme of the distribution of the water masses of the Gulf and Scotian Shelf.

Curiously different from all the other monographs in the report of the expedition is J.W. Sandström's contribution, 'The Hydrodynamics of Canadian Atlantic Waters.' E.E. Prince, who wrote the preface to the report, refers to this lengthy monograph on dynamic oceanography sparingly, with little apparent knowledge of its meaning, and mainly in relation to the static structure of the water column. But clearly Hjort had something else in mind in asking the principal Swedish disciple of

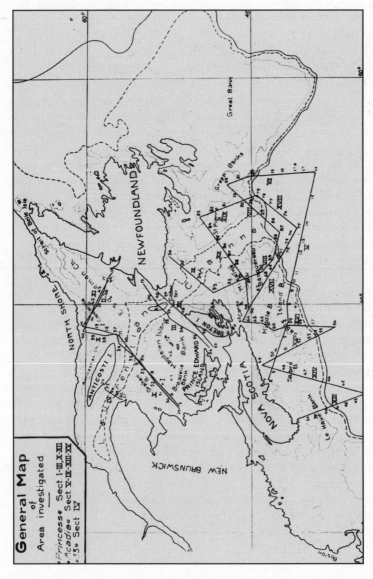

4.4 Cruise tracks and oceanographic stations of the CSS *Acadia* (on the Scotian Shelf and south of Newfoundland) and the Fisheries Protection Steamer *Princess* (in the Gulf of St Lawrence) during the Canadian Fisheries Expedition, spring and summer 1915. From Bjerkan 1919.

Vilhelm Bjerknes to provide a dynamic analysis of water movement.[59] Although nowhere does he explicitly spell out the relationship between the hydrographic (i.e., physical oceanographic) analyses and the biology of fishes, one can construct a reasonable idea of what he had in mind. Referring to the complexity of vertical structure in the Gulf of St Lawrence, Hjort called the variation of oceanographic properties 'of the highest importance to the study of all biological questions.' And making specific reference to the work of Bjerkan and Sandström, he states again the close link between biological features such as fish production and the physical properties of the sea:

> ... it would be well to ascertain what fluctuations may occur from the conditions found to prevail in 1915. What variation can take place, for instance, in the amount of fresh water discharged by the St. Lawrence river, in the Gaspé current, in the interchange of water between the gulf of St. Lawrence and the area outside, in the great cold intermediate water layer, in the Labrador current, and in the distance of the warm ocean water from the coastal banks? All of these questions will naturally be of the highest importance in the study of biological problems, chiefly, perhaps, the varying distribution of the cold water layers.

A little earlier, he had said that as the result of the expedition,

> ... we had now succeeded in procuring a material which enabled us to measure the thickness of the various water layers by sections taken transversely to their direction of movement, and this in the shortest possible time ... This furnished us with the basis for a theoretical treatment of the material, and an analysis of the influence exerted by the various factors; the earth's rotation, melting of the ice, specific gravity, temperature, etc. In Mr. Bjerkan's paper, the reader will find all the precise data concerning the values for salinity, temperature, and density. And Mr. Sandström has endeavoured, on a wide scale, to give a thorough analysis of the causes conducive to the circulation of the water as a whole, and its dynamics generally. This is, as far as my experience goes, the most thorough treatment of these questions which has yet appeared.[60]

Several years later, in an encomium to Otto Pettersson, Hjort made it clear what he was after in having Sandström do a dynamic analysis:

> ... I have advanced the hypothesis that the great fluctuations in the stock of

fish in northern waters, which the study of the yearclasses has revealed, are due to the variations from year to year in the hydrographical conditions, on which the development of the food of the larval fish and as a consequence the success of the years [*sic*] production of fish depends.[61]

No further wide-scale study of eastern Canadian waters took place during Hjort's lifetime, but had there been another, Hjort would certainly have used its results to further his aim of understanding year-class variation, that is, solving what he called 'the biological-hydrographical problem.'[62] But what of Sandström? What reasons can this physical scientist, working professionally as a meteorologist, have had to carry out a very time-consuming, highly detailed dynamic analysis, based on Bjerknes's theorem, of a remote and seemingly unimportant area of the North Atlantic Ocean?

The Hydrodynamics of Canadian Atlantic Waters

In introducing his analysis of the physical oceanography of eastern Canadian waters, Sandström complained that 'the scanty leisure left me by my official duties and other obligations'[63] had made it difficult for him to prepare the monograph. But the length, organization, and complexity of this monograph indicate that its author believed it was of high importance. That importance was certainly not its relevance to the fisheries of a small and distant nation, but, in Sandström's eyes, the opportunity to provide a full account of how Bjerknes's circulation theory (as he called it) could lead to striking, original, and frequently counterintuitive analyses of the effects of physical forces on oceanic circulation:

... the popular but erroneous theory of currents due to the action of wind, has here been abandoned ... These phenomena [for example, currents running against the wind], which are explained by Bjerknes' circulation theory, give the observer generally an impression that the water in the sea has a very remarkable and unexpected tendency to opposition against the forces striving to act upon it. The usual primitive conception as to the laws which govern movement in liquids will not suffice to explain such paradoxical phenomena; it will be necessary to introduce new ideas, for the proper comprehension of which a considerable amount of mathematical knowledge will be required ... [I]t seemed to me of the highest importance that the interested observers in the Canadian Atlantic waters, and also in other parts of the globe, should be able to familiarize themselves with these new principles, and I have therefore endeavoured to get around the mathematical difficulties as far as possible by extensive use of graphical illustrations, and by reference to well-known principles ... Such limitation and simplification will, I trust, render these complicated yet interesting theories accessible to a wider circle among those at all occupied with marine phenomena.[64]

In short, Sandström was presenting himself as the evangelist of the method originating with Bjerknes, using the opportunity provided by the Canadian Fisheries Expedition to write the first textbook of dynamic oceanography.

The opportunities presented by Hjort's data from eastern Canada were unparalleled. Summarizing the significance of his studies, Sandström concluded that

within the area of these investigations, from its boundary on the Gulf Stream side to the mouth of the St. Lawrence, a number of mostly [sic] interesting phenomena are encountered, which render the waters in question one of the most instructive fields on the face of the globe for hydrographical and hydrodynamic research.[65]

Not the least importance of the area was the chance to study an ocean current (the Gaspé Current) along its whole length, and the opportunity to see what happened when the Labrador Current and Gulf Stream met south of Newfoundland.[66] Hjort had welcomed the opportunity to test his year-class hypothesis in an area so different hydrographically from Norwegian waters; Sandström found the area equally significant in applying dynamic physical oceanography. Referring to the quality of the data, he said that 'the material collected by Dr. Hjort from the Canadian waters is ... beyond question the best that has ever been obtained from the atmosphere and the sea ... From a dynamic point of view, Hjort's measurements leave absolutely nothing to be desired.'[67] Here, then, was a magnificent opportunity to test and promote dynamic oceanography:

I purpose [sic], then ... to apply Bjerknes' circulation theory, as far as can possibly be done, to the present material, drawing such conclusions as thence be arrived at, and finally indicating what yet remains to be done in order that the system may be utilized to its fullest extent, and by the gradual adaptation of measuring instruments, methods of observation, etc., directed towards the solution of further oceanographical problems.[68]

How this was done in 'The Hydrodynamics of Canadian Atlantic Waters' reveals the state of dynamic oceanography in 1919, the need to promote quantitative methods to the uncomprehending, and some of the strains that existed in the small community of practitioners recognizing their importance.

The actual delineation of currents on the east coast of Canada plays a small part in the text of Sandström's lengthy monograph. The physical results are tabulated extensively but not discussed in great detail, and the currents are depicted mainly in two plates indicating differences from spring to summer in the strength of the Gaspé Current, the circulation through Cabot Strait, currents of the Scotian Shelf, and the details of the Slope Water south of Nova Scotia.[69] The seventy pages of text concentrate instead mainly on the physics of fluids, the principles lying behind the dynamics of ocean circulation, and the development of Bjerknes's theorem into a means of calculation, although it appears in detail only after forty pages of preliminaries.

At the outset, Sandström sets out to build a background of physical principles that apply to the sea as fluid in motion, starting with the effects of the rotating Earth, which cause currents in the northern hemisphere to appear to turn to the right. In addition to this basic feature of motions on a rotating sphere, the Earth, he took pains to emphasize the insignificance of the vertical dimension (compared to the horizontal) in the sea, and what he called 'the obstinacy with which sea water opposes the action of external forces.'[70] Then, using the data gathered by Hjort and his colleagues, he introduced the calculation of specific volume (the reciprocal of density, namely, volume per unit mass), the basic datum for the calculation of the density and pressure fields in the sea, and thus the current velocity. The distribution of specific volume, shown as isosteres,[71] is of the greatest importance, and he shows in simple examples how the redistribution of density (or its reciprocal, specific volume) within the water column can result in apparently counterintuitive motions when wind speed changes[72] as a result of imbalances in the buoyancy of the water. The most important feature in this dynamic adjustment of the water is the slope of the isosteric surfaces:

> The difference of level in the separating surfaces in the sea is of the same importance to the movement of sea-water as the varying level of the surface of a river to the movement of the latter. The force impelling a sea current may be calculated from the slope of the separating surface [between layers of different density or specific volume] in the same manner in which the force of a river's current is calculated from the slope of its surface water.

This was of the greatest importance, for, as he said, 'we can thus, from the form of the isosteric surfaces, discover what forces are acting upon the water.'[73] Always an experimentalist, Sandström then introduced the subject of how the distribution of density would be affected by the wind, using illustrations from his laboratory in which wind blew across small tanks of stratified seawater. Not only could the wind produce redistributions of density within the water column, resulting in complex currents, it also could produce large internal displacements along density surfaces, that is, it could produce internal waves.

With considerable insight, Sandström suggested that the widespread inability to understand the effect of the Earth's rotation 'is due to the fact that we have no senses for the direct perception of the rotation of the earth. All our direct perceptions indicate the earth as motionless.'[74] Nonetheless, it was essential to understand this effect, for from it came the general pattern of cyclonic and anticyclonic gyral circulations in the great ocean basins and the distribution of density surfaces in the water column characteristic of each.[75] And another factor with the potential to affect the motion of water significantly was melting ice, as his influential countryman Otto Pettersson claimed. This appeared to be a major factor causing the formation of the cold intermediate layer in the Gulf of St Lawrence, one of the most dramatic and puzzling discoveries of Hjort's expedition.[76] And all these motions were greatly affected by friction within the water, one of the most difficult factors to deal with because there were no direct means of measuring it; instead, it had to be inferred from the vertical distribution of velocities in moving water, or calculated from Bjerknes's equation when all the other variables were known.[77]

From this foundation of physical principles, Sandström then began to look at the big picture of the factors causing ocean currents. Both the wind and variations in the density of water were implicated:

> The causes which give rise to currents in the sea are either external forces, such as the action of the wind, or physical changes in the sea-water itself, occasioning an alteration of its specific gravity. We have thus two distinct categories of ocean currents, differing widely in character ...[78]

He was not reluctant to attribute important effects to the wind, such as the formation of the great anticyclonic gyre of the Sargasso Sea, in which a mass of warm and saline water extended to great depths as a result of wind stress and the rotation of the Earth (the Coriolis effect), but it is clear that he was more impressed by the effect of differences in density, and that he regarded the wind as a secondary geophysical factor compared to the great redistributions of water brought about by variations in specific volume, which he expressed as the number of solenoids (Bjerknes's term – see chapter 3):

> Currents arising from physical changes in sea-water ... are otherwise constituted [than wind-driven currents], and behave in a very different way. They are not restricted to a single layer, but may traverse several such [by convective and other motions], and thus have far greater freedom of movement than the wind currents. Consequently it is upon the former [physical differences in seawater – i.e., the number of solenoids or baroclinicity] that the task of bringing about interchange between the waters of different regions and different depths devolves.[79]

His interpretations of the distribution of density and its effect on baroclinicity, the results of laboratory experiments in which heat and cold were applied, and Sandström's unwavering commitment to the approach introduced by Bjerknes led him to an unequivocal conclusion: 'how insignificant is the part played by the wind' in producing the major systems of oceanic circulation, except in shallow surface layers. 'Bjerknes's circulation theory,' as he called it, provided all else.[80]

After this lengthy presentation of principles, Sandström finally presents 'Bjerknes' theory of circulation, with the clear and profound insight which it displays' into the causes of oceanic circulation, first describing the dynamic consequences of baroclinicity, that is, the intersection of specific volume and pressure surfaces (isosteres and isobars) in simple diagrams, and then in terms of the acceleration of the circulation of the resulting solenoids, as described by Bjerknes's theorem (see chapter 3), resulting in the equation

$$\frac{dC}{dt} = A - 2\varpi \frac{dS}{dt} - R$$

where C = circulation, A = the number of solenoids, ϖ = the angular velocity of the Earth, S = the area of a circle of latitude, and R = friction. He commented that 'this formula contains all that influences the circulation of the water in the sea,' and that 'Bjerknes' circulation theory will, it may safely be said, play a prominent part in future oceanographical investigations, on account of the ease with which it may be applied to hydrographical observation material, and the clearness and direct practical value of the results thereby obtained.'[81] From this development of the circulation formula, which in this presentation is rudimentary and must have left his mathematically unsophisticated readers confused, he went on then to show, quite clearly, how to calculate the number of solenoids (A in the equation above) in the waters of the Gulf of St Lawrence and Scotian Shelf.[82] This was the basis of current diagrams for spring and summer, indicating for the first time the overall current patterns off the east coast of Canada.[83]

Simplification of the basic equation was possible if one assumed that the currents were in a steady state, that is, were unaccelerated, and that the term for friction was small in comparison to the others. The result was an equation suitable for calculating current velocities, v, in which the only terms were the number of solenoids (determined from density and pressure in the water column) and the effect of the Earth's rotation upon water tending to circulate around them (the denominator, in which φ is the latitude):

$$v = \frac{A}{2\varpi \sin \varphi}$$

This was a result having similar form to, and the same purposes as, Helland-Hansen's simplification of Bjerknes's equation in 1905,[84] aptly characterized by Joseph Proudman decades later as bringing about 'a new epoch in physical oceanography.'[85]

Almost as an afterthought, Johan Sandström ended 'The Hydrodynamics of Canadian Atlantic Waters' with discussions of a few specialized topics that interested him, such as problems of calculating or measuring friction and its effects on circulation, the results of changes in acceleration (rather than assuming steady-state currents), the topography of the sea surface, and the distribution of pressure (closely similar to Mohn's analysis more than thirty years before, but without mention of Mohn), and calculations of the work involved in water movement. His summary finally brought together in only four pages the practical implications of his laborious calculations involving the interactions of the Gulf Stream and the Labrador Current, the land-hugging course of the Labrador Current (bringing ice close to shore along the east coast of Newfoundland), the dynamics of the coupled circulations of the Atlantic surface water entering the Gulf of St Lawrence through Cabot Strait and the formation of the Gaspé Current, the formation of the intermediate layer of the Gulf, and the significance of the deep Atlantic water flowing at great depth up the Laurentian Channel into the nearly enclosed Gulf. But these are only side issues, one feels, in a monograph devoted to spreading the word about a new, quantitative, and as yet unappreciated, way of investigating ocean circulation.

A Voice Crying in the Wilderness

My analogy of Johan Sandström with John the Baptist should not be overdone. In Europe just after the First World War, dynamic oceanography was recognized, if not fully understood or much used, by many in the marine science community. Bjerknes's accomplishment and its development by Helland-Hansen and Sandström were respected, but, admittedly, practised only in Scandinavia. It seems reasonable to see 'The Hydrodynamics of Canadian Atlantic Waters' as an attempt by one of those in the vanguard to build a greater understanding of a remarkable theoretical and practical advance in the physics of the ocean. Looked at only in this way, some aspects of this great monograph are hard to understand.

Sandström had been chosen by Johan Hjort to apply dynamic techniques to his data from eastern Canadian waters because Hjort and Helland-Hansen had fallen out. Sandström's text shows evidence of the rush that he felt to complete it: the absence of any references, the lack of explicit linkages between some of the sections, his rudimentary (even confusing) derivation of Bjerknes's theorem, and his sketchy account of the circulation of Canadian waters. More curious is the lack of any reference to his former colleague Helland-Hansen, whose simplification of Bjerknes's theorem provided the simplest way of using oceanographic data to calculate current velocity and direction. Nor is there even indirect mention of the only previous study of the same magnitude using dynamic methods, Helland-Hansen and Nansen's 'The Norwegian Sea,' published ten years before. But it seems characteristic of the self-sufficient Sandström that he took pains to develop the monograph on his own terms, rather than that he too had fallen out with Helland-Hansen. The result was that 'The Hydrodynamics of Canadian Atlantic Waters' lacks the clarity so characteristic of Helland-Hansen's 1905 monograph and many later publications, when Helland-Hansen had become the best-known exponent of the revolution begun by Vilhelm Bjerknes.[86]

Nonetheless, Sandström's monograph was an important one, increasing in influence for many years after publication.[87] But its presentation – and the circumstances of its production – make it unclear who were 'the interested observers in the Canadian Atlantic waters, and also on other parts of the globe'[88] mentioned in its introduction. He may have been positioning himself in relation to Helland-Hansen, who was well established in physical oceanography in Bergen, and who in 1919 became director of the new Geophysical Institute there. Or he may have been consolidating himself in Stockholm, where in 1919 he became the director of the Swedish Meteorological and Hydrographical Institute. Perhaps his intention was purely pedagogical, to carry the message about dynamical oceanography to the unconverted in Europe and the New World. If the last was the case, the effect in Canada was singularly unimpressive.

Although the opinion of the Canadian government's Editorial Committee that this 'abstruse and technical' report would be of no use to Canadians was overly pessimistic, even philistine, there was a grain of truth in this assessment.[89] Canadian science, even in the most sympathetic circles such as the Biological Board, was not prepared for a mathematical treatment of oceanographic problems. Canadian marine science in the early decades of the twentieth century was struggling out of the 'inventory' phase characteristic of all Canadian science during the previous century. This found its strength in providing an accounting of Canadian resources under practical, political, and more abstract intellectual pressures to strengthen an emerging nation.[90] E.E. Prince himself, although sympathetic to the expedition and its scientific aims, always took the high road politically, emphasizing the need (certainly real) to improve the lot of fishermen and the importance (also real, but politically loaded) of expanding the fisheries.[91] Within the universities, the only locales of pure research, the emphasis at least until the onset of the 1930s was largely on building fashionable experimental science – physics, chemistry, physiology, and biochemistry – imported actually or by example from Britain.[92]

The Biological Board was unusual in having had direct contact with one of the leaders of European marine science, Johan Hjort. But as an assemblage of individuals it was singularly unprepared to make a leap into quantitative physical oceanography. Until permanent employees resident in the biological stations were hired by the Board beginning in 1928, all its personnel were seasonal 'volunteers,' mainly from the ranks of university zoology or biology departments.[93] Most were mathematically unsophisticated, if not untutored, and certainly not able to handle the kind of analysis outlined by Sandström in 1919, if it had even seemed relevant. Until 1928 there was not a single physically trained person employed by the Board, and it was only then that the approach promoted by Hjort and exemplified in Sandström's monograph was applied to Canadian waters (as described in chapter 8).

An exception to these generalizations was A.G. Huntsman, who was profoundly affected by his summer with Hjort on the Canadian Fisheries Expedition. No more physically trained than any of his contemporaries, nonetheless Huntsman saw the potential of European survey methods for Canadian marine science. For nearly two decades, he patterned Board-sponsored research on the grand scale introduced by Hjort. In a series of summer expeditions – to southwestern Nova Scotia and New Brunswick in 1916, to the southern Gulf of St Lawrence in 1917, to the Miramichi River in 1918, to the Bay of Fundy coast of Nova Scotia in 1919, and to the Strait of Belle Isle (with physical oceanography specifically in mind) in 1923 – Huntsman attempted to follow the example set by Hjort and the Canadian Fisheries Expedition.[94] He failed, in large part, not for lack of ambition but because of the scarcity of scientific resources[95] and because as the 'one man band' of eastern Canadian marine science he could never bring any big project to full completion by himself. His resources – and his time – were totally inadequate to the kinds and magnitudes of the tasks introduced by Hjort. Even a promised dynamic analysis of the Bay of Fundy, based on six transects made during the summer of 1919, was never completed.[96]

Thus the results of the Canadian Fisheries Expedition, notably Johan Sandström's introduction of Scandinavian mathematical physical oceanographic techniques, lay fallow in Canada for more than a decade. The small scientific community in Canada of the early 1920s was simply unprepared in background, attitude, and resources for this kind of science. The sophistication of Canadian science increased rapidly in the years after the First World War, but its scientific aims were directed into research far from the geophysics of the atmosphere and ocean. This might be attributed to the lack of scientific sophistication in a small nation like Canada, but as the chapters to follow show, even nations regarded as scientific powerhouses between the wars, for example, Germany, France, and the United States, had their problems assimilating mathematical physical oceanography into the received canons of scientific practice. Canada was not alone.

5

'Physische Meereskunde': From Geography to Physical Oceanography in Berlin, 1900–1935

Oceanic circulation is a fundamental problem of marine science. Because it influences directly and decisively the climatically determined pattern of temperature and salt content and the distribution of gas content and thereby influences indirectly to a great degree the occurrence and abundance of the inhabitants of the sea, a solution of most of the problems of the sea without a complete understanding of oceanic circulation is inconceivable.

Alfred Merz and Georg Wüst 1922, p. 1.

The Birth of a New Discipline in Germany

During the first three decades of the twentieth century, quantitative physical oceanography was born. Its parents, applied mathematics and physical geography, were of European origin, but the new science, deriving its main energy from the work of Vilhelm Bjerknes, Bjørn Helland-Hansen, Johan Sandström, and Otto Pettersson in Norway and Sweden, found multiple foster homes in Western Europe and North America, especially Norway, Canada, Germany, and the United States. As the preceding chapter indicated, the reception of dynamical oceanography was not the beginning of a quick march to triumph, but rather the halting accommodation of established scientific practices to unfamiliar and unsettling new approaches to the oceans. In each national setting, older practices were not rapidly replaced, but often underwent slow evolutionary changes characteristic of the places where the changes occurred.

The early development of mathematical physical oceanography in Germany is of special interest because of the long tradition there of work on physical geography, including the configuration of ocean basins and the delineation of surface currents.[1] A very large number of technical and semi-popular books on the geography of the oceans, often called 'physische Meereskunde,' appeared in Germany from the mid nineteenth century well into the twentieth, only a few of which are Jilek's *Lehrbuch der Oceanographie* (1857), Gareis and Becker's *Zur Physiographie des Meeres* (1867), Kohl's *Geschichte des Golfstroms* (1868), Kayser's *Physik des Meeres* (1873), Witte's *Über Meeresströmungen* (1878), Attlmayr's *Handbuch der Oceanographie* (1883), Boguslawski's *Handbuch der Oceanographie* (1884), Walther's *Allgemeine Meereskunde* (1893), Janson's *Meeresforschung und Meeresleben* (1901), and the first edition of Schott's *Physische Meereskunde* (1903). Representative of a much longer list that indicates the German public's nearly insatiable interest in popular or technical but digestible science, these were descriptive and non-quantitative treatises, representing a long tradition of German travel, exploration, and geographical scholarship. But although some of these, and many not listed, were merely genre works, aimed at a curious public, few did not have reasonable scientific respectability and many were unashamedly didactic. Throughout this extensive literature, when the word 'Oceanographie' (later 'Ozeanographie')[2] was used, it represented what is now called in English 'hydrography,' the charting and tabulation of the properties of the oceans.

Of a different nature are the books by the great geographer (and later, unofficially, oceanographer) of the University of Kiel, Otto Krümmel (1854–1912): *Die aequatorialen Meeresströmungen des Atlantischen Oceans* (1879); *Versuch einer vergleichenden Morphologie des Meeresräume* (1879); *Der Ozean* (1886, second edition in 1902); and *Handbuch der Ozeanographie, Band II* (with Boguslawski, 1887). Krümmel's magisterial successors to Boguslawski's text of 1884, which appeared in 1907 and 1911, were of a different order too. They summarized very recent scientific work on the oceans, including the introduction of dynamic oceanography, marking the beginning of a transition from physical geography of the oceans to a physics-based ocean dynamics. No one had greater potential to introduce new ideas in oceanography to German physical geographers of the ocean than Krümmel, who prided himself on a synoptic knowledge of the most recent scientific literature. But his early death, as professor of geography in Marburg, ended his influence on an early succession of students and on his scientific peers.[3]

More than twenty years after Krümmel's introduction of Scandinavian dynamical oceanography to German readers, the physical study of the oceans was still in transition in Germany. Writing in 1928, Albert Defant

(1884–1974), newly appointed *Professor ordinarius* of oceanography and director of the Institut für Meereskunde in Berlin, described the situation:

> The scientific study of the oceans finds itself in a state of development that in the first place will be promoted by the provision of new observational material from modern geophysical viewpoints.[4]

In this chapter, I investigate the transition in Berlin from descriptive studies of the oceans toward (although not fully to) Defant's ideal of a geophysically based science. At the beginning of the twentieth century, German oceanographers found themselves in a well-defined intellectual realm in which there was agreement about the basic aims of geography, although often debate about the details, interpretations, and the weight given to conflicting theoretical approaches within that field as it applied to the oceans. By the 1930s, German oceanography had largely separated from its ancestral roots in physical geography. But this was in a hesitant and uncertain way as new shared assumptions, common approaches, and controversies over theory and practice developed, appropriate to a new discipline struggling to emerge from such well-constructed foundations.

Origin and Early Work of the Institut für Meereskunde in Berlin

When it began in 1900 and formally opened in 1901, the Institut für Meereskunde (Institute of Marine Sciences) of the University of Berlin represented an amalgam of scholarly, educational, and political aims,[5] a union, shifting with time, of the ideas of its founding director, Ferdinand von Richthofen (1833–1905),[6] and of the Imperial German Navy. In his lecture *Das Meer und die Kunde vom Meer*,[7] Richthofen viewed natural science as fruitfully removing the curbs that *Naturphilosophie* had placed on thought in Germany. Among its useful empirical results was an increasing interest in the sea, which was of importance even – perhaps especially – to inlanders like the nearly landlocked Germans. Sea commerce brought wealth, well-being, and power; these were the prerequisites of art, literature, science, and the spread of culture. Britain had dominated power over the oceans and knowledge of them, but as he spoke (said Richthofen) there were signs of a maritime awakening in Germany brought about through trade and naval power. The new Institut für Meereskunde had the goal of furthering popular knowledge of the sea and its significance to Germany. Although he did not live to see its opening in 1906, the Museum für Meereskunde planned by Richthofen, associated with the Institute, had an even greater role in spreading knowledge of German destiny on the oceans.[8] Writing of German expansionism at the turn of the century, Alexandra Richie tell us that

> it was during this period that the museums of Ethnology, Arts and Crafts, the Colonial Museum and the Natural History Museum grew most rapidly, and even the Maritime Museum, built in 1906, was little more than an excuse to present more nationalistic propaganda about the need for a large German navy to defend the new colonies or trade routes.[9]

But the reality was more complex, as the relation between these two branches of the University of Berlin bears out.

Richthofen's successor in 1906 as *Professor ordinarius* of geography and director of the Museum für Meereskunde was the distinguished geomorphologist and geographer Albrecht Penck (1858–1945).[10] Until his retirement in 1921, Penck promoted Richthofen's ideals of scientific study of the oceans and popular education. As he wrote in 1910, 'The point of view expressed in the founding of the Institute and Museum of Marine Science was in the first instance nationalistic.'[11] But to Penck, a distinguished scientist, a scientific program was equally important, and he promoted it with determination.

Nonetheless, even with the best will and the expenditure of much effort, Berlin's distance from the sea made the execution of the ideals of scientific research difficult in Berlin: 'It is not so difficult if one concerns oneself with the theoretical side [of research]. But the present status of oceanography demands before all else that observational material be broadened.'[12] Given the right person, however, the difficulties could be overcome. With the arrival in Berlin in 1906 of Alfred Grund (1874–1914), formerly Penck's student in Vienna, as *ausserordentliche Professor* and director of the Geographical–Natural Science Division of the Institute, teaching, field observation, and research in aquatic sciences began in earnest.[13] Grund established an oceanography laboratory, led students on trips to the Sakrower See near Potsdam[14] and to Norway and the North Sea, and proposed to the Biologische Anstalt Helgoland, a marine laboratory located off Germany's North Sea coast, a program of hydrographic investigations based on North Sea lightships. During the late summer of 1907, Grund studied oceanographic techniques in Norway as a participant in one of the annual courses taught at Bergens Museum by Bjørn Helland-Hansen, Johan Hjort, and others, beginning

5.1 Alfred Merz (centre) and Fritz Spiess (right), the captain of SMS *Meteor*, probably in 1925, before the beginning of the *Meteor* expedition. From Georg Wüst photographs, SIO Archives 91-1, Scripps Institution of Oceanography Archives, UC San Diego Libraries, with permission.

a link between the Institute and Helland-Hansen that lasted until the 1930s. But Grund's stay in Berlin was brief. In 1910 he took up the chair of geography in Prague. He was killed in action four years later during the early days of the Great War. His successor in the Institute was another of Albrecht Penck's ex-students, Alfred Merz.

Alfred Merz and Berlin Oceanography

During his student days in Vienna, Alfred Merz (1880–1925) (fig. 5.1) had visited the Adriatic Sea. There he began investigations of short-term variations in temperature and salinity, some of them due to internal waves. This work became his *Habilitationsschrift* in Berlin in 1910,[15] permitting his appointment as *Privatdozent* (equivalent to a university lectureship) in addition to his duties as a division director within the Institute.[16] Like Grund, Merz had an interest in the open sea, but no ready means of studying it during his early years in Berlin. He followed in Grund's footsteps, using the lakes near Berlin, including the Sakrower See, as substitutes for the oceans for teaching and research.

He soon took over Grund's studies based on lightships, expanding them into long-term work on tidal currents and hydrographic properties of the water column.[17] Using instruments developed for work by the International Council for the Exploration of the Sea, he developed refinements of thermometers and current meters for these studies. Merz's connections with the Imperial Navy prospered and strengthened too; his hydrographic work made increasing use of small naval vessels, and during the war he prepared tidal charts to aid submarine warfare.

Merz's aims were set out in an early publication, 'Berliner Seenstudien und Meeresforschung' (1912), and another early paper, 'Neue Anschauungen über das nordatlantische Stromsystem,'[18] reveals his scientific orientation and oceanographic knowledge. To Merz, although limnology might be an enforced replacement for open-ocean studies, lakes had their advantages because they were relatively simple systems uncomplicated by tides and salinity. They were equivalent, as he said, to simplification of a theorem by the mathematician or to experiment by the physicist. But lakes were not to be viewed as oversimple; he regarded them as systems within which water properties were linked closely with sediments and with organisms. As a result, to solve limnological problems (later he applied the same reasoning to the open ocean), it was necessary to have closely spaced time-series data on physical and chemical properties throughout the year and at all depths in addition to geological and biological information. Merz's first results, based on fifty days of work between April 1910 and May 1911 on lakes near Berlin, resulted in a reinterpretation of thermocline formation.[19] But his horizons were wider: even as his work on lakes and the coastal North Sea developed, Merz worked with students in the Berlin Aquarium's marine laboratory at Rovigno on the Adriatic,[20] continuing there the time-series observations that were the basis of his *Habilitationsschrift*.

It is clear too that Merz was paying close attention to theoretical physical oceanography during his first years in Berlin, especially the problem of the causation of ocean currents, stating that 'oceanographers have always actively concerned themselves with the problem of ocean currents, but without completely solving it and without conclusively resolving conflict of opinions.'[21] Directing his attention to Nansen's monograph 'The

Waters of the Northeastern North Atlantic' (1913), Merz reviewed the contributions of Nansen and especially of V.W. Ekman[22] to the mathematical formulation of wind-driven ocean currents, and Nansen's contention that oceanic deep water was formed by cooling and convection in the North Atlantic near the southern tip of Greenland.[23] He explored briefly the effect of the Earth's rotation and of internal pressure gradients on ocean currents, basing his comments on work by Vilhelm Bjerknes and especially upon the formulation of the dynamic method by Helland-Hansen and Sandström.[24] Reading widely in the English- and German-language literature on physical oceanography, Alfred Merz had extended his ideas well beyond the limits of the Sakrower See only a year or two after his arrival in Berlin.

Probably as early as 1911, certainly by 1912, Merz had begun to conceive of a major German oceanographic expedition to investigate the ideas suggested by Nansen and treated theoretically (on the basis of limited observations) by Ekman, Helland-Hansen, and Sandström. In 'Berliner Seenstudien,' concluding the scientific discussion with more general remarks, he stated that a major deep-sea expedition was 'the most urgent need today in oceanography.' Not only would such an expedition yield the richest scientific results, it would have a deeper significance for Germany: 'I believe that now the German people should grasp anew the opportunity to secure success and to add a great feat to its oceanographic accomplishments.'[25]

Oceanography and Germany's cultural and political roles on the world stage were never far from Alfred Merz's thoughts. In the scarcely unbiased eyes of Albrecht Penck,[26] Merz brought 'his own secure judgement' to the problems of nationality, nationhood, political boundaries, and the purported effect of geographical determinism on each. In this, he fitted well into the professorial culture of the University of Berlin, characterized by Alexandra Richie as a prestigious conservative class that

taught their students to admire the military men and the industrialists who ultimately directed the policy of nationalism and chauvinism from their offices a few blocks away from the university. By the turn of the century the dominant culture in Berlin stood for colonial expansion, naval supremacy, suppression of Polish separatism, a 'place in the sun,' national unity in the face of the Social Democratic menace; German *Kultur* should be passed on to other nations ...[27]

Well before the political, social, and scientific turmoil of the Weimar years,[28] Merz, like many of his colleagues,[29] had developed a firm view of Germany's cultural and scientific destiny. Reviewing Kjellen's *Die Grossmächte der Gegenwart* (1914) and Hettner's *Englands Herrschaft und der Krieg* (1915) in 1915, before the war had begun to go against Germany, Merz viewed the essence of national greatness, not in purely geographical boundaries, but in the production and development of German-influenced culture, which could and would lead to political unity. Where geographically discrete boundaries, strong cultural influences, and the will to power coincided, a great power might arise transcending mere narrow nationalism. England's power was based upon its natural resources (especially coal and iron) and on its worldwide shipping from many ports. But, according to Hettner's view – and Merz agreed – developments in transportation (not to speak of English parochialism and complacency) had swept the world (certainly Germany) past England into the twentieth century.[30] Although there is little or no evidence that Alfred Merz subscribed to the shriller versions of pan-Germanism, his general writings (and those of Albrecht Penck even more markedly) reveal a persistent strong belief that it was Germany's destiny to dominate a European cultural and scientific stage if not to expand into a middle European *Lebensraum*.[31]

By the end of the Great War, under the draconian provisions of the Treaty of Versailles, Merz, concentrating increasingly on his ambitions for a great German oceanographic expedition, appears to have accepted, along with many other German scientists, 'the proposition that scientific great power status could function as *substitute* for political great power status.'[32] Under the disastrous circumstances of post-war Germany,[33] little else was possible to Merz but a concerted attempt to place Germany in a pre-eminent position in oceanography.

Georg Wüst, Alfred Merz, and Oceanic Deep Circulation

Between 1910 and 1913, Georg Wüst (1890–1977) was a student of physics in the University of Berlin,[34] where, according to his own account,[35] he attended the lectures of several distinguished teachers, including Albrecht Penck, Alfred Merz, and Max Planck. He soon came under the wing of Merz, gathering data from German lightships and working at Rovigno. His doctoral thesis, on evaporation in the Atlantic, was presented in 1914.[36] For a few months, until beginning duty in a guard regiment in 1914, Wüst was a voluntary assistant to Merz. After further war service as a meteorologist, he returned to Merz as a voluntary assistant in February 1919. Later the same year, he was appointed *Assistent* in the Institut für Meereskunde, the first staff member educated in physics. Moreover, he had gone to Bergen in 1913 to learn dynamic oceanography from Helland-Hansen.[37]

Although Wüst's work for Merz between 1919 and 1924 involved frequent data-gathering trips on naval patrol vessels and hydrographic ships in the North Sea and the Baltic, it grew rapidly into collaboration on the deep circulation of the Atlantic. By 1921 Merz and Wüst had been compiling a card file of all the observations of temperature and salinity (along with other properties) for several years, in preparation for the long-awaited German oceanographic expedition to the Atlantic and Pacific Oceans. Their attention became more and more closely focused on meridional deep circulation of the Atlantic and the resolution of uncertainties about the directions and origins of deep currents. In that year, 1921, Wilhelm Brennecke's long-awaited summary of the oceanographic results of the German Antarctic Expedition on *Deutschland* in 1911–12 appeared, offering a verbal interpretation and a graphical model of the overall circulation of the Atlantic Ocean (fig. 5.2).[38]

Brennecke's model suggested that the bottom of the Atlantic was filled with cold dense water originating in the Antarctic and in northern latitudes (now called Antarctic Bottom Water, AABW, and North Atlantic Deep Water, NADW). At shallower depths were a variety of water masses in motion, including a low-salinity tongue extending from about 50°S to north of the equator (now called Antarctic Intermediate Water, AAIW), and a complex set of shallower currents, apparently wind-driven. Superimposed on the largely north-south pattern of deep circulation was a tendency for upward movement at the equator (indicated by the dotted arrows in fig 5.2), bringing cooler water to the surface and accounting for the puzzling shallowness of the thermocline there.

Brennecke's model was more than a summary – it was a hybrid of circulation models going back into the previous century.[39] Its immediate origin was the analysis made by his colleague at the Deutsche Seewarte (German Marine Observatory), Gerhard Schott, of the results of the German *Valdivia* expedition under Carl Chun down the length of the Atlantic into the Indian Ocean in 1898–9, published in 1902.[40] At that early date, Schott expressed the hope that some day dynamic analyses, of the kind attempted elsewhere by Mohn and Bjerknes, would be possible. But he was descriptive in his own work, invoking the principle of continuity to claim that deep-ocean currents were driven primarily by the wind-driven surface circulation, aided by increases in density at mid-

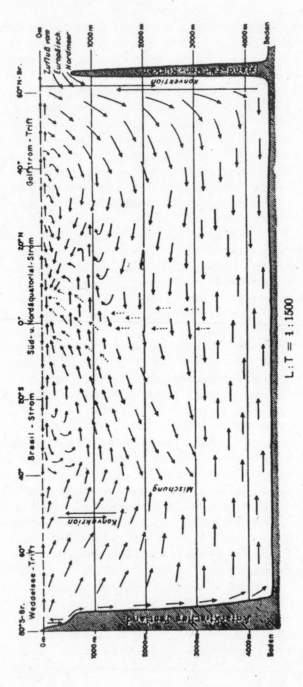

L : T = 1 : 1500

5.2 Meridional circulation of the Atlantic Ocean according to Brennecke 1921, p. 138.

latitudes that caused sinking and the formation of deep water. According to Schott, 'The driving force of the whole system of motion comes from the surface layer.'[41] The result, in Schott's depiction of the Atlantic, was a strikingly symmetrical circulation of the oceans, involving sinking at high latitudes and upward motion at and near the equator (fig. 5.3), closely resembling the pattern proposed by Emil von Lenz (1804–1865) based on his observations during Kotzebue's circumnavigation of the world during 1823–6 (fig. 2.1).[42]

Complexities began to arise even as Schott's report was being published. During the German South Polar Expedition of 1901–3 on *Gauss*, its leader, Erich von Drygalski (1865–1949), noted a salinity minimum associated with a temperature anomaly at 800–900 metres between Brazil and the Cape of Good Hope.[43] He did not attempt an explanation. Only four years later, Brennecke, accompanying *Planet* to the Southwest Pacific, confirmed the observation, noting that the low-salinity water at about 800 metres extended from the south to at least 13°N, beyond which its properties disappeared due to mixing.[44] Here was evidence for a north-to-south intermediate current (AAIW). In summarizing all his work, Brennecke did not depart much from Schott's model, but he added the complexity of mixing and intermediate water formation in the high latitudes of the South Atlantic and localized deep-water formation in higher latitudes of the North Atlantic (fig. 5.4). Astutely, Brennecke summarized the great problems still remaining: 'Where does the sinking from higher to low latitudes begin, what are the forces driving this water exchange, and how can one explain the fact that water under the surface layer in the tropics is colder than that in mid-latitudes?'[45] With far more information on density than was available to Schott, Brennecke invoked density-change and wind in oceanic circulation. In particular, he displaced deep-water formation to high latitudes, pointing out that the great depth of warm water at mid-latitudes (especially in the North Atlantic – the Sargasso Sea) need not result from sinking of dense surface water (caused by evaporation, according to Schott), but appeared anomalous only by contrast to the shallow equatorial surface layer. Winds drove the warm surface waters to high northern and southern latitudes; there cooling occurred, increasing their density, and deep-water formation took place by sinking.[46]

The most versatile, experienced, and insightful oceanographer of his generation, Brennecke accompanied the German Antarctic Expedition on *Deutschland* to the South Atlantic and the Weddell Sea from 1911 to 1913. His first reports on the oceanographic results[47] and a slightly

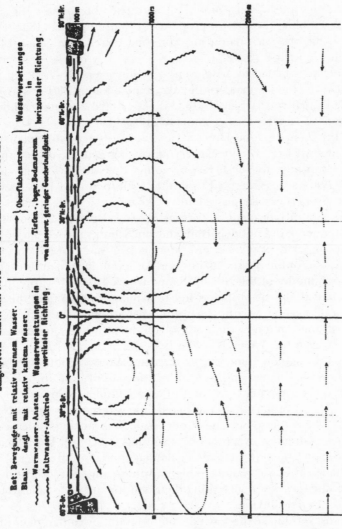

5.3 Meridional circulation of the Atlantic according to Schott 1902, p. 64. Note that sinking, in response to wind-forcing and increases in density, occurs across a broad band of latitude in the North and South Atlantic Ocean.

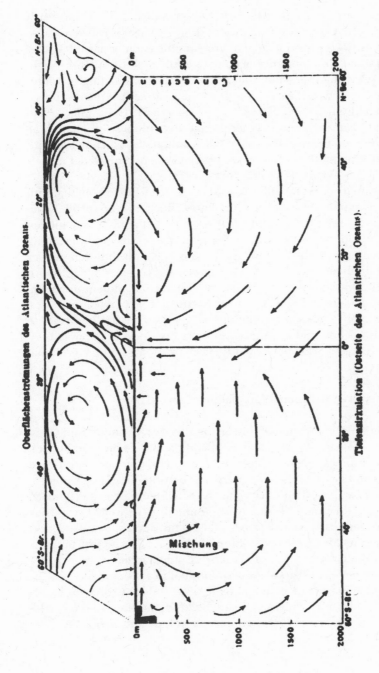

5.4 Atlantic circulation according to Brennecke 1909, p. 98. Note the differences from Schott's model in figure 5.3, especially the increased complexity in the south, including formation of Antarctic Intermediate Water ('Mischung') and the displacement of deep-water formation much farther north and south. But this scheme is still nearly symmetrical about the equator.

later review[48] added detail to his earlier observations from *Planet,* clearly delineating the AAIW (visible in the *Challenger* results thirty-five years before), a northward-directed current crossing the equator at mid-depths. They also showed the presence of cold water of far North Atlantic origin (NADW) overlying even colder and denser water originating by winter cooling in the Weddell Sea. Although his full treatment of the results was not published until 1921 (delayed by the war), by 1914 Brennecke had established the main pattern of meridional circulation in the Atlantic, albeit with debts to the past such as symmetry about the equator, and with uncertainties about the origin of NADW that were not solved definitively until the 1950s and after.[49]

In his obituary of Wilhelm Brennecke, Gerhard Schott wrote that with Brennecke's name was associated 'the knowledge that a powerful, hemispheric water-exchange in mid-depths [of the Atlantic] (and indeed all oceans) from north to south and the reverse occurs' and that Brennecke's results were 'the beginning of a substantial revolution in earlier views and the basis of later work.'[50] This can scarcely be denied. But in 1921, when Brennecke's synthesis of meridional circulation of the Atlantic appeared, Merz and Wüst saw *themselves* in the role of revolutionaries, about to present a new conception of Atlantic circulation. Conflict arose at once with Schott and Brennecke, not least because of the polemical way that Merz and Wüst presented their analysis of Atlantic meridional circulation in 1922.

Controversy over Meridional Circulation of the Atlantic

By 1922, Atlantic circulation had been on Alfred Merz's mind for at least ten years, in preparation for a major expedition that would promote Berlin oceanography and all of German science. His attention had turned from lakes and German coastal waters to the world ocean as a whole, as exemplified by the Atlantic. Despite the uncertainties of the political and financial situation in Germany beginning in 1919, in that year Merz began negotiations with the German Navy to provide a ship and support for what became the *Meteor* expedition of 1925–7. His scientific preparations, with the collaboration of Wüst, were summarized in their paper 'Die atlantische Vertikalzirkulation,' published in 1922.

Merz and Wüst provided an extensive historical study of ideas on Atlantic circulation, including the role of the wind, the effect of the Earth's rotation, and the controversy between Carpenter and Croll about the driving forces of oceanic circulation.[51] The gist of the argument that followed these preliminaries was a striking claim:

> ... observations show that the Atlantic meridional circulation actually consists of water exchange between northern and southern hemispheres. Only the equatorial surface circulation and the water exchange between the middle and polar regions of the North Atlantic are phenomena based on climatic differences intrinsic to the hemispheres concerned. Circulation of the South Atlantic high latitudes has a close relation to water exchange between the two hemispheres.[52]

This was depicted graphically in a diagram showing the meridional (i.e., south-north) patterns of circulation along a vertical section of the Atlantic from the south polar regions to high northern latitudes (fig. 5.5).

Merz and Wüst claimed new and revolutionary status for this model of the Atlantic circulation. Although Brennecke had recognized the significance of trans-equatorial deep circulation (of AAIW, AABW and NADW), he had retained much of the 'old circulation scheme' presented by Schott in 1902. This was of special significance because, according to Merz and Wüst, neither Schott nor Brennecke had paid attention to the 'revolutionary' results of the voyages of HMS *Challenger* (1872–6) and SMS *Gazelle* (1874–6) as outlined by J.Y. Buchanan in 1885 and by Alexander Buchan in 1895 and 1896.[53] Both, Merz and Wüst believed, had not only recorded the properties of AAIW, AABW, and NADW in their reports but *knew of their existence* and had indicated in their work an extensive meridional distribution of distinctive water layers. Thus, Merz and Wüst claimed, Brennecke's scheme, where it showed meridional circulation, was not new, and where it retained Schott's vertical circulation at the equator was wrong. Finally, bringing to bear new evidence on the reality of extensive meridional circulation, Merz and Wüst calculated the pressure distribution throughout the Atlantic, taking 1,200 metres depth as a level of no motion, to show that a general northward motion occurred above 1,000 metres and a southward one at 1,400–4,000 metres, with greatest velocities near the surface and below 2,500 metres.

Merz and Wüst's attack on Schott and Brennecke brought a rapid reply and a response in turn from Merz.[54] The emotional temperature rose quickly in their publications and correspondence. Brennecke and Schott's rejoinder to the first attack on their claims was temperate. They had known of the oceanographic results of the *Challenger* and *Gazelle*

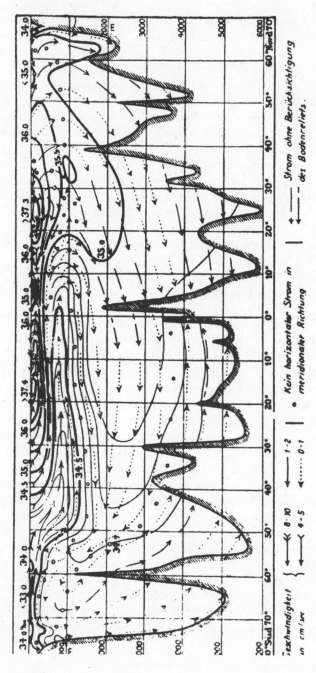

5.5 Meridional circulation of the Atlantic Ocean as described by Merz and Wüst in 1922. This diagram shows the distribution of salinity, and with arrows the pattern of circulation inferred from the distributions of temperature, salinity, and pressure. From Merz 1925, p. 567.

expeditions but had chosen to exclude them because of the possibility of errors in the data.[55] The scientists on *Challenger* and *Gazelle* had used maximum-minimum thermometers reading, at best, to 0.2°C; only on the German South Polar Expedition of 1901–3 had trustworthy modern reversing thermometers reading to 0.01°C been available, making reliable interpretations possible. As for the knowledge of circulation held by Buchanan and Buchan, it was clear to Brennecke and Schott that they knew of a cold northward flow (AABW) along the deep ocean bottom on the western side, but that neither had made an issue of intermediate depth motion from the south (AAIW) or of deep southward motion from the north (NADW). In short, Merz and Wüst were putting words in the mouths of Buchanan and Buchan rather than resurrecting 'revolutionary' but forgotten interpretations.[56] It was, according to Brennecke and Schott, 'grotesque' to read more into the old expedition results than was really there.[57] Moreover, it was unfair to accuse Brennecke of claiming the existence of strong vertical circulation at the equator – his interpretation (fig. 5.2) was that the circulation there was mainly meridional but with a slight upward tendency (indicated in his figure by the dotted arrows).

Analysis of the controversy makes it clear that any decision to accept or reject the data from *Challenger* or *Gazelle* was a matter of scientific judgment rather than a matter of fact. Brennecke and Schott were justifiably critical of the old data, but as Merz pointed out in a reply to their critique, his use of it was not arbitrary.[58] By carefully analysing the internal consistency of the *Challenger* temperature data and by estimating the magnitude of the errors, he claimed that the old data could be used reliably, providing suspect data were removed. This was just what he and Wüst had done.

At this point, late in 1922, with point and counter-point made by the Hamburg and Berlin authors, the controversy might have ended (the editor of the *Zeitschrift der Gesellschaft für Erdkunde* declared it closed, at least in his own journal) had it not been for the intervention by the grand old man of European oceanography, Otto Pettersson, in the *Annalen der Hydrographie und maritimen Meteorologie* (edited by Wilhelm Brennecke).[59] It took dead aim at Merz and Wüst's calculation of the pressure distribution along the length of the Atlantic, from which they had inferred meridional circulation in general agreement with that based on the distribution of temperature and salinity. There were other issues too. Merz and Wüst had criticized Pettersson and Sandström's old paper on Atlantic circulation, published only two years after Schott's monograph on the

Valdivia results, especially the dynamic calculations by Sandström using sparse data from Schott.[60] But to Pettersson there was a more important issue. Why had Merz and Wüst used only the distribution of pressure rather than employing the dynamic method of Bjerknes and Sandström? Referring to dynamical oceanography as 'a finely-polished tool which has put mathematical physics in the hands of oceanographers and cannot be replaced by calculations originating in static or isostatic principles,' he pointed out that oversimple interpretations of oceanic circulation not based upon dynamics were too likely to be wrong or incomplete: instead, 'the characteristics of oceanic circulation must be built upon dynamic principles. *Motu ignorato, natura ignoratur.*'[61] With gentle sarcasm, Pettersson then pointed out that rehabilitating the work of Buchanan and Buchan, based upon data from maximum-minimum thermometers that they had presented but not interpreted, 'would certainly have pleased – and surprised' those 'old masters.'[62]

Bad blood between Merz and Schott was probably not unexpected, given the generational, institutional, and scientific differences between them. Schott, on his side, was highly critical not just of Merz's interpretation of work originating in the Seewarte, but also of Merz's proposal that currents inferred on theoretical grounds be included in nautical charts.[63] Nor was it surprising that Merz saw underhanded activities in the actions of his opponent Brennecke as editor of the *Annalen der Hydrographie*.[64] But to fall afoul of a senior statesman of oceanography like Otto Pettersson was a serious matter and demanded attention. In a short reply, then in a much longer paper, Merz and Wüst defended and refined their use of pressure distribution as an indicator of circulation, concentrating their fire mainly on Johan Sandström's dynamic calculation in 1904 based upon only a few temperature sections and even fewer density values.[65] In this, in their attempt to defend themselves, they appear to have missed Pettersson's point that dynamic calculations took into account more than pressure within the oceans, specifically the motions imparted by baroclinicity and the Earth's rotation.[66] Until the end of his life only a few months later, Merz gave no evidence of understanding the difference between determining the pressure field of the ocean and dynamic calculations. In May1925 he wrote of Mohn's approach (interpreted only as determining the field of pressure)[67] and that of Bjerknes (dynamic calculations) in equal terms.[68] And yet he and Wüst had been shaken by Pettersson's criticism. During June 1923, Merz wrote in a conciliatory way to Pettersson and Sandström, expressing his respect for their work and papering over their differences.[69] And Wüst was impelled to approach dynamic oceanography anew, with a single question in mind: could it be tested empirically and thus be proved useful?

Georg Wüst, the *Meteor* Expedition, and Dynamic Oceanography

Georg Wüst was an anomaly in the Berlin Institute, for he had begun his university career by studying physics, in addition to geography with Penck and Merz. Bjørn Helland-Hansen had been a regular visitor to the Institut für Meereskunde beginning in the academic year 1910/1911. No doubt Wüst was introduced to him by Merz, who supervised Wüst's doctoral research, and in the autumn of 1913 Wüst travelled to Bergen to learn Scandinavian techniques. He returned to Berlin to prepare his doctoral thesis, presenting it in 1914, just before the outbreak of the Great War.

Probably it is not surprising that the pre-war lessons from Norway did not bulk large in Wüst's thought in the early post-war years. Returning to the Berlin Institute in 1919 after his war service, first as an unpaid assistant to Merz, then with a formal appointment as *Assistent,* his work increasingly involved the compilation and analysis of hydrographic data from the deep Atlantic Ocean, in preparation for the expedition that had become Merz's obsession. But as the Institute's only trained physicist, Wüst brought a new viewpoint and a more fully quantitative approach to the study of ocean circulation than his predecessor Grund or his senior colleague Merz. Otto Pettersson's criticism of Merz and Wüst's primitive approach to dynamic calculations in their 1922 paper was a special reproach to the physically trained Wüst. He set out to improve his knowledge of dynamic oceanography and its application by testing its validity through comparison with direct measurements of currents. The applicability and the field assessment of the dynamic method in Wüst's hands by 1924 became especially important as Merz's expedition began to take final shape, for the dynamic method could affect directly the sampling program that Merz was preparing.

Serious planning for the *Meteor* expedition began in 1919 with the designation of a ship and its modification for oceanographic work.[70] Refitting the ship and completion of a cruise plan were delayed by the German economic crisis of 1922, and it was only very late in 1923, after the crisis had ended, that the ship's conversion began and that a scientific program was agreed upon.[71] First, the initial plan to work in the Pacific had to be abandoned because of cost; it was too expensive to replace *Meteor*'s coal-fired engines with the diesel engines needed for very long cruises without fuelling stops. Thus the Atlantic alone, the arena of Merz and Wüst's work during the early 1920s, became the object of the expedition. According to Merz's evolving plan, the acquisition of primary data from the little-known Pacific would be replaced by a very detailed examination of the Atlantic from just north of the equator to the Antarctic (fig. 5.6). A new kind of open-ocean oceanography became possible because of his detailed knowledge of the Atlantic and his determination to use only the most modern instruments and techniques to verify and elaborate the model of circulation that he and Wüst were defending.

More of the same – that is, more hydrographic data adding detail to knowledge of the trans-equatorial circulation of the Atlantic – could readily be gathered during the expedition. But the interpretation of that data by qualitative means (as in all of the previous studies) was only part of Merz's plan. Dynamic oceanography could be used to provide confirming evidence for patterns of circulation and to give evidence of transports. The fourteen transects designed by Merz (see fig. 5.6), extending from the tropical Atlantic to Antarctic waters, were set up so that dynamic calculations could readily be carried out for the whole South Atlantic Ocean.[72] The only question remaining in 1924 – at least in Merz's mind – was whether or not the dynamic method was up to the job. Could it be shown to give the same results as conventional measurements of currents using current meters?

In a classic study in 1924, Wüst calculated the transport of the Florida Current[73] using data gathered between 1867 and 1889 by the United States Coast and Geodetic Survey officers C.D. Sigsbee, J.R. Bartlett, and J.E. Pillsbury. Pillsbury's work was particularly significant, for in 1885, 1886, and 1887 he had anchored the Coast Survey's ship *Blake* directly in the axis of the Florida Current to make direct measurements using current meters of the direction and rate of flow, a *tour de force* of late nineteenth-century deep-sea mooring and instrument use. Thus, for this special region, direct measurements of currents were available and the dynamic method could be used to calculate them indirectly. As Wüst showed, there was good agreement between the two methods, demonstrating what he called 'the utility of the dynamic method':[74]

> The results of calculating velocity from distribution of temperature and salinity under the prevailing conditions agree with results of direct current measurements and current variations. The agreement of results from three completely independent methods is proof of the validity of the data and of the utility of the dynamic method.[75]

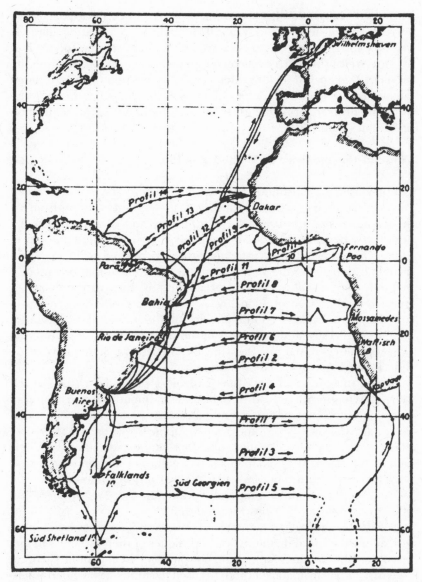

5.6 The sampling plan of the *Meteor* expedition (from Merz 1925, p. 576). The
east-west transects and regular stations (indicated by dots) were intended to add
detail to Merz and Wüst's model of deep Atlantic circulation, published in 1922,
and to allow dynamic calculation of currents.

This was of special significance to the projected work aboard *Meteor,* for currents in the deep ocean, while they might be inferred qualitatively from the distributions of temperature and salinity, were far too weak to be measured using current meters, especially from a ship swinging unpredictably at anchor in 4,000 metres or more of water.[76] As Wüst stated later, referring to current determinations in the deep ocean, 'The solution can only be to bring the general observations under the application of Bjerknes's tables.'[77]

Consolidating the Dynamic Method in Berlin

Georg Wüst's demonstration in 1924 of the close agreement between current measurements made with current meters and estimates from hydrographic properties using the dynamic method confirmed that the sampling scheme devised by Merz for the *Meteor* expedition was appropriate. On a larger stage, Wüst's conclusions were cited again and again in publications from the Berlin Institute as evidence that quantitative methods were useful in physical oceanography. But Wüst, only an *Assistent* in the Berlin Institute in 1924, was not in a position to make profound changes in the scientific practice of a hierarchical institution in a large conservative university, nor in the scientific practices of physical oceanographers raised in a descriptive rather than a quantitative tradition. This role was assumed by another.

During the course of the *Meteor* expedition, in August 1925, Alfred Merz, its scientific director, died in hospital in Buenos Aires. The overall direction of the expedition passed to *Meteor*'s captain, Fritz Spiess, while Wüst took over direction of the oceanographic work at sea.[78] Merz's position as director of the Institut für Meereskunde was filled by Albert Defant (1884–1974), a mathematical meteorologist from Innsbruck. Shortly after the return of *Meteor* in late May 1927, Defant was confirmed as Merz's replacement, and he soon became the first *Professor ordinarius* of oceanography in the University of Berlin.[79]

Defant, evangelically quantitative in his scientific work, was familiar with all the advanced mathematical techniques of fluid dynamics in meteorology and oceanography, notably Bjerknes's circulation theorem (the basis of the dynamic method), as used in studies of the atmosphere and the oceans. He regarded oceanography as being in a more primitive state than the more mathematically advanced meteorology, and viewed it as being in a time of transition to a truly quantitative geophysical science.[80] Defant's education and goals rapidly reinforced the move toward quantitative geophysical approaches begun haltingly by Merz and Wüst. In his monograph 'Physik des Meeres,' he presented the work of V.W. Ekman, Bjerknes, Sandström, and Helland-Hansen as modern standards (based in part on Wüst's test), not as speculative or peripheral.[81] To Defant, *Meteor* lay in a 'transitional' position in the study of the oceans, bridging older qualitative studies and newer systematic and mathematically based ones. The systematic sampling pioneered on *Meteor* would 'lead naturally to a more physical-mathematical treatment of individual problems of marine science, to an intensive application of the principles of modern hydrodynamics of heterogeneous media to the moving water masses of the oceans.'[82]

But Defant was more than a theoretician; he promoted knowledge of the overall structure ('Aufbau') of the oceans, the kind of work carried out before *Meteor* by Schott, Brennecke, Merz, Wüst, and others. This would give a framework for detailed work on smaller areas, or on the mixed layer and the deep waters of the oceans (the 'troposphere' and the 'stratosphere,' by analogy with the atmosphere).[83] His powerful and persuasive viewpoint was presented in a series of studies of the boundaries of water masses,[84] in reviews both technical and semi-popular of the theory of ocean currents,[85] and, of greatest significance, in the textbook *Dynamische Ozeanographie* (1929), the first modern treatment for students of mathematical physical oceanography. This major monograph in his newly adopted discipline expressed Defant's belief that oceanography could only prosper as a fully developed quantitative geophysical science.[86]

His reputation as an atmospheric scientist followed Defant into physical oceanography, transforming the approach to the oceans in Berlin. If the Institut für Meereskunde in Berlin gained the quantitative approach initially from Georg Wüst's test of the dynamic method, it was Defant who consolidated the transformation of physical oceanography there. Only slightly older than Wüst (Defant was forty-three in 1927, Wüst thirty-seven), Defant used the power of his position and the prestige of his reputation in a related discipline, meteorology, to transform permanently the practice of oceanography in Berlin.

In Berlin the dynamic method was an index of a change in approach – a mathematical instrument convenient in oceanographic calculations, and, initially, in promoting Merz's interest in analysing global circulation. By the mid-1930s, if not earlier, dynamic oceanography was accepted as a standard method in Berlin, judging by some of the doctoral theses of that period.[87] But descriptive and even geographically influenced studies, like Wüst's treatise on the origin of North Atlantic deep water, published in 1928, which placed a great deal of emphasis upon configuration of the bottom rather than upon hydrodynamics, were still part of the Berlin approach.[88] In fact, dynamic studies were a rather small component of the doctoral research and other published work that issued from Defant's Berlin Institute in the late 1920s and the 1930s: only four of twelve major studies used the dynamic method between 1923 and 1936. Defant's review in 1939 of German deep-sea oceanography between 1928 and 1938 makes it clear that detailed regional studies of the North Atlantic were the centre of attention. Even Wüst, who had pioneered the use of the method in Germany, used it rather little during his early career, preferring standard hydrographic techniques of water mass analysis (these included the development in the 1930s of his well-known core method).[89] Dynamic oceanography was only one convenient, although very powerful, addition to the *armamentum* of techniques and approaches used in Berlin.

It is not surprising that dynamic oceanography played a relatively small (though significant) part in the post-*Meteor* Institut für Meereskunde. The traditions of the Institute were geographical; long-held scientific attitudes are not cast aside readily. And the dynamic method was not a universal solution to oceanographic problems, merely a useful technique for the determination of currents in somewhat restricted circumstances. It became a part of the oceanographer's tool box, not a means of solving every problem. The real change was not in the acceptance and use of the dynamic method in Berlin (important as that was) but the acceptance of a new, quantitative approach to the oceans in place of the geographical and the solely descriptive approaches – what Defant called 'quantitative comprehension of the phenomena' of the oceans.[90] In this, Bjerknes, Helland-Hansen, and Sandström provided a technique, Merz the preconditions, Wüst a means, and Defant the mechanism (through power and prestige) for a change of practice in a science that had been taught in a well-established qualitative way in Berlin for nearly a quarter of a century.

In the course of these changes, physical oceanography began to change, both practically and conceptually, from a study of the distribution of water properties and currents, to a study of the dynamics of current production. Members of the Institut für Meereskunde in Berlin were not alone in being involved: other groups in Europe and North America had been there first or were going through similar processes of transition. Although each shared the use of a single mathematical approach, the pathways leading to change in physical oceanography were different in the case of each relatively isolated group. In Berlin the path leading to the use of the dynamic method began in Merz and Wüst's conflict with Schott and Brennecke at the Deutsche Seewarte in Hamburg. Had Helland-Hansen not become a regular visitor to Berlin, leading Wüst to learn dynamic oceanography (however imperfectly at first) from its Norwegian master, no one in the Berlin of the 1920s would have been prepared to try it. Alfred Merz, promoting the most up-to-date techniques as necessary for the advancement of German science, was unprepared by background to make the attempt, but he gave latitude and support to Wüst in an academic setting that appeared conservative and restrictive. Finally, Defant's advocacy was essential to consolidation of the change of orientation that had begun with Merz and Wüst's studies of the deep circulation of the Atlantic.

Dynamic oceanography was first used in Berlin as a tool to promote a particular view of the deep circulation of the oceans against opponents. Once accepted, the approach that the method exemplified, quantitative analysis of data and modelling, began to exert its influence on the kinds of people who studied physical oceanography, favouring mathematically trained physicists over physical geographers. Berlin was not alone in this. The use of the dynamic method in other places, especially North America, soon became a stimulus to changes in the whole practice of physical oceanography as small, isolated groups of practitioners turned from qualitative science to mathematical physics. But in neighbouring France, certainly as scientifically advanced in most fields as Germany, the situation was different.

6

'Découverte de l'océan': Monaco and the Failure of French Oceanography

I am full of hope that France, by the renewed efforts of its mariners, its engineers, and its scientists, will not delay much longer in following the fruitful route in which, it must be admitted, she has been preceded by mariners and scientists of the United States, of England, and of Norway.

Julien Thoulet 1887, p. 430

As for the hydrodynamic theory of M. Thoulet, it is an anachronism. We don't live in the stone age of hydrography. To offer oceanographers of our era such a foolish imple-ment after the generalizations of Kelvin and Bjerknes! What would a biologist say if one offered to exchange his modern microtome for a stone knife?

Otto Pettersson to Jules Richard, 22 May 1905
(letter in Archives of the Musée océanographique de Monaco)

A French Paradox

Second scientifically only to Germany in the mid and late nineteenth century, France had a formidable system of scientific teaching, centred in the *Grandes écoles,* notably the École polytechnique and increasingly the École normale supérieure, as well as in newly reformed universities in Paris and the provinces.[1] French physics and chemistry were widely admired, and in marine biology France was in the vanguard, having been far ahead of much of the rest of Europe in establishing marine stations: Concarneau in 1859, Arcachon in 1867, Roscoff in 1872, Wimereux in 1873, and, among many others through the years, even a fisheries sta-tion, the Station aquicole of the national Service des pêches maritimes in

Boulogne-sur-Mer in 1884.[2] And yet, despite the power of French physics and chemistry in the universities, early in the twentieth century French marine science was mainly biological. The American Charles Kofoid, surveying the biological stations of Europe, commented that the success of French science was due to 'the highly centralized national system of education, with its practically coequal subdivisions' and 'principally, the large share which men of scientific interests and training have had in shaping educational policies and practice,' combined with 'large maritime interests and fisheries.'[3] Despite this, it was only marine biology, and a largely non-overlapping discipline, fisheries, that dominated the marine sciences in France well into the twentieth century, not any physical science of the sea.

The scientific output and quality of science done in France in the late nineteenth and early twentieth centuries have sometimes been questioned, on the basis that French science was over-centralized, had structural problems (especially the lack of an institutional hierarchy – that is, too much local independence), weak universities (because power lay primarily in their faculties), and narrow academic preparation (especially the *baccalauréat* and *agrégation*).[4] But this can hardly be so if marine biology is considered, for it sprang from exactly the conditions that seemed to lead to lack of success in other areas. So we must look to other factors to attribute responsibility for the failure of French scientists to develop a new non-biological science of the sea at a time when change was occurring rapidly outside France.

This is the paradox of French marine science: the virtual invisibility of France at a time when dynamic oceanography was making inroads into the way that physical oceanography was being practised in Scandinavia, Germany, and the United States. During the growth of a new brand of oceanography in most of Europe and in North America between the 1890s and the 1930s, France was never a player. In fact, one is hard-pressed to name even a single distinguished French physical oceanographer until the 1960s, and there were no world-class texts in that field until the same time. Paradoxically, though, the best-known exponent of oceanography in Europe during the late nineteenth and early twentieth centuries was French-speaking – Albert I[er], Prince Souverain de Monaco. His influence on the practice of oceanography and on its practitioners was great, but never on the physics of the ocean, even though one of his followers, Julien Thoulet, made every attempt to create such a role for France. Their failure requires exploration and explanation.

'Chef et propagateur de l'océanographie': Albert Ier, Monaco, and Exploration of the Oceans

Albert-Honoré-Charles Grimaldi (1848–1922) (fig. 6.1), who succeeded his father Charles III as Sovereign Prince of Monaco in 1889, was born in Paris and schooled in France.[5] He was soon sent to sea, as an ensign in the Spanish Navy in 1866, and joined the French Navy in 1869, staying long enough to see action in the Franco-Prussian War of 1870. Ever a sailor, Albert bought a British yacht, *Pleiad,* in 1873, renaming it *Hirondelle* to give it his own stamp. In this 200-ton, 32-metre schooner, he ranged the North Atlantic, until, in 1884, his attention turned to science at sea, apparently as a result of his acquaintance with the Paris zoologist Alphonse Milne-Edwards and the example of French scientific cruises on *Travailleur* (1880–2) and *Talisman* (1883).[6] He began to take plankton hauls and deployed floats to trace surface currents; as a result of the latter, he published a preliminary map of North Atlantic surface currents in 1892.[7]

For thirty years after 1885, Albert and a number of scientific colleagues spent several weeks at sea each year in a series of increasingly larger and more powerful yachts: *Princesse-Alice* (1891–7), *Princesse-Alice II* (1898–1910) (fig. 6.2), and *Hirondelle II* (1911–15). Each was equipped with scientific instruments and oceanographic gear designed by the Prince himself or by members of his scientific and naval staff.[8] The range of activities was impressive, including, along with current studies, mid-water and benthic trawling at great depths, the use of huge traps for fish and invertebrate animals, capture of whales and dolphins (primarily to reveal their stomach contents – which, in the case of a sperm whale, included a giant squid), and meteorological work using kites and balloons.

It was the demands of hard work at sea and long cruises that led to the replacement of the worn-out little *Hirondelle* by the 650-ton, 53-metre, auxiliary steam yacht *Princesse-Alice* in 1891. The new ship was fitted out with the most modern of oceanographic gear, including steam winches (*Hirondelle* had a manual capstan), a sounding machine, freezer, purpose-built laboratories, and electric lighting.[9] Mainly working from Monaco to the Azores, along the coasts of northwestern Europe, and in the Mediterranean, Albert and his colleagues carried out a full range of biological, chemical, geological, and meteorological studies at least as varied as on the much larger *Challenger* a generation earlier. However, in my judgment, the peak of Albert's career at sea was reached with the use of *Princesse-Alice*'s successor, the 1,420-ton, 73-metre *Princesse-Alice II,* for more than a decade beginning in 1898.

6.1   Albert Ier about 1889, when he became Sovereign Prince of Monaco. From J. Carpine-Lancre 1998b, p. 27; original photograph from Walter Lewis, Bath, England. With permission of Jacqueline Carpine-Lancre.

6.2 Albert Ier's third yacht, *Princesse-Alice II*, in service from 1898 to 1910, the most productive period of his scientific career. Collections personelles de S.A.S. le Prince Souverain de Monaco; artist De Simone, 1909. With permission of H.S.H. Prince Albert II of Monaco.

The beautiful *Princesse-Alice II* spent its first season afloat at Spitsbergen under the direction of Albert, with a scientific staff including the Scottish polar explorer W.S. Bruce, *Challenger*'s physicist John Young Buchanan, and the German zoologist Karl Brandt. It returned to the Arctic in 1899, 1906, and 1907. Throughout its sea-going career, *Princesse-Alice II*, like its predecessor, under the general direction of Albert and the day-to-day supervision of the chief of his scientific staff, Jules Richard (1863–1945),[10] was used for a very wide range of scientific studies. These included deep dredging and trawling, mid-water plankton and fish collections, and studies of marine bacteria. *Princesse-Alice II*'s deep benthic trawling at 6,035 metres southwest of the Cape Verde Islands in August 1901 was the deepest sampling of the oceans until the work of the Swedish *Albatross* in the late 1940s.[11] And as on his previous vessels, new equipment was continuously in use or being developed. As Albert said, '... one exploits all the imaginable means to study the fauna of the seas.'[12] Shortly after the turn of the century, he was working mainly on intensive studies of restricted areas, much like H.B. Bigelow's 'intensive area studies' in United States Atlantic waters a decade later,[13] including in 1909 a five-day station in 5,940 metres of water west of Portugal, which was the most complete study in open-sea oceanography to that date.[14]

By 1910, Albert was planning even more extensive cruises and studies using his new 1,600-ton, 89-metre yacht *Hirondelle II*. Although he was able to begin work on deep-water plankton using huge nets that could be towed very rapidly, his ambitious plans were cut short by the outbreak of the First World War in 1914. Despite a further cruise in 1915, the ship and the Prince never resumed full-scale oceanographic work thereafter, and after a period of ill-health, Albert died in Paris in June 1922.

Although Albert's researches and travels at sea were very influential among scientists and were noted in his obituaries, there was another side of his life that has received less attention, his social conscience and the responsibility that he felt to modernize his tiny Principality. Within a decade of becoming Sovereign Prince, Albert had begun the creation of a harbour suitable for large ships, and was busy encouraging businesses to establish themselves in Monaco. He was responsible for the building of a modern hospital and the establishment of a high school, and in 1911 established a new constitution and legal system. Encouraging the arts, particularly music, was an expression of his love of Massenet and Saint-Saëns – and brought paying guests to the Principality, as did congresses on all the subjects popular in his era.[15] Behind this lay a deep belief in the ability of science to provide a bulwark for the best in European civilization and for the advance of civilized ideals. In 1904, at the Sorbonne, Albert told his audience that 'as for me, I have contributed with all my strength, for eighteen years, to the conquest of scientific truth, sole emancipator of thought and conscience, infallible guide to a strong and generous civilization.'[16] Speaking to an international audience in 1910, he said,

> Here, gentlemen, you see that monégasque soil has given rise to a proud, inviolable temple, devoted to the new divinity that reigns over understand-ing. I have devoted all my strength of mind, of my conscience and of my sov-ereignty to the enlargement of scientific truth, the only ground on which the elements of a stable civilization, safeguarded against the foibles of hu-man law, can mature.[17]

His 'inviolable temple' was the Musée océanographique de Monaco, be-gun in 1898 and inaugurated in 1910, a terrestrial focus of Albert's social and scientific ideals.

Shortly after he began work at sea, in 1885, Albert had the idea of founding a zoological station in Monaco. The time was ripe, for in the 1880s Europe was in the midst of an explosion of new marine biologi-cal stations, including Villefranche (near Monaco) and Banyuls-sur-Mer. Intrigued by the revolution in zoology going on around him, Albert had plans drawn up for a marine station to be built at Fontvieille, on the western edge of Monaco, involving a public aquarium, a variety of spe-cialized laboratories, a library, and working space for an international clientele of scientists. For some reason, his plans for the laboratory hit a roadblock and were replaced by the idea of a much more grand es-tablishment, an oceanographic museum. Encouraged by the public and scientific response to his displays at the Exposition universelle of 1889 in Paris, Albert's plans were reinforced for a place where his collections, stuffed into his Paris hotel rooms, could be displayed in a permanent and educationally powerful way.[18]

During the late 1880s, Albert was occupied with the modernization of Monaco, the building of *Princesse-Alice* and her successor, *Princesse-Alice II*, his first current chart of the Atlantic, and with the deterioration of his relationship with his first scientific assistant, Jules Malotau, Baron de Guerne (1855–1931), whose responsibilities included the work at sea and editing the lengthening series of published monographs on the cruises.[19] As a result, the planning of the new museum went slowly, and it was 1898 before construction began on the cliffs of Monaco, in the vicin-ity of the Jardins Saint-Martin (fig. 6.3). In 1887 a young zoologist from

6.3 The Musée océanographique de Monaco, built during 1898–1909 and inaugurated in 1910. Photo by Eric Mills, 1989.

the Auvergne, Jules Richard, had become Albert's principal scientific assistant, eventually taking on the major load of planning and supervising construction.

When the Musée océanographique was inaugurated in 1910, only two display halls plus library, auditorium, and the laboratories were complete. The last display hall was complete in 1913, when the Ninth International Congress of Zoology met in Monaco. The Prince said of it then:

> ... I wanted to give to Oceanography an international temple in accordance
> with the wide span of its position in the progress of human knowledge, and
> to create at the same time a rallying point where the servants of Science,
> that is of Truth, can rally their forces, finding new weapons for combatting
> together the obstacles of ignorance and superstition from the past, as well
> as the blindness that brutal revolutions bring to the progress of thought.[20]

He was also reported to have likened the Museum to 'a ship anchored on the coast with the riches extracted from all the abysses' and to have claimed that he had given it to scientists as 'an arch of alliance.'[21]

It was by no means sure at first that the funding or legal status of the Museum was permanent. During construction, in 1906, after negotiations with the French government, Albert arranged to have the Museum incorporated as part of a private foundation, an integral part of a newly founded Institut océanographique, which was provided with a building in the university quarter of Paris and inaugurated formally in 1911.[22] The foundation received an endowment of four million francs (there was a further million when Albert died in 1922). The Institute's function was to be the teaching arm of the foundation, while the Museum was to house the Prince's collections and be a centre of research. In Paris, three professorships were established, one in each of biological oceanography, physical oceanography, and the physiology of marine organisms.

Teaching had been on the Prince's calendar since 1903, when he first arranged a series of lectures on oceanography in Paris. More followed from 1904 through the spring of 1906, involving Albert himself, along with two of the Paris academics who later took up professorships in the Institut océanographique, Paul Portier (1866–1962) and Louis Joubin (1861–1935). Beginning in 1904–5, these lectures became a formal course in oceanography, which, upon the establishment of the Institute, were the foundation of its academic program. A third professorial appointment was necessary, in physical oceanography; the physicist

Alphonse Berget (1860–1933) was Albert's choice.[23] Curiously, and significantly, the chair was not filled by the scientist widely acknowledged as France's most active oceanographer (apart from Prince Albert himself), the man later called 'the patriarch of French oceanography,'[24] the mineralogist and self-educated physical oceanographer from the Faculty of Sciences in Nancy, Julien Thoulet, who had lectured in Paris and elsewhere.[25]

Julien Thoulet – a Convert to Oceanography

If life moved in straight and logical lines, Marie Julien Olivier Thoulet (1843–1936) (fig. 6.4) would have gone from being a protegé of Albert Ier to the professorship of physical oceanography in Paris and, in that enviable near-sinecure, begun to work with the dynamic method, after studying the works of Bjerknes and Helland-Hansen. Thereby French oceanographers would have taken their place in the succession of apostles learning a new approach to ocean circulation. But nothing like this happened. Instead we get a glimpse of the role of contingencies, personal eccentricities, and even narrow nationalism, in scientific change – or, in this case, scientific stasis.

Thoulet was born in the French colony of Algeria.[26] He left Algiers for schooling in Paris, in the course of which he studied mathematics, developed interests in cartography and mineralogy, and aspired to enter the École polytechnique. Rejected there twice, in 1864 Thoulet left France for jobs in mining and engineering in France and Spain, then in the United States, where for a time he was an engineer during the construction of part of the Northern Pacific Railroad near Lake Superior. Back in France by 1872, he took a preliminary degree in the Paris Faculté des sciences and prepared a doctorate in mineralogy in the Collège de France, awarded in 1880 for studies on the physical properties of minerals. His academic career began as *maître de conférences*[27] in Montpellier and then in the highly patriotic, anti-German University of Nancy in Lorraine.[28] Appointed to a chair in 1884, he appeared set to live out his scientific life as a teacher of mineralogy in the provinces.

In 1886, Thoulet arranged to spend several months at sea around Newfoundland on a French naval ship, *La Clorinde*, which was supporting the fishery.[29] This was the event precipitating his entry into oceanography, leading to the bulk of his scientific work, which extended well beyond his retirement from Nancy in 1913. Two years after his return from Newfoundland, Thoulet lectured on oceanography to naval offi-

6.4 Julien Thoulet (1843–1936), professor of mineralogy in the University of Nancy from 1884 to 1913 and 'the patriarch of French oceanography' according to Camille Vallaux. From Vallaux 1936.

cers at the Observatoire de Montsouris, and visited marine laboratories in Scotland and Norway on a mission from the Ministry of Public Instruction to assess the state of the marine sciences there in comparison with France.[30] In 1886 and 1887, Thoulet met Prince Albert, perhaps through de Guerne, and later took part in at least three cruises on the Prince's ships as well as carrying out sediment and water analyses of samples taken on many of the Prince's other cruises. His texts *Océanographie (statique), Guide d'océanographie pratique,* and *Océanographie (dynamique)* were published in 1890, 1895, and 1896, cementing his reputation as the premier physical scientist of the sea in France. Lecturing in prestigious venues, including his free course on oceanography at the Sorbonne in 1891, in the École des hautes études maritimes during 1897–9, and in the Conservatoire nationale des arts et métiers in 1903 (and elsewhere in Paris through 1905),[31] gave him high visibility as teacher as well as researcher. Why then was Thoulet not appointed professor of physical oceanography in the Paris Institut océanographique in 1906?

The last three decades of the nineteenth century saw a great deal of attention to maps of the oceans, especially bathymetric charts incorporating the many soundings that were beginning to accumulate from cable surveys and for a variety of other reasons.[32] By the time of the Seventh International Geographical Congress in Berlin in 1899 there was much discussion of the need to standardize the terminology applied to features of the sea floor and for a new, synoptic set of charts of the ocean depths. The outcome was a commission given the task of 'instigating the preparation and publication of a bathymetrical map of the oceans.'[33] Its members, all distinguished geographers or marine scientists, included Julien Thoulet, and Albert I[er] (fig. 6.5).

Despite a paper by Thoulet, written in 1901, outlining the principles on which such a chart should be based, and some discussion between Thoulet and Albert in 1902, there was little action by the commission until, at the instigation of Ferdinand von Richthofen, it met formally in Wiesbaden in April 1903. At that time, Prince Albert, appointed chairman of the commission, agreed to take over the production and financing of the new chart. One of Albert's staff, Charles Sauerwein (1876–1913), a French naval officer, was given immediate responsibility for organizing the production of the chart on principles that had been spelled out by Thoulet. Work soon began in Paris.[34] By June 1904 a manuscript had been prepared, and in September, Thoulet travelled to the United States on behalf of the Prince to the Eighth International Geographical Congress, where he lectured on the new bathymetric chart and presented

6.5 The Commission on Sub-Oceanic Nomenclature appointed by the Seventh International Geographical Congress of 1899 meeting in Wiesbaden in April 1903. From left to right: Charles Sauerwein, Otto Krümmel, Julien Thoulet, Albert Ier of Monaco, Alexander Supan, Otto Pettersson, and Hugh Robert Mill. From J. Carpine-Lancre 1998b, p. 121; photographer C.H. Schiffer. With permission of Jacqueline Carpine-Lancre.

drafts for inspection by the delegates. Between February and May of 1905, the final printing was completed, and Thoulet and Sauerwein congratulated themselves on an outstanding accomplishment, the production of the first Carte générale bathymétrique des océans.[35]

Their satisfaction was short-lived. Despite the initial enthusiasm in Washington and Paris, a critical examination of the Carte générale by the French geographer Emmanuel de Margerie showed that there were major inconsistencies and inaccuracies, including the projections, transcriptions of errors from the original sources, inconsistent terminology, and non-standard geographical names. He, quite correctly, attributed the problems to lack of proofreading by a competent cartographer. Here Thoulet might have played a role, but he had been excluded from the final production of the Carte générale by overwork, and perhaps by Sauerwein's belief that perfect accuracy was not necessary in a first edition. The two fought it out by letter, and in July 1905 they were brought together with Albert to discuss the imbroglio. As the historian of these events, Jacqueline Carpine-Lancre, points out, no written record of this meeting has been found.[36] Nonetheless, although Sauerwein came out of the events of mid-1905 unscathed, Thoulet, who appears to have been blameless, did not. After 1905 his relations with Albert were distant, he was dropped from the list of participants on the Prince's cruises, and although he was the ranking candidate to fill the third professorship in the Paris Institut océanographique, the physicist Alphonse Berget was given the position.

It is not surprising that for a time after 1905, Thoulet appears to have given more time to mineralogy than to oceanography, publishing more than thirty papers on sediments and other subjects between 1905 and 1911, although as the next section will show, he never fully divorced himself from marine science. When he retired from Nancy in 1913, Thoulet returned to his childhood home, Algiers, where he hoped to carry on research. As this proved to be a vain hope, in 1914 he went back to his laboratory in Nancy. But with the outbreak of war it was impossible to remain in Nancy, close to the German border, and so he moved to Brittany, then in 1919 to Paris, where he was given an office in the Institut océanographique. There he renewed his writing on oceanography, revisiting subjects that had occupied him since early in the century, especially deep-ocean currents, and, despite his location, becoming more and more distant intellectually from a science that had left him behind. After his last publication in 1933, and his death in 1936, Julien Thoulet was forgotten by all except for a few French scientists with antiquarian interests.

The Oceanography of Julien Thoulet

From the beginning of his work on the sea, in the Western North Atlantic off Newfoundland in 1886, there was a powerful nationalistic element in Thoulet's approach to oceanography. Convinced, even before his investigation of the state of marine science in Scotland and Norway in 1888, that France was falling behind the other major scientific nations of western Europe, particularly Britain and Germany, a constant theme in his public lectures was the dire consequence of being bested in science:

> ... bad luck to the nations that let themselves be passed and who do not take advantage of all the resources of their continental and maritime territory; their populations will fail to be augmented, and thanks to the ease of transport, will be reduced to the point of death by flight, emigration, establishing themselves among other peoples, and, as a nation, disappearing from the face of the earth. Speaking only of the study of the ocean, England, Germany, Austria, Sweden, Russia, the United States, have paid the expenses of expeditions intended to explore the ocean, this field of exploitation for the future. Despite these warnings, France remains outside this movement: may it not pay too dearly for its ignorance or its disdain![37]

Along with this apocalyptic vision, expressed in many of his publications, came a distinctive and narrow view of how oceanography was to be defined. With surprising vehemence, he claimed that oceanography must be separate from biology, especially zoology: '... the zoologist is incapable of arriving at firm results ...' because of his ignorance of physical laws, the result being a blind search for discoveries.[38] Oceanography was, to Thoulet, 'application of the principles of physics, chemistry and mechanics to the study of the sea,' and he claimed it was 'essentially an exact science, of numbers and experimentation.'[39] At greater length, and in telling order, he elaborated that

> oceanography is the study of the sea. Static oceanography deals with salt water considered independently of the movements which are manifested in it; it treats successively of the topography of the ocean beds and of their formation, their lithology. It analyses the waters, their composition and their influence on the nature of the depths, their numerous physical properties, the effect on them caused by changes of temperature, their density, their compressibility, the way in which light is diffused throughout the superposed strata, and the different optical phenomena ...

In dynamic oceanography the ocean is studied in motion. We study the
waves, which move the surface under the influence of the wind, and the
currents, which ... traverse its mass to a certain depth, and result from the simultaneous actions of heat, evaporation, the topography of the sea bottom,
the geographic configuration of the surrounding continents, the climate,
the force of the winds; in a word, from the total of exterior causes which ...
exert some influence ...[40]

But if oceanography was 'the ensemble of all the laws applicable to the
sea ... not only in the domain of chemistry and physics but also in that
of mathematics,'[41] he was distrustful of mathematics – 'it is necessary to
show prudence in the use of mathematics in natural science' because

a natural phenomenon being a unique equation with a great number of
unknowns, to want to apply to it mathematics in all its rigour seems to be
the affirmation that this equation can be resolved mathematically, which is
mathematically absurd.[42]

He clarified his viewpoint later:

... in the majority of cases, mathematics is a precarious aid for the elucidation of natural phenomena. It is the ideal and not the reality; it is simple
and nature is not so. The equality sign which is the basis of its reasoning is a
material absurdity because there do not exist on the whole earth two leaves
of a tree that are mathematically equal, nor in the depths of the ocean two
grains of sand rigorously equal, nor in the whole mass of oceanic water two
absolutely identical drops.[43]

As a result, although he was not averse to quantitative science, he regarded mathematical treatment as oversimplification of a natural world that
was ineluctably complex and not capable of simplification. It is clear that
to Thoulet mathematical modelling, even of the simple kind resorted to
by the exponents of the dynamic method, lay outside the bounds of acceptable scientific practice.

Oceanic circulation, in particular, was a complex phenomenon, the
resultant of processes going on from the coasts into the shallow depths
of the sea. In his paper 'Les courants de la mer' (1893) and in many publications thereafter, Thoulet described how the winds, sun, evaporation,
salinity, the Earth's rotation, the form and depth of the bottom, and the
geographical configuration of the continents, all played interacting roles

in producing ocean currents.[44] He believed that the problem was, in principle, explicable, not through new physical approaches but through the accumulation of new information about the distribution of properties in the oceans. In this, temperature and salinity played large roles as the main determinants of the density of the water, a physical property that his hero, John Young Buchanan, the physicist aboard *Challenger,* had recognized early.[45] Such properties enabled currents to be traced, but it was primarily the wind that caused ocean circulation. But how deep was the surface circulation? He cited the calculations of Karl Zöppritz[46] to the effect that surface wind-driven circulation could never extend very deep because of the variability of the winds and the inability of motion to be transferred through a fluid (seawater) with such a low coefficient of viscosity. Thus, below the moving surface waters lay abyssal water in eternal repose, as indicated by the stable density structure of the water, which increased constantly with depth: '... the deep waters are therefore immobile; they are veritable fossil waters.'[47]

The evidence against abyssal circulation dominated Thoulet's thoughts on ocean circulation for many years,[48] and although there is mention, indirectly, of the Carpenter-Croll controversy (see chapter 2) in his writings on abyssal circulation, he never makes direct reference to it.[49] Because of their remoteness, it was not possible to make direct observations on deep waters; instead evidence from chemistry and sedimentology had to suffice. The evidence they gave, in combination, was unequivocal according to Thoulet's interpretations. First, he believed that deep cold water would only form in high latitudes until the sea froze, forming a cap that would prevent further convection. In any case, the cold water found everywhere in the abyss need not have moved there along the bottom – instead, it might have been there since the origin of the oceans. Such was suggested by the fact that the coldest bottom water appeared to be located in isolated basins not in direct contact with the open floor of the deep ocean. Furthermore, in some regions of the oceans there were temperature inversions, regions where colder water lay over slightly warmer water; this showed that direct vertical motion of water could not be occurring. In most places, however, once corrections were made for pressure, there was a steady increase of density with depth, just as would be expected in a perfectly stable and immobile water column.[50] Thoulet also examined the hypothesis that high levels of oxygen in deep waters could only have been maintained if there were sinking of oxygenated water at high latitudes (a suggestion made by William Dittmar, who had analysed the seawater samples from *Challenger*), claiming that if deep ocean sediments were highly oxygenated (as he believed, due to the sinking of sediment particles bearing oxygen) the deep waters would take up oxygen from the sediments – eliminating the necessity that high oxygen content was the result of the abyssal circulation of water that had been at the surface originally. He claimed too that the stability of deep bottom waters would be enhanced as they dissolved material from the sediments – a hypothesis implying that there would be local variations in the chemical composition of the bottom waters.[51] Further evidence for the stability of the bottom waters lay in the nature of the bottom itself: in many places, it was far too rugged and steep to allow the steady flow of bottom water from high latitudes throughout the abyssal basins.

Variation in the composition of deep oceanic waters lay at the heart of Thoulet's belief that the deep waters of the oceans were stable and immobile:

> On the bottom, at places very close together, the chemical composition of the water can be very different. And, in the same region, the waters of the deepest depressions do not always have the greatest absolute density. Thus it happens that in the real ocean, from the special point of view considered here, deposits are localized; their distribution, their dimensions, their thickness, whether extensive, weak or non-existent, depends less on time than on internal conditions.[52]

His support of the idea that seawater was of variable composition, expressed in a series of papers,[53] soon involved Thoulet in an international controversy centring around the adequacy of salinity determinations in chemical oceanography. If, as Thoulet believed, there were small but significant variations in the ionic composition of seawater leading to variations on small scales in density, it would be necessary to make many chemical determinations before understanding the movements of the water. And yet chemical oceanography had just been presented, in 1901, with a carefully prepared set of tables from the laboratory of Martin Knudsen in Copenhagen relating temperature, salinity, and density, and making calculation of salinity and density (at given temperatures) a simple matter of measuring the chloride ion content of the water. Behind Knudsen's tables lay the long-held belief that the relative composition of seawater was unvarying, that is, that the ratios of the abundances of the various ions were constant.[54] Thoulet's beliefs about the small-scale variation of ionic composition of seawater, and soon the evidence for it based on analyses in his laboratory in Nancy,[55] contradicted the emerging orthodoxy of Knudsen's Tables, which were soon adopted by the International Council for the Exploration of the Sea as an international standard.[56] If Thoulet was correct, density of seawater and chlorinity, which could be readily measured chemically, did not have a constant relationship – a situation having serious implications for any calculations, like dynamic ones, that had used Knudsen's shortcut to density.

After several years of controversy over Thoulet's results, the results of Knudsen's Tables were borne out, and Thoulet's claims discredited as being due to a combination of inaccurate analyses and special interests. Otto Pettersson commented a little later that 'practically Thoulet has exaggerated the differences in the chemical composition of sea-water existing in nature tenfold. This is a practical question – a question of limits not of principles.'[57] A curious outcome of Thoulet's claims and the ensuing debate over the constancy of composition of seawater was his careful examination of the relationships between salinity, temperature, and density in all their combinations. Pairs of characteristics – salinity and temperature, salinity and density, density and temperature – could be used to characterize particular water types, giving what he called the 'personality' of water in particular locations.[58] Although Thoulet did not make more of this in print, it is an early use of bivariate relationships that found constant use after Bjørn Helland-Hansen introduced the T-S diagram to the oceanographic world in 1916, almost certainly without knowing of Thoulet's work.[59] Thoulet himself never gave up the idea of local variations in the composition of ocean waters, returning to the *Challenger* results, always his holy writ of oceanography, in the late 1920s to claim that abyssal vulcanism was responsible for local variations in composition and temperature of the deep water and might even result in local circulation, a 'microcirculation abyssale générale.'[60] In one of his last publications, he gave abyssal vulcanism profound powers:

Abyssal vulcanism is ... the powerful agent of the eternal restoration of chemical and physical equilibrium, the continual conservator of the unity of composition of oceanic waters [which is] perpetually destroyed by the eternal cycle of water leaving the sea as fresh water, permitting life and materials on land ['la vie des êtres et des choses'], and ceaselessly returning to maintain constant salt content in the vast ocean.[61]

But this was only a variation on the theme that dominated Thoulet's thought for forty years.

Thoulet's Stone Knife – an Approach to Dynamic Oceanography

At almost exactly the same time that dynamic oceanography was being developed in the hands of Vilhelm Bjerknes, Bjørn Helland-Hansen, and Johan Sandström from the 1890s to 1905, Thoulet was working on his own version of what may loosely be called 'dynamic' calculations of ocean currents. Much later, he wrote to Jules Richard, of his belief that 'science in general and oceanography in particular ought to be as much as possible above all an affair of common sense and my theory of circulation is ... based on measurements interpreted through simple common sense.'[62] The measurements were, in line with Thoulet's life-long interest in density of ocean waters, based on calculations of density and on their use in determining the topography of the sea surface, or of specific density surfaces within the water column of the deep ocean. The inspiration came not from Scandinavia, but from the example set earlier by French naval engineers, including Urbain Dortet de Tessan (1804–1879), and particularly Anatole Bouquet de la Grye (1827–1909), who had encouraged Thoulet's teaching of French naval officers between 1888 and 1890.[63] As Thoulet expressed the basis of their approach, which he appropriated as his own, 'the key to oceanic circulation is a levelling.'[64]

Dortet, who had accompanied Abel du Petit Thouars during his circumnavigation in 1836–9 on *Vénus* (see chapter 2), suggested that mathematical methods could be applied to the circulation of the oceans, which had as its ultimate causes solar heating and the rotation of the Earth, modified by friction and other factors. However, in his journal for October 1838, ignoring the Earth's rotation, he wrote that '... in a basin filled with liquid, if there exists a constant cause of a difference of level between two points, this difference must produce a current from the highest level to the lowest level.' Because the temperature difference between the poles and the equator was a 'constant cause' resulting in a difference of sea level, there should be permanent currents from one to the other.[65]

Bouquet de la Grye accompanied a French transit of Venus expedition to Campbell Island, near New Zealand, in 1874–5. He occupied himself at sea by measuring the density of seawater and testing the utility of measuring the chloride ion as an index of salinity (it proved to differ only very slightly from measurements using hydrometers, a result that could have saved Thoulet much wasted effort thirty years later).[66] Included in his monograph is a nomogram relating density, volume, and chloride ion content to the temperature of seawater. He noted that because of differences of salinity, and thus of density, the ocean surface is not at a constant level but varies from place to place.[67] The result is motion:

> Each molecule of water ... runs continuously toward the precise point its density would indicate, following the route with the greatest slope, which is that of the greatest differences of relative density.[68]

Commenting that direct measurement of the level of the ocean surface is so difficult that it is surprising that no one has attempted to use indirect means, he then set out, using density measurements from *Challenger*, to calculate the level of the Atlantic Ocean based on the premise that very deep water, down to about 4,000 metres was of such constant temperature and density that it could be considered a stable surface. The result showed that the surface level of the North Atlantic fell from its western margin, the axis of the Gulf Stream, to a depression centred in the eastern Sargasso Sea and then rose again along the West African coast (fig. 6.6). He did not comment on the implications for oceanic circulation, and never returned to the subject after publishing the outcome of his calculations in 1882.[69]

Thoulet, who was alert to new publications on the oceans, particularly from French sources, began to find his way into ocean surface levelling about 1890, shortly after his lectures to naval officers of the Observatoire de Montsouris and his contacts with Bouquet de la Grye. In his *Océanographie (statique)* of 1890, Thoulet quoted extensively remarks on ocean levelling by Bouquet de la Grye, discussed the relationship between density and the height of the sea surface, and used some of his own data (from his work on *La Clorinde* off the east coast of North America in 1886) along with *Challenger* data to claim that the Gulf Stream 'descended' from a topographic high off the coast of North America 'to the plains represented by the waters of the central regions of the Atlantic.'[70] Not only had Bouquet de la Grye demonstrated a depression in the central North Atlantic, but this appeared to be a general feature of oceans, as shown in Henrik Mohn's studies of the Norwegian Sea, which were summarized in his monograph of 1887, 'The North Ocean.' Based on the two, Thoulet concluded that the normal shape of the open ocean was to be depressed in the centre, due to the presence there of the most dense water, whereas along the coasts of the continents lighter (less saline and thus less dense) water stood at higher levels. Again, basing his discussion closely on Mohn, he claimed that such a structure of the oceanic water column (depressed) and coastal water (at a higher level) must result

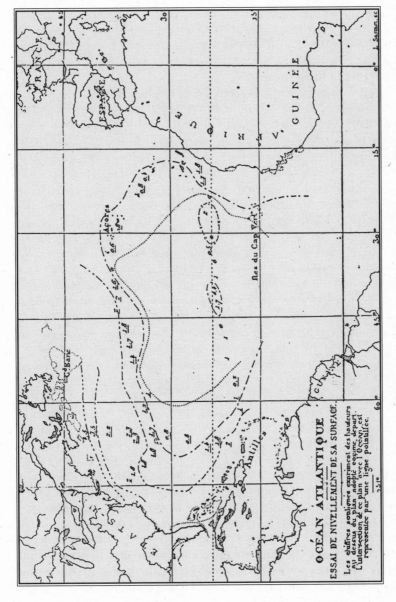

6.6 The height of the Atlantic Ocean, based on data from the *Challenger* expedition, 1872–5, calculated using the density of seawater by the French hydrographic engineer Anatole Bouquet de la Grye. From Bouquet de la Grye 1882, p. 474.

in downwelling in the centres of oceans, and corresponding upwelling along the coasts, completing a cycle of circulation from and to the central waters of the oceans.[71] In theory, the Earth's rotation should affect such motions, as he pointed out in a discussion of currents driven by ocean slope in his 1896 text *Océanographie (dynamique)*, but despite giving the Coriolis equation, he claimed that 'it is practically zero for ocean currents because of their low velocity.'[72]

By 1904, Thoulet's 'common sense' had led him to a technique that he believed could be widely applied to determine ocean currents. The primary data were measurements of density, leading as he said, to 'measurement of ocean currents by means of the physical and chemical analysis of water samples collected in series.'[73] The essence of the technique, described in a series of papers in 1904 and 1905, and taken up by at least one of his students, was to use measurements of density to calculate the slopes of density surfaces whether at the ocean surface, or in the ocean interior.[74] The current direction and velocity were determined by the slope of the surface, and could be calculated in absolute terms if there were a way of getting a direct measurement by some means. At sea, seawater samples were taken in a triangular pattern at the surface and below. Then measurements of density, preferably from the use of hydrometers, not by calculation using titration for chloride and application of Knudsen's Tables,[75] could be used in conjunction with the known relations between the density, temperature, and volume of seawater to give the height of the column at each corner of the triangle, and, by calculation, the slope of the triangular surface. Water ran downhill along these surfaces, following a track and with velocity that were the resultant of the heights at the corners of the triangle (fig. 6.7).

Thoulet's 'dynamic method,' based on frequent, simple measurements at sea, was necessary because of the variability inherent in nature. In 1905, in the most formal description of the technique, he wrote that

> it is the individuality of each drop of water, an individuality established by direct measurement and not by reasoning, and consequently indisputable, which allows one to approach with certainty and to succeed in the discovery of the secrets, for the most part still unknown, of oceanic circulation.[76]

What he called 'the identity, the personality, of a seawater' was determined by a number of variables such as its density, halogen, and sulfate content, which he regarded as variable from sample to sample, characteristic of it, but 'static.' But it was necessary to go further:

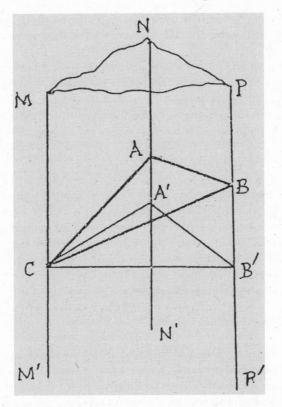

6.7 A graphical representation of Thoulet's method of calculating ocean currents, likened by Otto Pettersson to using a stone knife. Measurements of density at sea throughout the water column at each corner of a triangular area were used to calculate the height of the water column at each location, A, B, and C. The slope of the triangle indicated the direction and velocity of the current. From Thoulet 1925c, p. 4.

> ... if the static characteristics define the sample perfectly, they remain silent on the role of each in the heart of oceanic waters. To obtain this information, it is necessary to have recourse to its dynamic characteristic, to the number represented by the symbol S^{θ}_{4} or $n\,S^{\theta}_{4}$, weight per litre of water, at the time when, shallow or deep, it plays a role in nature ...[77]

Then, with this dynamic number at hand (actually, as many as could be obtained under the difficulties of working at sea), one proceeded to build up sloping, triangular density surfaces throughout the water column that indicated the direction and velocity of the currents:

> The intensity of the currents will be, whatever the locality and depth considered, measured as a function of the gradient of densities and consequently comparable among themselves. If one wants to evaluate them in units of velocity, it is necessary by previous experience to establish the relation existing between this velocity and a certain number of known gradients.[78]

But there were qualifications:

> This measurement procedure should only be applied with the greatest rigour to deep currents. The facts described are true near the surface and at the surface even where the water flows from the continental margins, a region of low density, toward the centre of the oceans, where owing to evaporation and non-influx of continental fresh waters they have a high density. However, in superficial regions [that is, at the surface], atmospheric meteorological phenomena, climate, season, wind, ice, the geographical configuration of the continents and still other factors, will exert a greatly predominating disturbing influence.[79]

These reprises of Thoulet's method from his major monograph of 1905, in the *Résultats des campagnes scientifiques accomplies sur son yacht par Albert Ier, Prince souverain de Monaco*, are given here because it is in all likelihood this monograph that so offended Otto Pettersson (see the opening of this chapter). Calling Thoulet's method 'an anachronism' suitable for a 'stone age of hydrography,' a stone knife rather than a microtome, was typical of the outspoken Pettersson, who knew and supported Scandinavian dynamic oceanography as developed and propounded by Bjerknes, Helland-Hansen, and Sandström. What is not clear, for no evidence seems to exist bearing on it, is how Thoulet reacted to Pettersson's devastating critique.

By 1905, Thoulet and Pettersson had been in regular contact for several years over the adequacy of Knudsen's Tables, the chlorinity method of determining salinity, and the hypothesis of the constant relative composition of seawater that were their foundation. Letters in the Archives of the Musée océanographique de Monaco between the two show a genteel exchange of views at the time when tests of Knudsen's method were being organized by means of intercomparisons in several laboratories.[80] Pettersson's personal correspondence shows a somewhat less genteel side to the debate. At almost the same time that Pettersson wrote disparagingly to Jules Richard about Thoulet's method of determining ocean currents, he was writing to Martin Knudsen about Thoulet's claim to have found fatal flaws in the salinity technique:

> I am beginning to think that I can foresee hydrographical troubles. Now our friend Thoulet has been out, and ... this old stubborn and conceited man does not stop denying the basis of our observation system until he has got a public answer. I knew the fellow and predicted that a controversy was unavoidable.
>
> I can inform you that on my proposal Thoulet 2 years ago sent a control sample to the Central Laboratory which was found to be absolutely wrong. Nansen who communicated this to him got a rude letter as answer. Nansen is down on the old Thoulet and will, I think, readily join issue with him.[81]

Thoulet was forced, eventually, to yield on the question of salinity (although he never accepted the inherent validity of Knudsen's Tables), but evidence is lacking that he knew of Pettersson's criticism of his current method. Because it was characteristic of the outspoken Pettersson to deliver his criticisms directly to the authors of papers he could not accept, it is likely that Thoulet got a forceful letter or two from Pettersson. Whatever may have happened between the two, with dogged determination, ignoring Pettersson and other critics,[82] Thoulet stuck with his method for the rest of his life, making it an important component of a well-developed, if highly idiosyncratic, theory of ocean circulation.

When the aging Thoulet returned to work in Paris after his unhappy sojourn in Algiers and after losing his home and laboratory in Nancy as a result of the war, he did not hesitate long to return to themes that had dominated his work before retirement and the conflagration:

> The physical circulation of the total oceanic mass is the resultant of a densimetric circulation which is more or less constant the whole year in the warm regions, semi-annual in glacial maritime regions where it corresponds to the season of freezing and to that of the fusion of ice ... Densimetric phenomena explain the continuity of the water cycle on the globe, the permanence of sea level, the existence of ocean currents at moderate depths, ascent of cold waters within warmer ones, warm waters in the midst of colder ones ...[83]

In a lengthy series of papers and monographs, Thoulet visited and revisited all the scientific themes of his early career, centring all on the utility of using potential density to determine the circulation of the oceans.[84]

In this highly repetitive *oeuvre* there are new elements, such as the increasing use of the hypothesis of vulcanism to account for the variable composition of seawater from place to place and its overall constancy of composition. And there is evidence that Thoulet was reading new literature within a restricted range of subjects, although the basic currency of his work was the reworking of density results from the *Challenger* expedition, summarized mainly by Buchanan in the 1880s.[85] To this could be added some new information, for example, the temperature sections made by Murray and Hjort in the North Atlantic during the *Michael Sars* expedition of 1910.[86]

Differentiating what Thoulet knew from what he expounded presents difficulties. For example, he rapidly picked up the concept of potential temperature (temperature of seawater *in situ* corrected for warming due to compression at depth), which was introduced by Helland-Hansen in his chapter on physical oceanography in Murray and Hjort's *The Depths of the Ocean*, published in 1912,[87] but does not mention its source. In a letter to Jules Richard early in the 1920s, he mentions some of the work of V.W. Ekman on the compressibility of seawater as 'a magnificent exercise in absurdities.' The precipitating factor was apparently 'a four-line equation!!' and an overly enthusiastic approach to the use of mathematics in oceanography that he attributed to Germans, specifically Otto Krümmel.[88] Elsewhere, he commented approvingly on the experimental work of J.W. Sandström, partially for his use of experiment, partly, apparently, because Sandström's results provided evidence for the strong separation of shallow and deep waters long claimed by Thoulet.[89]

The outcome of Thoulet's life work in oceanography, despite a torrent of publications from the 1880s into the early 1930s, was an oceanography that could all have been written before the Scandinavian contribution to the fluid dynamics of the ocean had occurred. Others ignored his work, notably the French fisheries oceanographer Edouard Le Danois in treatises on the circulation of the North Atlantic beginning in the 1920s,[90] as did even the sympathetic obituarist of Thoulet, Camille Vallaux.[91] But Thoulet was not deterred; he had found the true way and he never abandoned it. Despite, or because of, his dogged espousal of a few ideas, he had been forgotten long before his death in 1936.

The Failure of French Oceanography

Was Julien Thoulet's oceanography an individual instance of science that did not adapt to change? Or was he an element in a system of thought and practice larger than himself that limited the ability of French marine scientists to make changes in their approach to physical oceanography throughout the first decades of the twentieth century? Why was dynamic oceanography of the Scandinavian kind, so widely accepted elsewhere, rejected (by Thoulet) or ignored (by his successors – with the exception of Le Danois) for so long? I cannot claim to provide definitive answers to these difficult questions, but some clues are worth following up.

The first feature of the story is Thoulet himself. His belief in the nearly infinite heterogeneity of oceanic properties, and the disdain he felt for mathematics (as opposed to quantitative experimentation), crippled his ability to look for ways of simplifying oceanic processes like circulation. To simplify, according to his scientific beliefs, was to leave nature behind, rather than to make it tractable. His holistic approach was quantitative according to his own understanding, which meant gathering numerical data to develop generalizations about the ocean – but quantitative certainly did not mean physical modelling as used, for example, by Mohn, Bjerknes, and the other mathematical physicists who were working on the same problems. Thoulet's claim that it was mathematically absurd to believe that complex natural phenomena could be resolved in equations removed him from the mainstream of mathematical physics and certainly from the main currents of physical oceanography in the early twentieth century. He could never be an apostle of Scandinavian dynamic oceanography; instead, he came to personify what had to be overcome.

It may be significant that Thoulet represented no school of thought in France, and did not develop one. He may have been called the patriarch of oceanography in France,[92] but the appellation was honorary, not a mark of the abundance and influence of his descendants. In fact, there were none, at least in France.[93] Certainly the system in which he worked, an independent laboratory of mineralogy in a provincial university, represented some of the worst aspects of French scientific life at the turn of the nineteenth century – a fiefdom with no necessary relation to any other institution. When he was cast adrift by Prince Albert in 1905, he had no scientific resources at his command except his own and no peer group within which to test and refine his ideas.

In this, Thoulet was not greatly different from a host of other French scientists, particularly in the marine sciences. In comparison with Germany, where professors were state employees and could readily link themselves with state enterprises (such as the need for information on fisheries), in France there was no such link. Thoulet was well aware of the significance of the foundation of the International Council for the

Exploration of the Sea in 1902, but for complex reasons, France was not one of the early members of ICES.[94] This isolated its marine scientists from the scientific stimulus, the development of new techniques, the sampling programs, and the regular scientific meetings that were so important to scientists from the nations participating in ICES.

The degree of isolation of French scientists other than Thoulet from currents of scientific thought outside the country is striking, judging by the oceanography textbooks of Thoulet's time and after. This was partly a consequence of the educational division within France that limited scientific training to narrow faculties and institutes, and probably partly to the lack of teaching in fluid dynamics except in the engineering context of the École polytechnique. If even physicists were unlikely to get an introduction to the behaviour of real fluids in an oceanic or meteorological context, what hope for biologists, chemists, or geologists? To his credit, the biologist Jules Richard wrote a textbook, *L'océanographie* (1907), that attempted to survey much of the field, although mainly centred on instruments. Thoulet's method is mentioned very briefly, without comment,[95] Significantly, so too is the 'méthode de Bjerknes et Sandström,' without Helland-Hansen's simplification, but no details are given and the reader is referred to V.W. Ekman's difficult series of papers published in 1906.[96] The prudent Richard, knowing of Thoulet's difficulties, and the criticisms of his method by Otto Pettersson, was sitting on the fence. Fifteen years later, the mainly descriptive *Manuel d'océanographie physique* of the French naval officer and writer Jules Rouch (1884–1973), who was professor of physical oceanography in the Paris Institut océanographique from 1937 to 1945, referred to the 'classic' works of Thoulet and Richard,[97] repeated the conclusions of Bouquet de la Grye on the topography of the North Atlantic, and described the meridional circulation just as Schott had in 1902.[98] To give him credit, the busy Rouch, who gave up his professorship in Paris after he became director of the Musée océanographique in 1945, brought his later text (written under wartime conditions), the *Traité d'océanographie physique,* at least up to the 1920s, including short descriptions of the work of Bjerknes, Helland-Hansen, Sandström, Ekman, and E.H. Smith.[99]

The most significant work to appear late in Thoulet's life was the two-part 'Leçons d'océanographie physique' (1930–1) by Alphonse Berget, who had taken the physical oceanography chair in Paris denied to Thoulet. Both volumes contained thorough physical descriptions of the atmosphere and oceans, but practically no up-to-date oceanography, and no mention of standard instruments except those designed and used in France. The *Meteor* results (see chapter 5), hot topics in the 1930s, were not mentioned. Although Zöppritz's results were included, Berget completely ignored the twenty-five-year-old work of V.W. Ekman, which vitiated Zöppritz's conclusions about the effect of the wind on deep circulation.[100] In a section on measuring currents, Berget referred to Thoulet's technique as 'a method of rare scientific elegance,' claiming that 'this method will be *in the first rank* of those that one can apply to the study of water movement in the great depths of the sea,'[101] but he wrote only a few perfunctory sentences on the 'method of Bjerknes and Sandström' without any discussion of its significance and use.[102] There is no mention, in the context of ocean circulation, of any dynamical oceanographic results. The Norwegian Mohn is referred to as an Austrian, and Defant's *Dynamische Ozeanographie* is misspelled. In short, Berget was woefully out of touch with the content of the field and with the changes in physical oceanography that had occurred between 1900 and 1930 – while teaching in France's premier institution for the study of the sea. Students who wanted to be at the cutting edge of physical oceanography would have to go elsewhere, and this remained the case until the 1960s, when the first mathematically based text in French burst upon the scene.[103]

The remarkable accomplishments at sea of Prince Albert, and the self-absorbed concentration on a single aspect of the sea by Julien Thoulet, when placed in larger context, occurred when oceanography in Germany was undergoing a rapid transition from physical geography into a branch of mathematical geophysics. The same techniques and approaches were about to leap the Atlantic, into the United States, and to be permanently incorporated into physical oceanography there and in Europe. France was set apart partly by the individualism of these two men and those they influenced, but perhaps even more by the absence of any direct way of incorporating new mathematical techniques into the practices of a scientific community singularly unprepared – for personal and institutional reasons alike – to accept such a change.

7

Slipping away from Norway: Dynamic Oceanography Comes to the United States

There is one other undercurrent ... which I have noticed, but which Rossby feels strongly about. As you know geophysics has been dominated for many years by the Norwegians. There are nine of them here and it is evident that they are beginning to slip. This would be alright if they would take things more gracefully, but they continue to point out endlessly that nobody except the Norwegians understand [sic] the problems, even though it is clear that they have now been surpassed in some questions by other countries. Rossby is very pleased that he will train the new hydrologist for Bermuda. He counts this as the beginning of the fall of Bergen.
Columbus Iselin to H.B. Bigelow, 22 September 1936[1]

Dynamic Oceanography in a New World

In the previous chapters, I have made the point that the introduction of mathematical techniques – 'dynamical oceanography' – into the practice of sea-going geographers and physicists was not a monolithic takeover of one scientific approach by another, but a more subtle shift, an evolution rather a revolution. And in the changes, the outcome was governed in a major way by the environments into which dynamic oceanography was introduced. Scientific contexts, previous training, local needs, and often the personality and scientific preferences – the 'scientific style' – of the new practitioners had very important influences on the way the new technique, a vanguard of a different physical style, was applied and to what degree it was successful. While it may be true that, as Columbus Iselin claimed in 1936, the Bergen proponents of dynamic oceanography were 'slipping,' it seems more likely to me that they were merely being overtaken by the success of their own methods and approaches. Others had taken the examples given in Scandinavia and the techniques pioneered there and run with them.

The United States in the first three decades of the twentieth century provides an important test of this hypothesis. Dynamic oceanography was introduced to the country – although not to the continent of North America (see chapter 4) – by scientists working on opposite coasts, although in very similar settings, namely, small laboratories or research groups with a need to make the study of the sea quantitative. One was marginalized and others entered the developing mainstream. Out of each experience came the science of physical oceanography that was consolidated, made canonical, and entered the Second World War to be transformed further in scale and goals. Modern physical oceanography came out of this era. The foundations of this change were established, not by individuals building empires, but by solving the problems they faced as scientific seafarers, working scientists, and managers of men.

'Useful in Many Capacities': George McEwen Finds the Sea

In April 1908, W.E. Ritter, the director of the Marine Biological Association of San Diego's laboratory at La Jolla, California, wrote to his friend and patron Dr Fred Baker,

> I believe I have found exactly the man for the hydrographic work: it is George F. McEwen, an assistant in physics at Stanford. He is a candidate for a doctor's degree there and is willing to take hold of our problems for his chief research work. He will consequently stay with us for two or three years at least. He appears to be a very intelligent young man and is taking hold of it with genuine interest ... With the rest he has some knowledge of surveying and navigation and so should be like Jas G. Blaine, 'useful in many capacities.'[2]

George McEwen (1882–1972) (fig. 7.1) stayed in La Jolla long beyond his nominal two or three years, retiring in 1952 as professor of physical oceanography at the Scripps Institution of Oceanography (SIO), the descendant of Ritter's early laboratory.[3] Upon his part-time appointment in 1908, McEwen became the first professional physical oceanographer in North America. By the time his part-time appointment in La Jolla became a full-time job in 1912, McEwen had begun to transform himself into a mathematical physical oceanographer, basing his work upon the European pioneers of the field like V.W. Ekman and Vilhelm Bjerknes

7.1 George F. McEwen (1882–1972) in his office, Scripps Institution of Ocean-ography, about 1935. Photo by Eugene Lafond, Lafond Papers, with permission of Scripps Institution of Oceanography Archives, UC San Diego Libraries.

and applying European ideas as well as his own techniques to calculate current speeds and directions off the west coast of the United States.

Largely by chance, the young McEwen found himself working in the rapidly developing new field of marine science. With his abilities as a physicist, he was well equipped to comprehend the mathematical work of the Scandinavians. His notes show that while he was still working on a Ph.D at Stanford, certainly no later than January 1910, McEwen had translated V.W. Ekman's papers of 1906 on the wind-driven circulation, and that a year later, while he was teaching mathematics at the University of Illinois, he read the works of Bjerknes, Ekman, and Nansen, though probably not at that time the papers of Sandström and Helland-Hansen.[4] No other North American at that time could match the combination of ability and opportunity that met in McEwen, allowing him to apply mathematical physical oceanography to North American waters.

Despite his advantages of time, place, and ability, McEwen did not found a North American school of dynamical physical oceanography. His lengthy career in La Jolla took place outside the mainstream of physical oceanography in the New World, and the opportunity passed to others on the East Coast of the United States. Outlining McEwen's work and ideas reveals why this was so. McEwen's personality, the intellectual setting, and the physical environment in which his work was done played roles in what fairly can be called his failure to develop dynamical oceanography in California. But such a judgment is also unfair, even anachronistic, because when McEwen began work in La Jolla, it was not clear what a physical oceanographer was expected to do, nor in which direction the field was headed. His career shows how loosely drawn were the limits of physical oceanography early in the twentieth century, and how one consistent line of work, like McEwen's, could be left behind or become irrelevant as the new discipline evolved into the mainstream of mathematical physical oceanography. George McEwen, a young man from Iowa, had a different, ultimately unsuccessful, vision of the physics of the sea compared to members of the East Coast elite who began studies similar to his nearly a decade later.

Dr Ritter's Laboratory at La Jolla

The Marine Biological Association of San Diego, and the laboratory in which George McEwen found himself in 1908, came about as a result of W.E. Ritter's wish to found a biological station on the coast of California from which a 'biological survey' of the coast between the Mexican border and Point Conception, based upon new principles, could be carried out.[5] Ritter (1856–1944) came to California in 1866 from Wisconsin, where he had been a schoolteacher, to study zoology at Berkeley. After graduate work at Harvard, he returned to Berkeley as lecturer and chairman of the Zoology Department in 1891. Scientifically prolific, persuasive and energetic, he rose quickly to assistant professor in 1894, associate professor in 1898, and professor in 1902. By the time he chanced upon San Diego as the site for a marine station, he had already spent a year's leave (1894–5) in Liverpool, Naples, and Berlin, and had been a member of the Harriman Alaska Expedition.

Ritter's laboratory in San Diego began its life in the Coronado Hotel in 1903.[6] Between 1905 and 1909 it was housed in a small building in the La Jolla city park, but at the urging of the newspaper magnate E.W. Scripps, who with his half-sister Ellen Browning Scripps befriended Ritter and became his patron, the Association bought land north of La Jolla, where a laboratory building was completed in 1910 (fig. 7.2). Ritter and his wife soon moved to La Jolla and took up residence in the new laboratory. Within a few years, the laboratory became part of the University of California, becoming the Scripps Institution for Biological Research in 1912 (fig. 7.3), and upon Ritter's retirement (in 1923) and his replacement as director by the marine geologist T. Wayland Vaughan (1870–1952) in 1924, it was renamed the Scripps Institution of Oceanography, with the aim, agreed upon by Ritter and Vaughan, of becoming a full-fledged oceanographic institution.

From the beginning, Ritter's intention of conducting a 'biological survey' based at the Marine Biological Station in San Diego was more complex than the words suggest. He was aware that the marine biota of the California coast was unexplored, but he was not content merely to list and catalogue the organisms available along the shore and in the coastal waters. Morphological or taxonomic studies for their own sakes were worthwhile but not intellectually satisfying aims for Ritter. He wrote of 'the vast scale on which things are done in the ocean and the literally infinite complexity of cause and law there in operation.'[7] The biologist needed help to understand the complexity of conditions under which marine organisms lived. That help must come from other specialists, working together in a group, much as teams of specialists collaborated in an astronomical observatory.[8] Specialists, carefully selected for their ability to contribute to common goals, would find it advantageous to contribute because they would be able to do more as members of a group than they could as individuals with less extensive resources.[9]

7.2 The George H. Scripps Building (left-most, closest to the sea), with other buildings of the Scripps Institution of Ocean-ography in 1925. With permission of Scripps Institution of Oceanography Archives, UC San Diego Libraries.

7.3 The staff of the Scripps Institution for Biological Research (later the Scripps Institution of Oceanography) in 1916. From the left, W.E. Ritter (director), C.O. Esterley, George F. McEwen, Percy S. Barnhart, F.B. Sumner, Ellis H. Michael, H.H. Collins, and Wesley C. Crandall. Printed with permission of the San Diego Historical Society – Ticor Collection.

As Ritter's early ideas consolidated into a full-fledged philosophy of biology,[10] he became convinced that there were dangers in an overemphasis on laboratory science. While laboratory work was essential for the study of biological processes under controlled conditions, or to show the potential of organisms under conditions they would never meet in nature, he was aware that frequently it was substituted for field observation and was regarded more highly than field studies. Laboratory science was essential to understand nature, but only field observation could direct the laboratory investigator to ask questions that had reality in the natural world. He asked, '... how far can the differentials between two kinds of organism be correlated with differentials in the environments which they respectively occupy?'[11] To what extent did currents, temperature, the specific gravity, and chemistry of the water affect the abundance of organisms? The question raised the issue of what Ritter called 'organic adaptation,' that is, the degree to which animals were matched to their

physical and biological environments, and the degree to which their evolution resulted in, or was caused by, cooperation in nature, as in human
society.

Ritter was not blind to the virtues of mathematics, which he saw as
having a two fold role in marine biology. First, statistical tests (corre-
lations) would reveal associations between the morphological and be-
havioural traits of organisms and their physical environments. And the
cyclical and periodic phenomena of individual development (ontogeny)
could be expressed and examined using mathematical techniques. Ritter was unusual for his time in believing 'that mathematics called to the
service of biology and kept strictly in its place as an assistant, is not only
enormously important, but for many of the deepest problems absolutely
indispensable.'[12] Mathematical analysis was a tool, among many, useful
for discerning the complex of factors that interacted to give an organism
distinctiveness and a distinctive environment.

An early expression of Ritter's scientific philosophy appeared in 1912,
the year that George McEwen became a full-time employee of Scripps
Institution for Biological Research. Ten years after he had begun the
search for a site for a seaside laboratory, Ritter took the opportunity to
outline what he called 'a sort of confession of faith as to the larger meaning of science, of biology in particular.'[13] This reveals the sort of intellectual environment in which the young McEwen found himself at the
beginning of his career as an oceanographer.

Science, to Ritter, was not merely an expression of human imagination
but also a source of morality and value. He stated clearly his 'belief ... that
science must justify its right to live and flourish, not alone in its ministra-
tions to physical well-being, but also to the higher and highest reaches of
man's nature.'[14] Scientific institutions such as his laboratory had a three-
fold duty: to study nature; to communicate the results of those studies to
the public; and to provide thereby an objective and humane support for
religion, philosophy, ethics, art, and other non-scientific aspects of human culture. To restrict science to the practical side of life alone would
be wrong, and to deny the cultural values of science, as he frequently
exhorted teachers, was to deny the lessons that nature held for human
society as well as for the higher morality of human intellectual life.[15]

Ritter later called his philosophy of science the organismal conception
of life.[16] This had both operational and theoretical aspects. His laboratory in 1912 was based upon the belief that 'the widest generalizations
which any science is capable of reaching can never be reached until the
whole range of phenomena touched by such generalizations has been

examined.'[17] Once such an examination had taken place, scientific value lay in the synthesis of information, not in the mere analysis that had taken place. The reasons for this lay deep within Ritter himself, for he hoped to achieve a synthesis of his own tendencies toward mysticism as well as analytical thought and action. As long as mind was sustained by nature, all its manifestations had to be taken seriously, to be examined analytically, and to be synthesized into a union of man's intellect with the natural world.[18] The reality of nature lay in its fundamental diversity, not in the kind of reductionism that saw ultimate reality in matter alone and would exclude the complexity of human experience, the view he credited to his then colleague at Berkeley, Jacques Loeb.[19] Thus, in 1912 and after, Ritter's laboratory had to work 'in both the small *here* and the vast *yonder.*'[20] His laboratory was in a position to make unique contributions, Ritter believed, because of its concentration on the diversity of nature and experience. Other marine laboratories, many of them larger and better equipped, could concentrate on the analysis of nature; the Scripps laboratory could carve out its niche with modest tools because of its attention to the synthesis of diversity. Francis B. Sumner, who came to Ritter's laboratory in La Jolla from Woods Hole in 1913, summarized its approach wryly but accurately:

> The Scripps Institution had a creed, which its members have repeated with child-like faith, following the words of their father-confessor. One of the articles of this creed has been the importance of studying the relations between the organism and its environment. Another has been the recognition of the one-sidedness of either field or laboratory study, considered by itself, and the consequent need of combining the two for a proper understanding of vital phenomena. Still another has been the necessity of employing rigidly quantitative methods, as far as these may be applicable. Finally, the organism itself has been wholeheartedly recognized as having a real existence, in its own right, and not merely allowed a provisional existence, pending its analysis into chemical, morphological or genetic elements.[21]

After four summers of work in La Jolla, George McEwen found himself a full-time member of Ritter's laboratory, in day-to-day contact with the California environment at all seasons and with Dr Ritter himself. In 1908, Ritter had announced McEwen's initial appointment with the remark that 'the particular satisfaction in having a physical laboratory operating in conjunction with the biological work lies in the fact that whenever a special biological question comes along requiring information from the physical side, the physicist can be appealed to *then and there*.'[22] But it was an open question how well McEwen, viewing himself as a mathematician and physicist, would adapt himself to Ritter's grand view of science.

The Work of a Physicist

During his first four summers in La Jolla, while he began Ph.D. studies at Stanford, McEwen began to do hydrographic work, collaborating with the station naturalist, Ellis Michael, to measure temperature and density at many locations off the Southern California coast. These rapidly demonstrated that the water lying along the California coast was cooler than water farther offshore. Hermann Thorade (1881–1945), a German schoolteacher who worked part-time at the Deutsche Seewarte in Hamburg, had noticed the same phenomenon when he compiled surface water temperatures recorded by ships en route to and from the Panama Canal or in the eastern Pacific between 1898 and 1906. The close relation between low water temperatures near the coast and the strength of northwest winds could be explained using V.W. Ekman's mathematical theory of the drift of surface water at an angle to the downwind direction, which would result in coastal upwelling. By early in 1910, McEwen had read Thorade's paper, was translating Ekman's works, and was rapidly exploring the possibilities that Ekman's work offered for work off California.[23]

With considerable originality, McEwen reasoned in 1911 that it would be possible to test Ekman's theory provided one assumed that without upwelling the sea temperature along the coast would be the same as that of water far from shore, where the temperature was established due to the sun's radiation, conduction, and evaporation. This he called the 'normal temperature,' defining it later as the temperature 'which would prevail in the absence of a general drift or flow of the water as a whole, either vertical or horizontal.'[24] The measured departure of the observed sea temperature from the 'normal temperature' would indicate the amount of upwelled water that had reached the surface. Testing his method with sea temperature data from four locations along the California coast from San Diego northward to Cape Mendocino, McEwen found that he could predict the temperature reduction with fair accuracy, thus giving strong support to Ekman's theory. Any departures of the calculated temperatures from the observed ones could be explained by variations in the direction of the coast relative to the wind, variable winds, and the slope of the bottom, all factors known to affect current direction in Ekman's theory.[25] He noted too that the coastal climate was linked to water temperature through upwelling and wind, suggesting the possibility that sea temperature might be used in some way in weather forecasting.[26]

No later than 1912, McEwen had touched upon the two topics that were to occupy him for the rest of his career, the determination of ocean currents using temperature and problems of long-term weather forecasting based upon linkages between marine and terrestrial climates. In his thinking, the two were closely related; moreover, the two required the application of a particular method, which Ritter referred to as 'the search after physical formulae by the use of which the relatively few field observations that can be made at any one time shall give more information concerning the phenomena in question than the observations alone afford.'[27] Although McEwen was never comfortable with scientific work not grounded firmly in causally linked phenomena, a category that included long-term weather forecasting, his early work undoubtedly satisfied Ritter's wish to have all the research in La Jolla directed toward complex, large-scale problems.

After 1912, encouraged by his success in testing Ekman's theory with simple physics, McEwen began to use mathematical physics more extensively to describe and explain the distribution of ocean temperatures in both the horizontal and vertical directions. As he summarized his approach, 'Such considerations led to a general investigation of the relation of temperature to the ever-present factors, radiation, evaporation, and the resulting alternation or mixing motion of small water masses, and the effect of a given drift, horizontal or vertical, upon the normal distribution of any property of the water.'[28] The contrast with the dynamical method of the Scandinavians Bjerknes, Sandström, and Helland-Hansen was great. Rather than an estimation of ocean currents based upon temperature and salinity, which entered into a calculating equation based upon a number of untested assumptions (like the level of no net motion – see chapter 3), or unmeasurable physical properties (such as the eddy viscosity of the water), McEwen preferred to work with a suite of measurable physical properties affected by the movement of the water which were quantitative indices of the water's motion. In a series of publications between 1915 and 1938, he doggedly amplified and refined mathematical models expressing the horizontal and vertical distribution of temperature and other properties in the ocean,[29] taking the approach embodied in the title of his lengthy paper in 1919, 'Quantitative Comparisons of Certain Empirical Results with Those Deduced by Principles and Methods of Mathematical Physics.'

After several years of work, interspersed with other activities such as routine data gathering and work on climate, McEwen had by 1918 arrived at two generalized differential equations for the calculation of normal temperatures. Thereafter, most of his work on horizontal currents lay in refining this approach, occasionally using dynamical methods (what he called 'Bjerknes's circulation theory') to check the plausibility of his calculations. The most troublesome problem with the use of his differential equations was knowing the appropriate constants to express cooling and the absorption of heat. Through the years, he slowly revised and corrected the constants and the equations based upon new information on the distribution of temperature in the Pacific and on evaporation, which his students studied in the laboratory and in lakes beginning in 1919. By 1938, after much labour, McEwen had arrived at what he considered the definitive equation expressing theoretical sea temperature in the top thirty-five metres in terms of heating, cooling, and water motion.[30]

In this work, the problem of vertical motion, which McEwen had tackled first in 1912, was most challenging, for vertical motions, being relatively slow, involved turbulence, which was not readily dealt with using classical equations of motion or diffusion. Zöppritz, in his early attempts to deduce the effect of the wind upon oceanic circulation, had used a laboratory value of the viscosity of water; the result was that the wind could cause significant currents at any depth only if it blew for hundreds or thousands of years.[31] By the time of V.W. Ekman's first paper, in 1905, hydrodynamicists were well aware that because of small-scale turbulent motions in the sea, some much larger coefficient of eddy viscosity was appropriate. As Ekman showed, once a realistic coefficient was used, the wind readily caused currents, and realistic current velocities could be calculated. In addition, the sub-surface ocean actually responded rapidly to the wind, contrary to Zöppritz's results.[32] Before Ekman and McEwen, only the Dane J.P. Jacobsen had investigated the coefficients of eddy viscosity in the sea, basing his calculations on work in the northeast Atlantic.[33]

McEwen's preoccupation for many years was what he called 'a single mathematical theory of the mixing motion.' His mature theory of the vertical distribution of temperature and salinity was published in 1929. He solved the problem of linking the quiescent deep waters with the well-mixed surface waters using the physical approach of statistical mechanics, in which parcels of water moved up and down, redistributing temperature, salinity, and other properties in the top hundred metres or so; below that, the classical equations of motion or diffusion applied. In this way, the 'normal' distribution of properties such as temperature, salinity, and nutrients, could be calculated, as well as the effects of turbulence, upwelling, and the biological uptake of nutrients.[34] The summit of McEwen's achievement was what he termed 'a pair of simultaneous differential equations expressing causal relations between turbulence, rate of surface cooling, rate at which solar radiation penetrates the surface, vertical distribution of temperature and salinity, and the rate of vertical flow of water.'[35] There were, however, two drawbacks: the large number of constants involved in the equations, each of which had to be determined; and the fact that the equations could not be solved analytically, only by graphical or numerical approximations.

Despite McEwen's repeated claims that he was close to having suitable 'approximate' solutions of the equations, these were never published. And although he continued tabulating coefficients and constants to increasingly greater accuracy, he had reached the end of the line with this work on currents after the mid-1930s. The complexity of his equations, which required lengthy computation, the unfashionability of his approach, and the environment in which his work took place, combined to push his work into obscurity. Late in his career, he found himself marginalized by an intellectual descendant of the Scandinavian mathematical physical oceanographers who had established the method that he hoped to replace.

Harald Sverdrup Takes the Helm

W.E. Ritter regarded George McEwen as one of his brightest stars, and McEwen's stock remained high when T. Wayland Vaughan succeeded Ritter in 1924 to preside over the newly-named Scripps Institution of Oceanography (fig. 7.4).[36] In fact, under Vaughan, who had the connections to arrange that large amounts of new temperature data be sent to McEwen, McEwen's long-range weather forecasting program at SIO had its best years, and his work on ocean circulation developed its complexity.[37] Even during the early years of the Great Depression, Vaughan's enthusiasm and the demand of power companies for McEwen's forecasts enabled his program to expand so that the labour of calculation could be shared with a group of assistants.

Vaughan's successor was Harald U. Sverdrup (1888–1957) (fig. 7.5), regarded as one of the world's leading physical oceanographers.[38] Sverdrup had been both student of and assistant to Vilhelm Bjerknes in Oslo

7.4 Staff of the Scripps Institution of Oceanography in 1935, during the visit of Bjørn Helland-Hansen. From the left, F.B. Sumner, George F. McEwen, Roger Revelle, Blodwyn Lloyd, Richard H. Fleming, Helland-Hansen, T. Wayland Vaughan (director), Erik Moberg, and W.E. Allen. From Lafond Papers, with permission of Scripps Institution of Oceanography Archives, UC San Diego Libraries.

and Leipzig, where he worked on meteorological problems. In 1917, he joined Amundsen's north polar expedition on *Maud*, remaining with the ship until its return to Norway in 1925.[39] By 1926, when he succeeded Bjerknes as professor of geophysics in Bergen, Sverdrup had published authoritatively on meteorology, atmospheric physics, marine chemistry, magnetics, tides, oceanographic instruments, and even Arctic ethnology. Like Bjerknes, he was well known at the Carnegie Institution of Washington, where he had reduced the magnetic data from the *Maud* expedition while the ship was in Seattle in 1922. Using data from the oceanographic cruise of the Institution's vessel *Carnegie* in 1928/1929, Sverdrup made new, important contributions to the understanding of upwelling in the Peru Current and the origin of Pacific deep waters.[40] His appointment was a clear signal that Scripps was expected to develop a world-class reputation in physical oceanography.

When Sverdrup succeeded Vaughan in September 1936, McEwen's

7.5 Harald U. Sverdrup (1888–1957), director of the Scripps Institution of Oceanography, 1936–48, in his office in October 1941. With permission of Scripps Institution of Oceanography Archives, UC San Diego Libraries.

long-range forecasting program was still alive, though greatly reduced in scope, and McEwen was continuing the refinement of his temperature-based method of calculating ocean currents. Sverdrup found, in addition to McEwen, six other faculty members, only two of them young (Roger Revelle and Richard Fleming), and a very small group of students. He set out to reform the entire program at Scripps, beginning with its teaching and its overall approach to research. Nearly from the beginning, partly as a result of his reforms and partly because of a clash of personalities, there was strain between Sverdrup and McEwen.[41] The result was to make McEwen's work less and less characteristic of what went on around him after 1936.

Long before Sverdrup's arrival, Vaughan had set out to extend and formalize the teaching program begun by Ritter by instituting a general seminar in oceanography given by all the staff, along with a number of other courses, including dynamic, geological, and chemical oceanography, microplankton, and marine ecology. Later, mathematics was added to this. McEwen's seminar course on 'dynamic oceanography' changed little during the Vaughan era. Based mainly on Bjerknes and Sandström's *Dynamic Meteorology and Hydrography* (1910), later partly on E.H. Smith's account of the dynamic method (1926), it included sections on units, the properties of seawater, hydrostatics, and the dynamic method. Like many of the other courses, it was relatively informal and probably given irregularly, especially in the early years.[42]

Sverdrup's reform of teaching at Scripps was realized in June of 1938. It included a short-lived course for undergraduates, and a series of courses required for graduate study, including a general seminar, physical oceanography (Sverdrup, McEwen, and Fleming), marine meteorology (Sverdrup and McEwen), marine geology, chemical oceanography, marine microbiology, phytoplankton, marine invertebrates, marine biochemistry, and biology of fishes.[43] The young faculty members were now involved in the Institution's expanded, formalized program, and McEwen found himself giving way to Sverdrup in teaching physical oceanography. This appears to have been Sverdrup's goal, for to Vaughan in 1936, before arriving in La Jolla, he wrote of not wanting 'to disappoint those who think me able to carry your plans further, and, according to these, help making Scripps Institution a center also of dynamic oceanography.'[44] Clearly McEwen had not done so.

Research in Oceanography – the McEwen Style

Some time in 1929, George McEwen prepared an entry for *Who's Who in America*. Describing his profession, he wrote:

> Professor, Physical Oceanography, engaged mainly in scientific research involving applications of mathematics to 'field' problems in oceanography. Research in geophysics, {oceanography / meteorology} and statistical methods. Temperature; salinity and circulation in the North Pacific Ocean. Statistical methods, measurement of viscosity.[45]

Quite accurately, McEwen saw himself as a mathematical physicist. It was this approach that governed his thinking, his work, and ultimately his failure to play a significant role in the development of physical oceanography in North America.

As a physicist, McEwen always emphasized the primacy of method and measurement. He approvingly quoted the French oceanographer Julien Thoulet's definition of oceanography as 'the application to the natural phenomena of the sea [of] the precise methods of the exact sciences, mathematics, mechanics, physics and chemistry.'[46] This was in striking contrast to the approach taken by both Vaughan and Sverdrup. Vaughan's approach was synthetic; oceanography 'comprehends all studies of oceans themselves and of the interrelations between the oceans and other parts of the earth.'[47] Sverdrup took an equally broad view, beginning the great text *The Oceans* with the statement that 'oceanography embraces all studies pertaining to the sea and integrates the knowledge gained in the marine sciences that deal with such subjects as the ocean boundaries and bottom topography, the physics and chemistry of sea water, the types of currents, and the many phases of marine biology.'[48] Late in his career, McEwen wrote that the 'essential unity' of oceanography had brought together many lines of work at Scripps,[49] but a close examination of his professed methods of science shows that his work was dominated by one narrowly drawn approach to the physics of the sea.

Although McEwen referred to physical oceanography as 'a synthetic science,' to him its first task was one that any physicist would recognize, namely 'the accurate measurement, throughout the ocean, of the physical and chemical properties of the water, such as its temperature, motion, salinity, gas content, and acidity.' Only when these tasks of measurement and description had been accomplished was the second goal to be tackled, that of 'coordinating and interpreting the direct objective knowledge, and thereby discovering general laws of the phenomena of the sea.'[50] The 'general laws,' however, were always secondary to the precise measurement of physical conditions in the sea, and true to his training as a physicist and mathematician, McEwen was never satisfied with a physics on which the boundary conditions were not closely drawn, or in which the physical conditions had not been characterized with precision. His career shows his constant search for ever better calculated constants, or more accurately solved differential equations expressing physical phenomena in seawater.

McEwen's attempt to reduce the complexity of nature to 'true and useful principles' was and is one that most physicists and physical oceanographers would find congenial. Most recent mathematical modelling is aimed at exactly what McEwen hoped to achieve, the match between physical models and observations from nature. However, what is unusual about his work, in retrospect, is how it stalled in mathematics, the precise evaluation of constants, and the solution of differential equations. McEwen never overcame the mathematician in his nature to achieve what gave success to less able mathematicians like Sverdrup – physical insight, to which mathematics was secondary. He was an intelligent, skilful mathematical physicist who was never willing to make the compromises in precision and physical reality that are necessary to understand, provisionally and subject to revision, the complexity of the sea.

Given this knowledge of his scientific philosophy, it is easier to understand why George McEwen paid so little attention to dynamical oceanography as it was developed and practised in Europe even though he knew the European methods intimately after only a few years in La Jolla. By 1911, when he was teaching in Illinois, he was reading Bjerknes's papers, and when he taught his first course on physical oceanography, about 1917, Bjerknes and Sandström's *Dynamic Meteorology and Hydrography* was part of his canon (although he did not refer directly to the work of Helland-Hansen and Sandström, which converted Bjerknes's theorem into a practical method of calculating currents). No later than 1919, McEwen reported that he was working on application of 'Bjerknes' Theory.' Once his own methods of current determination, using 'normal temperature,' had been worked out, he proposed, but never tried, combining his technique to predict changes in current patterns.[51] And during the time he was actively developing his method, no one else applied dynamic oceanography to eastern North Pacific waters.

It was the theoretical basis of the dynamic method that McEwen found objectionable. Its unwarranted physical assumptions, its use of physical shortcuts (the reference level of no net motion was one) rather than empirically based measurements, made 'Bjerknes' method' philosophically incorrect and 'speculative,' as well as too prone to error, for his satisfaction. Admittedly, McEwen became more sympathetic to the dynamic method in 1924 when Vilhelm Bjerknes visited the Scripps Institution, lecturing on oceanographic methods and ocean dynamics.[52] Within a year, he was using the dynamic method to calculate currents off Southern California;[53] four years later, he used it to calculate currents off Alaska using temperature and salinity data gathered for W.F. Thompson of the International North Pacific Fisheries Commission.[54] Then and later, his lecture outlines and general reviews of physical oceanography reveal that he was closely familiar with dynamic methods and their background, partly through his conversations with Bjerknes, partly through keeping up with the published literature.[55]

But the dynamic method was useful only under certain circumstances (especially when only temperature and salinity were available) and as a check on his own, very differently based method. McEwen viewed his own, original techniques as preferable to dynamic methods only partly because they comprehended more phenomena, including the thermodynamics of the ocean surface. The real problem, according to McEwen, was the difficulty of working with techniques like the dynamic ones that involved physical causes: 'Investigations of ocean currents may be approached from the dynamical side in which case, the currents are regarded as the effect of a set of causes or they may be regarded as causes producing certain effects.' The dynamic method of Bjerknes, Sandström, and Helland-Hansen fell into the first category; his own 'normal temperature' method into the second. He had developed his own method because of his belief that causes were frequently unapproachable in physics, that 'the theoretical determination of currents from the causes, even though the latter may be known, is in general, extremely difficult.' He concluded that 'the subject of the dynamics of ocean currents is far from satisfactory' because so little was known of internal friction, eddy motions, and similar phenomena.[56] It was for this reason that McEwen followed his own path into the physics of the ocean.

Although by 1940 McEwen was teaching mainly dynamic methods at Scripps, his heart still lay in his temperature method. He never discussed its grave defects, although they were clear, and certainly obvious to Harald Sverdrup. Chief among them was the problem of averages: McEwen's normal and calculated temperatures were based upon monthly (occasionally weekly) averages of ocean temperatures. Although he understood the necessity of understanding variations in the waters he studied, such as the California Current, where his work began, his approach to calculating currents obscured the very variation that should be determined. Fleming and Sverdrup's cruises showed that rapid changes occurred in the California Current, and Sverdrup's application of dynamic methods gave rapid and precise estimates of short-term changes in the California Current upwelling zones for which McEwen had first developed his 'normal temperature' method.[57] Moreover, McEwen's methods used a mathematics that was patently too complex and inappropriate to the quality of his data, bogging down repeatedly in differential equations whose analytical solutions were very difficult or impossible. Even if approximate solutions were obtained, they were frequently restricted to specific latitudes or depths, such as subtropical surface currents, and required far more prior knowledge than was necessary for dynamic calculations. No doubt McEwen understood the significance of many of these problems, but he persisted with his non-causal methods because they represented what he considered the proper scientific approach to the problems of the sea.

When McEwen left the Mathematics Department at Illinois in 1912 to join W.E. Ritter in La Jolla, he was assured that there was no such field as oceanography.[58] Vaughan's and Sverdrup's attempts to define oceanography during the 1930s show that many still held this view two decades later. McEwen suffered directly for his decision to enter a non-classical scientific field, for, as Vaughan's correspondence shows, when McEwen was proposed for membership in the National Academy of Sciences there was no support from basic scientists, even outright rejection from R.A. Millikan. Vaughan abandoned his campaign for McEwen after nearly two years of politicking.[59] It was ironical that McEwen, who chose to make a career in oceanography, should have ended up the odd man out in his profession. But such was the case; in 1912 it was impossible to forecast what the physics of the sea would become. On scientific and personal grounds, McEwen chose one route into his chosen field. The rest of the oceanographic world took another.

'Uplift, Propaganda and Scientific Gymnastics' on the East Coast of the United States

It is hard to judge how much scientific isolation contributed to the direction that McEwen's work took. It is clear, however, that the Scripps Institution of Oceanography *was* isolated both physically and intellectually from the rest of science in the United States until the 1940s. Francis Sumner, who was at Scripps nearly as long as McEwen, commented on the situation in the early years:

> Rarely did we visit other centers of learning, even Los Angeles or Berkeley; and rarely did we see scientific guests as transients, making us a few hours visit. And so we tended to become more and more introverted, and more and more lacking in perspective. The rigid limitations of the Scripps Institution, and its component members, were often forgotten. At times we seemed to feel ourselves divinely commissioned to expose the errors of our less favored scientific colleagues elsewhere.[60]

Even late in the 1930s, the letters of Richard Fleming's wife, Alice, to Roger and Ellen Revelle in Europe show how small Scripps still was and how distant the connection with the outside world remained even into the Sverdrup years.[61] Isolated on its hillside north of San Diego, the Institution was a small village unto itself in physical location and ethos.

Roger Revelle quotes one staff member of Woods Hole Oceanographic Institution, which was founded in 1930 on southern Cape Cod, to the effect that 'before World War II the Oceanographic (that is, Woods Hole Oceanographic Institution) was a place where about ten professors could come in the summer and not lose money. Scripps was an institution without much of a ship; Woods Hole had a ship but it wasn't much of an institution.'[62] How each institution played its role was significant, for as Susan Schlee has shown, only 10 per cent of the papers published by the Scripps Institution staff were on open ocean processes or organisms, whereas at Woods Hole fully 72 per cent were on the open ocean.[63] This was a mark of the isolation of Scripps, but equally of the different situations under which oceanography was developing on each coast.

In 1928, when oceanography in New England was beginning to take off, Henry Bigelow (1879–1967) (fig. 7.6), associate professor of zoology at Harvard and the newly appointed secretary of the National Academy of Sciences Committee on Oceanography, wrote from Cambridge, Massachusetts, to Wayland Vaughan in La Jolla that 'at least you will be several thousand miles away from all centres of uplift, propaganda and scientific gymnastics, a picture into which I find I do not in the least fit.'[64] In truth, Bigelow fitted very well into the expansion of oceanography on the East Coast that was leaving Scripps in its wake, and he was reaping the fruits of his own efforts, which dated from early in the first decade of the century.

At Harvard, the young Henry Bigelow came into contact early with Alexander Agassiz (1835–1910), the wealthy director of the Museum of Comparative Zoology (MCZ), accompanying Agassiz to the Maldive Islands in 1901, the year of his graduation. In 1904, as a promising young zoologist, he went with Agassiz to the eastern tropical Pacific on the U.S. Fish Commission steamer *Albatross,* and in 1906 he completed a Ph.D. on coelenterate biology at Harvard and joined the staff of the MCZ. When Alexander Agassiz died in 1910, Bigelow's taste for work at and on the oceans was established, but the opportunity to work at sea became a problem. Agassiz had established a working relationship with the U.S. Bureau of Fisheries (before 1902 named the U.S. Fish Commission), through which he hired research vessels, notably the magnificent and able *Albatross,*[65] and Bigelow had investigated deep-water pelagic fauna of the Gulf of Maine for Agassiz on the Bureau's schooner *Grampus* in 1908, but ambitious expeditions on the Agassiz scale were beyond his means.

It was apparently a visit of Sir John Murray to the United States in

7.6 Henry Bryant Bigelow (1879–1967) at the helm of the U.S. Fish Commission schooner *Grampus*, perhaps in the period 1912–16. From H.B. Bigelow Issue, *Oceanus* 14(2), July 1968, p. 2. With permission of Woods Hole Oceanographic Institution Archives.

1910 that stimulated Bigelow to link his work to the needs of the Bureau of Fisheries and in 1912 to begin a lengthy study of the Gulf of Maine, aimed at learning the environmental setting of what was regarded as the onset of overfishing.[66] By the time his close association with the Bureau, and especially the ability to use Bureau vessels, came to an end in 1924, Bigelow had amassed a large amount of information on the fish, plankton, and circulation of the Gulf of Maine, and was well on his way to producing major monographs that would establish his reputation worldwide in the marine sciences.[67] Among these, his 'Physical Oceanography of the Gulf of Maine' was a largely descriptive account of the entire region from southwestern Nova Scotia to Martha's Vineyard and Nantucket, with a great deal of detail on the seasonal and regional changes in temperature and salinity. About a hundred and twenty pages (just over 10 per cent of the total) dealt with the circulation of the area, beginning with a cautionary statement:

> Study of the circulation that dominates any part of the sea can be attacked in two different ways: (1) Directly, by observation with current meters or drift bottles, by ships' log books, and by interpreting the distribution of salinity and temperature, or (2) indirectly, by calculation of the hydrostatic forces tending to set the water in motion. The second method has greatly concerned oceanographers of late, and its value can hardly be overestimated in the study of ocean currents in the open sea; but its application to the Gulf of Maine is complicated by the disturbing factors introduced by the irregular contour of the bottom, the limiting coast line, and the strong tides ... It is fortunate, therefore, that the following account can be based on the more direct methods of observation, supported by consideration of hydrodynamic forces as causative agents.[68]

But Bigelow, trained as a zoologist, was never comfortable with the mathematics involved in dynamic oceanography, and frequently turned the calculations over to others. In the case of Gulf of Maine study, he states in the first paragraphs, that 'the chapter on hydrodynamics has been made possible by Lieut. Commander E.H. Smith's collaboration.'[69] A year later, his analysis of the physical and biological oceanography of Monterey Bay, California, may owe its dynamic analysis in part to George McEwen, but gives the credit for the method used to a monograph by Smith, who had been contributing to Bigelow's knowledge of mathematical oceanography since the early 1920s:[70]

The U.S. Coast Guard, Edward H. Smith, and Oceanography for the Ice Patrol

Edward H. Smith (1889–1961) (fig. 7.7), a native of Martha's Vineyard, Massachusetts, became a commissioned officer of the U.S. Coast Guard in 1910 upon graduation from the U.S. Coast Guard Academy. He came to know Bigelow about 1920, when he was assigned to duty on the cruises run by the USCG as participants in the International Ice Patrol, dating from just after the *Titanic* disaster of April 1912. Sometime late in 1920 or early in 1921, Smith was taken to sea off Gloucester, Massachusetts, with Bigelow and was, as he wrote to Bigelow, 'introduced ... to my first oceanographic station.'[71] During 1922–3 Smith took classes toward a master's degree, mainly under Bigelow (the degree was awarded in 1924), and thereafter clearly became a good candidate, in Bigelow's eyes, for a Ph.D. in oceanography, a subject that existed in name at Harvard, but without students or institutional structure.[72]

Bigelow's connection with the Coast Guard, and specifically the Ice Patrol, began early. After the disastrous sinking of *Titanic* in 1912, the nations involved in North Atlantic shipping met to consider how the risk of collision with icebergs could be reduced. During the summer of 1913, the Scottish research vessel *Scotia,* with the British oceanographer D.J. Matthews aboard, surveyed the southern Labrador Current area for ice and made some preliminary collections of oceanographic data.[73] But the United States was the only nation with the resources and sufficient interest to continue the work, and, after preliminary cruises in the summer of 1913 (before the Ice Patrol was established), became the major participant in the newly constituted International Ice Patrol when it was inaugurated in 1914.[74] A series of summer cruises, mainly into the Grand Banks area and the southern end of the Labrador Current, continued until the Second World War, after which surveillance by air began.

Shortly after its foundation, Henry Bigelow was appointed a technical advisor to the United States' committee of the Ice Patrol.[75] He, among others, was convinced that more than patrols and the tracking of ice were needed; *prediction* of the tracks of icebergs should be possible if there were sufficient knowledge of the currents of the area. The difficulty lay in finding skilled oceanographers who could contribute to the work of the Ice Patrol. Smith outlined the problem facing the Ice Patrol into the late 1920s:

7.7 Edward H. Smith (1889–1961) during the U.S. Coast Guard's *Marion* expedition to Baffin Bay and West Greenland, summer 1928. From Ricketts 1932, p. vii.

... the Patrol has had difficulty in attracting individuals, in this country at least, who possessed the particular qualifications requisite to head active investigations. The Coast Guard, therefore, during the past few years has encouraged its own officers – one especially – to improve themselves in a scientific way, wisely realizing that the Ice Patrol will give the most efficient economic service to shipping only when useful, scientific methods are employed to support the practical work.[76]

That 'one especially' was, of course, Smith himself. And early in his involvement with the scientific side of ice drift, it is clear that he was thinking and reading independently. Reporting by letter to Bigelow on an Ice Patrol cruise that began in February 1922, he wrote from Halifax that 'I have been reading Sandström's article on "Hydrodynamics" in the Canadian fisheries report.[77] Is that theory of his generally accepted by scientist's [sic]? I can make use of it, but I'd like to know, first, if it is considered sound.'[78] Bigelow's answer must have encouraged Smith, for within four years he became an authority on the use of the dynamic method, promoting it at every opportunity to his Coast Guard superiors and to physical scientists.[79]

During the inter-war years, Bjørn Helland-Hansen's Geophysical Institute in Bergen was the Mecca of aspiring young physical oceanographers, who went there to learn the dynamic method from one of its founders. Helland-Hansen welcomed overseas visitors, one of whom was E.H. Smith in 1924–5. Smith went with a sense of mission, visiting hydrographic offices in Britain and on the Continent while en route, encouraging major shipping lines in France and Holland to stick to predetermined (and thus presumably safer) courses to and from the New World, and attending a meeting of the International Council for the Exploration of the Sea in Copenhagen, where he spoke of the work of the Ice Patrol and met most of the big names in European marine science. Then, in late September 1924, he arrived in Bergen to attend Helland-Hansen's lectures and work with data from the Ice Patrol cruises. His aim was clear:

If the Patrol had knowledge of the drift tracks which bergs would follow after arrival at the gateway of the Atlantic [the Tail of the Grand Banks, where the Labrador Current and Gulf Stream met], much more detailed information could be furnished to approaching vessels, especially during the protracted periods when fog enshrouds this cold water region. It has been found that the drift of nearly all bergs is primarily controlled by relatively deep-seated currents and that the effect of the winds, such as those which usually prevail from April to July in this region, is practically nil. A monthly current map (for oceanic circulation does not fluctuate so rapidly as the atmosphere) of this critical area where the Labrador Current and the Gulf Stream meet, is an indicator of the courses the ice will follow. It is just here-in that modern methods of oceanography as taught by Helland-Hansen, and a few other European oceanographers, have a practical application for this particular problem confronting the Ice Patrol.[80]

A little more than a month after his arrival in Bergen, Smith wrote to Bigelow, 'H.H. has already given four lectures on the circulation theory which lies at the basis of all the computations and which was really first developed by Bjerknes. Solenoids, specific volume, and isosteres, do not have the terror they once had when I used to puzzle over Sandström's article in the Canadian Fisheries Expedition.'[81] By early December, he was beginning to convert the Ice Patrol temperature and salinity data from the Tail of the Grand Banks area for 1921 through 1923 into dynamic charts, and in early January 1925 he reported that the 'charts give the stream lines of the current around the Tail of the Bank and reveal vastly more regarding the scheme of circulation than we have ever been able to secure heretofore. It is encouraging to find that the stream lines and the drift tracks followed by the bergs in the critical area at the time, agree very closely.'[82] Just before he left Bergen in August 1925, Smith had completed most of the dynamic analyses, and in addition was about to send to Bigelow a lengthy paper based on Helland-Hansen's lectures describing the use of the dynamic method.[83] The practical outcome for the Coast Guard, according to Smith, was that the Ice Patrol should in-corporate permanent scientific programs into its routine work based upon dynamic oceanography and statistical meteorology. In his report to the commandant of the Ice Patrol, he spelled out an agenda for the future based on dynamic oceanography:

These new methods the application of which is virtually unknown to hy-drographers of America today, is not only recommended as scientific but also as extremely practical, for example, leading to advance information re-garding the drift of icebergs. Briefly stated a vessel equipped with the usual oceanographic water-bottles may be cruised over given region from which the temperature and salinity of the water be collected at frequent intervals. These data, when treated by the aforementioned computations result in an accurate chart showing the direction and velocity of the prevailing currents. It seems fitting that the Ice Patrol should lead in an attempt to give increased economic value to a work which it is believed will take place as an extensive ocean survey in the not far distant future.[84]

Dynamic analysis was now routine for Smith, who, as he prepared to resume graduate studies at Harvard in the spring of 1926,[85] sent Bigelow a sketch of the latest results he had calculated while on board an Ice Patrol vessel on the Grand Banks (fig. 7.8), emphasizing that 'this is the first time that the Ice Patrol, and to my knowledge any other expedition, has calculated the movement of ocean masses immediately for use on board of a ship. This work represents the practical application of the scientific knowledge acquired abroad.' He continued: 'We are now in a position to predict the courses which bergs will follow as they drift southward into the so-called "critical area" – "the gateway to the Atlantic."'[86]

The *Marion* Expedition – and a Practical Method for Determining Ocean Currents

During E.H. Smith's busy, lengthy trip to Bergen in the early autumn of 1924, he stopped in Copenhagen, and on the recommendation of the ICES hydrographer Martin Knudsen stayed there long enough to attend the annual meeting of ICES. Following his talk to the ICES hydrographers (physical oceanographers), Smith recorded the reaction of the powerful and influential Johan Hjort:

> Among those who spoke was Dr. Johan Hjort ... who had just returned from an expedition in his ship, the Michael Sars, to Davis Strait and Baffin Bay, a region of particular interest to the Ice Patrol, since it is from here that the Arctic ice really begins its drift – a journey which often does not terminate until such ice menaces even the extra-southerly steamship tracks. Dr. Hjort pointed out some of the important results which might be attained for the Ice Patrol by a pioneer expedition to this unknown region of the north ... Hjort believes that important information for the Patrol would be revealed by ... vertical sections made at critical points from the coast between Newfoundland and Baffin Land. I spoke with several Danes who had cruised in recent years in waters around Greenland, and from such conversations I have concluded that we have a vast deal to learn yet regarding the currents and the general ice movements in the far north.[87]

Hjort's remarks and Smith's growing interest in where the icebergs were

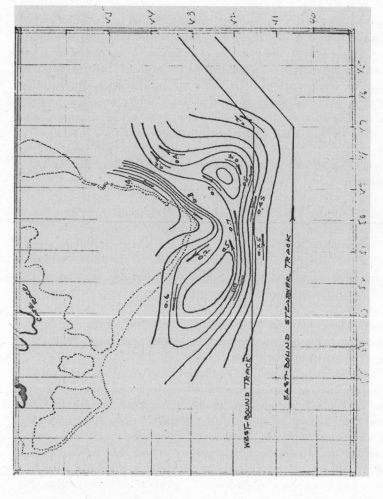

7.8A Currents around the Tail of the Grand Banks, Newfoundland, as calculated by E.H. Smith while at sea on an Ice Patrol cruise during the first week of May 1926. From Harvard University Archives, H.B. Bigelow Correspondence, Smith to H.B. Bigelow, 8 May 1926. With permission of Harvard University Archives.

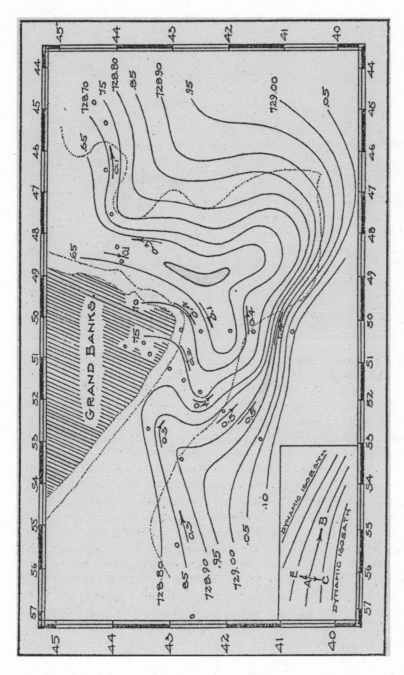

7.8B Currents in the same area four years earlier, 5–7 May 1922, calculated from Ice Patrol data by E.H. Smith while he was studying in the Geophysical Institute, Bergen, in 1924–5. From Smith 1926b, p. 39.

7.9 The U.S. Coast Guard cutter *Marion* off West Greenland during the summer of 1928. From Ricketts 1932, p. 24.

born, as well as where they ended up, resulted in a major U.S. Coast Guard expedition to Baffin Bay and West Greenland, the calving ground of Labrador Current icebergs, during the summer of 1928.[88]

The U.S. Coast Guard patrol boat *Marion* left New London and Boston in mid-July for Newfoundland, Labrador, and West Greenland (fig. 7.9) with two officers (Smith, now a lieutenant-commander, was in command) and twenty sailors. Smith's contacts in Scandinavia paid off now, for the expedition was organized in conjunction with the Royal Danish Navy, which had also been planning an oceanographic expedition to Baffin Bay to learn more of the marine resources available to Greenlanders. By pre-arrangement, the *Marion* expedition worked from southeastern Newfoundland into southern Baffin Bay along the coast of West Greenland to Disko Island (approximately 70°N). The Danes agreed to work to the north, beginning their survey area in Baffin Bay at Disko Island, working north to Smith Sound (78.5°N) on the naval barkentine *Godthaab*.[89] Throughout the summer, the two ships kept in touch, coordinating their work by radio.[90] By mid-September the area had been extensively covered – *Marion* alone made more than 190 complete oceanographic stations (including 2,800 temperature and salinity observations) along 16 transects concentrated mainly along the Labrador coast, off southeastern Baffin Island in Davis Strait, and off southwest Greenland north from Cape Farewell.

Such an unprecedented amount of data took several years to prepare for publication,[91] complicated in the case of the *Marion* results by the addition of new extensive sets of data by summer expeditions in 1931 and 1933–5 on the USCG cutter *General Greene,* mainly off Newfoundland and Labrador. By now Smith had assistance, for the Coast Guard and Ice Patrol made their first appointment of a full-time civilian physical oceanographer in 1930. This was Olav Mosby (1896–1956), graduate of the University of Oslo in oceanography, sometime assistant to Fridtjof Nansen, and, at the time of his appointment, an oceanographer in the Geophysical Institute in Bergen, closely associated with Helland-Hansen.[92] When Mosby returned to Norway in 1932, he was replaced (in 1933) by Floyd Soule (1901–1968), based, like Mosby, at Woods Hole Oceanographic Institution, who made a long career as physical oceanographer with the Coast Guard, retiring in 1965.[93]

The results of the *Marion* and *General Greene* expeditions, when they appeared in 1937 under the names of Smith, Soule, and Mosby, were the culmination of E.H. Smith's career as an oceanographer and a major contribution to regional oceanography of the Northwest Atlantic. More than a hundred figures illustrated the dynamically calculated current directions and velocities from the Tail of the Grand Banks to Disko Island, making the region one of the best known in the world, matching the Norwegian Sea and the eastern North Atlantic.[94] It was the culmination, too, of Smith's attempts to make ocean science not just palatable but positively necessary to the Coast Guard.

In a more subtle way, though, Smith's influence was just as great, perhaps even greater, because of earlier work, dating from his year in Bergen. In 1937, Smith, Soule, and Mosby, drily described their methods, stating that 'the dynamic computations for the stations occupied in 1928 and 1934 have been made by means of anomaly tables published by Sverdrup ...; and for the years 1931 and 1933 after the manner described by Smith (1926).' The latter was a substantial publication of the U.S. Coast Guard titled 'A Practical Method for Determining Ocean Currents,' first conceived and drafted in Bergen in 1925. It was, according to Smith's comments in his doctoral thesis, the first publication incorporating in print his dynamic calculations of currents. But it was more too, for the 1926 monograph quickly found its way onto oceanographers' desks, becoming a classic cookbook of the dynamic method along with Johan Sandström's 'The Hydrodynamics of Canadian Atlantic Waters' of 1919. It was one inspiration of George McEwen's more theoretical review of 1932, 'A Summary of Basic Principles Underlying Modern Methods of Dynamical Oceanography,' in which McEwen used four of Smith's illustrations and cited the monograph as an important aid in calculating baroclinic circulation. It is among the few publications of its kind mentioned in the first (some would say the only) great textbook of oceanography, Sverdrup, Johnson, and Fleming's *The Oceans* (1942), which is discussed in the next chapter.

The success of Smith's 'Practical Method' lay in his aim to make dynamic oceanography accessible to a wide range of readers. As he wrote, 'The writer has tried to present a rather technical scientific subject in such a manner that it may easily be understood by the ordinary student.'[95] Always the Ice Patrol officer, he added that 'always there has been the hope that the methods elucidated herein would serve some practical purpose.' That was likely to be true for the ensuing work of the Ice Patrol from 1928 through the 1930s, but the increase of understanding that Smith aimed at, rather than the practicalities, proved to be more important in the long run. The contribution of Smith's 'Practical Method' was important to oceanographers in many places, and most of all to Henry Bigelow, his students, and young associates at Harvard and in the newly developing Woods Hole Oceanographic Institution (WHOI) during the 1930s.[96]

Columbus Iselin, the Gulf Stream, and the Oceanography of the Western Atlantic

Columbus O'Donnell Iselin II (1904–1971) (fig. 7.10) is the only professional oceanographer to have appeared on the cover of *Time* magazine.[97] At that time, 1959, Iselin was long beyond a sea-going career and had not made major scientific contributions for nearly a decade, but his influence on oceanography and oceanographers was immense. He was a transitional figure in the transformation of oceanography from the scale characteristic of Bergen, Scripps, and WHOI in the 1930s to that of the post-war world.[98] He was in part the cause, and in no small part a victim, of the transformation of oceanography from the small-scale and descrip-

7.10 Columbus O'Donnell Iselin (1904–1971) as director of Woods Hole Oceanographic Institution about 1940. From *Oceanus* 16(1), June 1971, facing p. 1. With permission of Woods Hole Oceanographic Institution Archives.

tive to the large-scale, mathematically driven science that it became during and after the Second World War. Oceanography did slip away from Bergen,[99] but it slipped away too from all the early practitioners who found themselves in a new world of practice and organization. Iselin was both a cause and a legatee of this change.

Columbus Iselin was an unlikely recruit to oceanography. Scion of a wealthy Swiss-American banking family from New Rochelle, New York, he went to Harvard as an undergraduate to study mathematics. Perhaps because of his yachting experience, he found his way into the office of Henry Bigelow at the Museum of Comparative Zoology (MCZ), and by the time of his graduation in 1926 he had shifted fields and was becoming an oceanographer. Shortly after his graduation, as a new graduate student under Bigelow, he was on his way up the coast of Labrador with a group of Harvard friends in his seventy-eight–foot schooner *Chance*.[100] A year later, he made several sections of the Gulf Stream from Nova Scotia to Cape Hatteras, and in the summer of 1928 a long section across the extension of the Stream north of the Azores.[101] Upon the completion of a Harvard M.S. degree in 1928, Iselin was appointed assistant curator of oceanography in the MCZ, formalizing a relationship with Bigelow that lasted for several decades.

In Iselin, Bigelow had captured a student who had a vessel and the aptitude to advance his studies of the waters offshore of New England. Iselin, for his part, found a scientific focus for his fascination with the sea and his outstanding ability as a sailor. And when Bigelow was appointed director of Woods Hole Oceanographic Institution in 1930, Iselin was designated the master of the Institution's first research vessel, the ketch *Atlantis*, sailed it back from the shipyard in Copenhagen in 1931, and was its master for its first few cruises into the Gulf Stream.[102] He was assistant to Bigelow as director of WHOI during the first few years, and succeeded him as director in 1940. For much of the 1930s, especially late in the decade, as his papers make clear, he was *de facto* director of the Institution and certainly one of its most thoughtful chroniclers.[103]

In mid-summer 1929, Bigelow wrote to Bjørn Helland-Hansen, who was expecting a visit from Iselin as he made his way to Copenhagen:

> With this summer's work, he will have accumulated a considerable mass of data in the shape of profiles fit for dynamic calculations as to the flow of the western and northern margins of the Gulf Stream draft which it is now time for him to work up. And I feel sure that whatever advice and counsel you are willing to give him will help him far more than I now can, for on the physical side, he has rather outgrown me.[104]

Iselin, under Bigelow's stimulus, and no doubt with the help of E.H. Smith, was at work on dynamic calculations based on the data from his 1926 Labrador cruise within a few months of his return, and on similar calculations for the 1927/1928 Gulf Stream cruises shortly afterwards. This involved nothing much more complicated than computation of dynamic heights and inferences about current directions and velocities based on the horizontal contour charts and vertical sections.[105] Later, as his studies of the Gulf Stream system developed, showing the variations of current velocity and volume, the presence of multiple filaments rather than a river-like flow,[106] and the prevalence of eddies inshore of its axis,

Iselin often depended more on physical intuition than on mathematical analysis. His publications and personal papers reveal an underlying unease about the applicability of the dynamic method. Writing for general readers about advances in oceanography, he commented on 'the usefulness of Bjerknes's equations':

> His theory of calculating the velocity of currents by the slope of the surfaces of equal density demands that the circulation has assumed a steady character. There is some reason to believe that ocean currents are usually either slowing up or increasing their velocity and often break down altogether. The immediate question before physical oceanographers is how far it is advisable to use for dynamic calculations stations from different expeditions or even from the same expedition, but made over a period of a month or more. It may be that in certain regions the ocean circulation is disturbed by eddies corresponding to the extra-tropical cyclones of meteorology, in which case nothing but simultaneous stations will give a true picture of the circulation.[107]

Iselin's reservations about the utility of dynamic methods were not decreased by the results of the *Meteor* expedition (1925–7), being worked up by Albert Defant and Georg Wüst. Defant and Wüst had proposed a two-layered division of the oceans, using the 8°C isotherm as the major dividing feature and as a level of no motion for dynamic calculations.[108] Even after a revealing and helpful visit to the Institut für Meereskunde in Berlin in 1930, Iselin was convinced, based on his knowledge of the Gulf Stream, that this had been an unwise decision, and that dynamic methods were often inapplicable either because of too much variability of motion or because deep-water currents were too slow.[109]

Columbus Iselin's best-known paper, published in 1936, 'A Study of the Circulation of the Western North Atlantic,' based upon fifty-four cruises of *Atlantis* between North America and Bermuda, is a purely descriptive study, including diagnosis of the slope water, and an introduction of the idea of lateral isopycnal or isentropic mixing in the Sargasso Sea–Gulf Stream area.[110] As more information accumulated, mainly from *Atlantis* cruises through the late 1930s, he extended his interests to geographically larger areas along with the climatic and fisheries implications of the North Atlantic circulation, especially long-term variations of the Gulf Stream system.[111] And in summarizing and synthesizing a decade's work on the Gulf Stream in 1940, including collaborative work with the oceanographer E.F. Thompson of the Bermuda Biological Station since 1937,

Iselin presented relative velocities and volume transports calculated using 'Bjerknes's circulation theorem' to show that the presence of eddies was an important feature of the Gulf Stream, and that its seasonal variations were great, often greater than year-to-year variation. Unlike many of his previous papers, and in his talks to the WHOI staff, there is no editorializing on dynamic oceanography – in the 1940 paper, it is used simply as a necessary tool to get the requisite results, the velocity and transport of the Gulf Stream. It is taken out of the tool kit, used, and put away.[112]

In an obituary of this complex man, Henry Stommel wrote tellingly of Iselin's approach to oceanography as seen in his syntheses of information on the Gulf Stream:

> As a report of the data obtained, this paper [Iselin 1940] is useful ... However, the structure of these sections [of the Gulf Stream] is complex, with evidence of multiple streams, or waves, or eddies, and Iselin was able to offer no satisfactory objective criterion for defining 'the transport' of the stream. He does not make clear why he decided to choose the extremes of dynamic height on two sides of the stream to define the instantaneous transport. This is especially surprising because he shows quite clearly that a previous overly optimistic suggestion that variations in transport could be determined by time series at only two stations – one in the slope water and the other near Bermuda – could not be used. And then he proceeds to choose two stations on each section in an arbitrary way ... One senses that as a scientist he was beyond his depth. Some obituaries mention that Iselin studied mathematics as an undergraduate, but there is no evidence of it in his work – in fact he had a distrust of any formal ideas and theoretical work that sometimes surfaced in the annual reports that he later wrote as director.[113]

This is accurate and fair, but may miss some of the complexities of time and approach that make Columbus Iselin such an interesting and important transitional figure in the transformation of physical oceanography.[114] His notes on the state of the science tell us much about Iselin and about a science in transition in the last years before the Second World War.

'The Present Day Status of the Circulation Problem in the Sea'

In 1939 and again in 1940, Columbus Iselin gave talks to the staff of the Woods Hole Oceanographic Institution, the first while he was, in effect, assistant director under Bigelow, the second in the early days of his ten-year term as director.[115] They show his search for a unified approach to the physical understanding of the sea, and for a unified mission for the Institution. It was the latter, especially, that made Iselin's musings as director seem unfavourable to theoretical oceanography, rather than some antipathy to mathematical physical oceanography. He found it necessary during the years just before the war to develop a *modus operandi* that could keep a scientific staff occupied and happy on their own projects while also turning its efforts toward oceanographic and marine biological problems that could be financed from scarce private or governmental resources.

In 1939, according to Iselin, the major contributions of the Institution since 1931 had been studies of long-term variation of ocean currents, investigation of the nature and scale of mixing processes and the oxygen minimum layer, the discovery and investigation of large eddies and internal waves, development of new instruments, work on air-sea interactions, and, as he phrased the problem, concern with 'the limitations of the Circulation Theorem.' The last was of importance on purely practical grounds (in addition to theoretical ones) because, if dynamic analyses were to be applied full-scale to the Gulf Stream sections, it would require not only detailed sampling at sea but particularly the work of a technician to do the calculations. This was a matter of time and money.

Iselin claimed, as he said to the WHOI staff in 1939, that the study of lateral mixing (isentropic or isopycnal analysis) 'probably represents the most important single contribution to physical oceanography that has come from the Institution, ' largely because it could be used to explain properties of ocean waters, such as the distribution of oxygen, salt, and temperature, that were difficult to explain by vertical mixing.[116] It impinged closely on dynamic calculations, for as he pointed out,

> in the circulation theorem the importance of frictional forces had to be minimized. In the principle of isentropic mixing and in the wake stream theory, friction is emphasized. On the other hand, the Circulation Theorem is able to give special importance to the stability of the water column and to the effect of the earth's rotation. No theoretical analysis has been able to take all three factors equally into consideration. Through a program of observations designed to explore the limitations of the Circulation theorem, it should be possible to gain a good idea whether or not frictional forces have the importance we now think they have.

The investigation of problems such as this required not only the 'special qualifications' of the various scientific staff members, but also a background of common knowledge. WHOI had evolved in a loosely *ad hoc* way, but in the investigation of complex problems, Iselin claimed, 'it will become necessary that each investigator acquires the same background' because '... there now exists [*sic*] in our staff considerable differences in fundamental oceanographic knowledge which should gradually be smoothed out. Such knowledge comes through study rather than specialized research.' Iselin's call for collaboration and a common canon of knowledge was built on a different basis than the isolated contribution of equals that Ritter had promoted thirty years before, and was necessary now that the magnitude of oceanographic problems made it clear that they could only be solved by large-scale collaboration rather than by individuals occasionally seeking advice.

Ocean circulation, whether attacked through dynamic calculations, isentropic analysis, or descriptive study of eddies and currents, was still at the heart of physical oceanography. But Iselin, as responsible administrator, saw that the Institution, perhaps too the profession as a whole, was too small to tackle everything, particularly in tough financial times:[117]

> In attempting to be critical of our work in physical oceanography, certain changes have been suggested. Automatically it is implied that mistakes have been made in the past. This is not at all true. If the various investigators had not been free to work as they pleased, it would not now be possible to present this report. Having gone through a period of individual exploration, we seem now to be in a position where it is possible to decide which of these lines of work shall be given emphasis during the next few years. Having made a general attack on the circulation problem, we now know that it is extremely complex and that some approaches are more hopeful than others. Our main point is that now it would perhaps be wise to consolidate our gains and to undertake a definite program to support them and to demonstrate their exact relations to the circulation problem. Only in this way can we prepare the ground well for another advance.

His solution was to back Bigelow's proposal for a concerted study of the fisheries and circulation of Georges Bank, where a major haddock fishery likely depended in an intimate way on the physical circulation, and especially on its variations from year to year:[118]

The main difficulty with oceanography is that the Lord made the ocean too big and this is the chief obstacle, which we must use our collective ingenuity to overcome. All the major oceanographic problems require the best sort of team-work. There are many questions which we can only hope to solve as an institution and not as individuals.

Iselin's advice to his colleagues was to begin to reign in their ambitions, although not their imaginations, for in the financial climate of the late 1930s it was too easy to try to do too much and end up doing less. As he said, '... no research institution can stand still for long. It either develops or it deteriorates. Once you begin to slow down, you are soon likely to start going backwards.' By the end of 1940, under Iselin's direction, WHOI oceanographers had completed six cruises to Georges Bank and had headed in what appeared to be a totally new direction.[119]

By 1940, Columbus Iselin could be called, quite aptly, 'one of the creators of the modern multi-disciplinary oceanographic institution.'[120] Beginning as a yachtsman and inspired by Bigelow, he acquired some of the mathematical tools of physical oceanography that had been developed in Norway and were applied and taught at Bergen. He used those tools in a selective way, and soon came to recognize their limitations because of the scale and complexity of Gulf Stream circulation. His imagination was large and synoptic, and although not a skilled mathematician, he sought out tests of dynamic oceanography. His ability as an administrator and inspirer of research led him away from ocean-going science into the management of science, where he was one of the engineers of a new structure of science, the multi-disciplinary and highly specialized oceanographic institution.[121] Just as in the case of Helland-Hansen in Bergen, who was never able to live up to the potential of the scientific innovations he helped to introduce and promote, Iselin too had science slip away from him even as he was contributing to its metamorphosis. Edward Smith, as Coast Guard officer, learned and made use of dynamic methods and then moved on. But Bjørn Helland-Hansen, George Mc-Ewen, and Columbus Iselin, all physical oceanographers, were left behind, in their individual ways, by a mode of doing science that each had contributed to creating and that each practised, but over which each, in the long run, had no control.

8

Facing the Atlantic and the Pacific: Dynamic Oceanography Re-emerges in Canada, 1930–1950

> *The Biological Board was originally constituted ... for the development of scientific knowledge related to the fisheries with a view to their conservation. Biological stations were accordingly established as bases for this work ... The general organization of the Board appears well adapted for the purposes it has to serve, although as the work develops minor modifications will from time to time doubtless be necessary ... There is a particular need for a pathologist, a biological statistician, and a biological hydrographer ... The biological hydrographer would have charge of the collection of hydrographic data such as temperature and salinities, from which to work on the seasonal and other changes in the water climate of the various regions, on which the distribution and abundance of the various fishes and their food is so directly dependent.*
>
> Report of the Royal Commission Investigating the Fisheries of the Maritime Provinces and the Magdalen Islands, 1928, pp. 78–9

Coming In from the Wilderness

New methods of doing oceanography had been introduced in Canada by Johan Hjort during the Canadian Fisheries Expedition of 1915, especially in the section of its report written by Johan Sandström on currents of the Canadian east coast determined using dynamic oceanographic methods. As I pointed out in chapter 4, this attempted transplantation of European techniques to the New World was unsuccessful, at least in the short run, and it was in the United States, on both East and West Coasts, that dynamic oceanography took root for reasons endemic to each location (as outlined in chapter 7). Canadian science was not ready in 1919 for the application of mathematical physics to the oceans, not least because of the very small size of the marine science community and the concentration of the Biological Board of Canada, through its biological stations on each coast, on the very practical problems of aiding underdeveloped and little understood fishery resources rather than taking up seemingly abstract, and certainly abstruse, analyses of water movement.

However, one person had taken the lessons taught by Hjort very seriously. This was A.G. Huntsman (1883–1972), a young marine biologist at the University of Toronto, who had accompanied Hjort on the expedition of 1915, and who before that had been Hjort's host at the University of Toronto in the winter preceding the expedition.[1] Huntsman had been appointed part-time curator of the Biological Board of Canada's biological station in St Andrews, New Brunswick, in 1911, and his position became a directorship in 1918, beginning a career with the Biological Board (and its successor, the Fisheries Research Board of Canada) that lasted until the end of his life. Huntsman was the first marine scientist in Canada to see the 'need to make the study of the sea quantitative,' as I described the imperative that drove his colleagues south of the Canadian-U.S. border at almost exactly the same time. Hjort's example was a driving force to Huntsman, leading him to direct a series of marine expeditions (table 8.1) on the Canadian east coast from 1916 until 1923, when lack of resources and competing programs ended his series of attempts to emulate the Canadian Fisheries Expedition.[2]

Huntsman's early attempts to modernize Canadian marine science along European lines were unsuccessful. When dynamic oceanography did come permanently to Canada, it was as a result of the transfer of ideas and influences from upstream in the United States, as the result, indirectly, of the influence of Huntsman's friend Henry Bigelow of Harvard and Woods Hole. Later, when mathematical physical oceanography had taken root in Canada, its two active practitioners found their inspiration and closest scientific links in the United States, largely as a result of the pervading influence of Bigelow, and later Columbus Iselin and other physical oceanographers in the rapidly expanding American oceanographic community. The result was, seemingly paradoxically, 'an independent progress'[3] of quantitative marine science in Canada that was distinctly different from that in the United States, although having many of its sources there.

Hjort's Legacy: Canadian Marine Science up to 1930

Writing sometime in 1925, A.G. Huntsman described the reasons behind one of the early, and most successful, expeditions that followed the Canadian Fisheries Expedition, the Biological Board's Cheticamp Expedi-

TABLE 8.1   Early expeditions of the Biological Board of Canada, beginning with the Canadian Fisheries Expedition, led by Johan Hjort. Later expeditions, directed by A.G. Huntsman, were attempts to apply the science brought to Canada by Hjort in 1915.

1915   Canadian Fisheries Expedition
 To the Gulf of St Lawrence and Scotian Shelf. Oceanography and fisheries biology under Johan Hjort
1916   St Mary's Bay & Annapolis Basin, Nova Scotia, and Kennebecasis River, New Brunswick
 Hydrology and biology
1917   Cheticamp Expedition
 From the west coast of Cape Breton Island, Nova Scotia, to the Magdalen Islands, Quebec. Hydrology and biology of the western Gulf of St Lawrence
1918   Miramichi River and Bay, New Brunswick, and the adjacent Gulf of St Lawrence
 Hydrology and biology
1919   St Mary's Bay & Annapolis Basin, Nova Scotia, and Kennebecasis River, New Brunswick
 Hydrology and biology
1921   Shelburne Expedition
 SW Nova Scotia fisheries and hydrography based at Barrington Passage
1923   Strait of Belle Isle Expedition
 Concentrating especially on the circulation of the Strait and on drift-bottle studies of currents along the west and east coasts of Newfoundland

tion of 1917 to the southern Gulf of St Lawrence (table 8.1):

> In 1915, under the direction of Dr. Johan Hjort, a general survey was made of the Atlantic waters of Canada, and the results of this survey have been published ... In the following year the Biological Board approved of plans for intensive studies of selected regions, which studies would form a natural extension of the knowledge gained in the survey of 1915. It was planned to choose for each year a region possessing features of particular interest, either as being representative of a larger area, or as exhibiting certain conditions in an extreme form. As the hydrographical and biological conditions were followed in some detail for an entire season from spring to autumn – the more critical part of the year biologically – a firm basis would be laid for the interpretation of the varied conditions existing in our waters, and this would make possible more rational and successful exploitation, administration, and conservation of our fisheries.[4]

In the case of the Cheticamp Expedition, Huntsman's aim, directly related to the work of the Canadian Fisheries Expedition, was to study the oceanographic and biological characteristics of the warm, rather fresh water that left the southern Gulf of St Lawrence to enter the open Atlantic via Cabot Strait.

Edward E. Prince, Dominion commissioner of fisheries and chairman of the Biological Board of Canada, wrote of the Cheticamp Expedition that 'it is expected that this work will increase the value of Dr. Hjort's discoveries and be of permanent value to the fishing industry of the Gulf of St. Lawrence.'[5] Such, indeed, was the motivation of much oceanographic work done on both East and West Coasts of Canada under the aegis of the Biological Board well before and after the great expedition of 1915.

Hydrographic work – that is, measurements of temperature and salinity – had started under Huntsman's stimulus in 1914, just before the organization of the Canadian Fisheries Expedition, in the area around St Andrews, New Brunswick, in the hands of his colleague J.W. Mavor. Mavor's assistant, E.H. Craigie, wrote that the purpose of this work was

to obtain as much information as possible not only about the actual temperatures and densities of the water, but also about the nature of the currents of warm and cold water ... Such observations, besides being of interest and importance in themselves, are valuable on account of their bearing upon the haunts and habits of fish frequenting the waters studied, or passing through these waters in their migrations.[6]

For this initial work at sea, Mavor and Craigie had only limited equipment, including at least one Richter reversing thermometer and a couple of types of reversing water bottles. They determined the density (and thus salinity) of the water with a hydrometer (a colleague, the Rev. Alexandre Vachon, used the internationally accepted and more sophisticated titrimetric determination of salinity in similar studies in Passamaquoddy Bay and the Bay of Fundy region in 1916). In August of the same year, they went farther to sea, then did so again in the summer of 1915, carrying out a temperature and density section of the Bay of Fundy from near St Andrews to Long Island, Nova Scotia. This was no inconsiderable feat considering the small amount of equipment available and that their work was only accessory to the main project, collecting plankton and dredging for bottom animals.[7] In 1919 Mavor expanded the work, dropping more than 300 drift bottles at strategic locations in the outer Bay of Fundy, from which the general circulation, inferred earlier from the tidal studies and the general distribution of temperature and salinity, could be demonstrated.[8] Huntsman's motivation in supporting this work was to be able to relate the distribution of the planktonic food of herring, and of the fish themselves, to water circulation at the entrance to the Bay of Fundy. Mavor, however, seems to have taken an interest in the water itself, for in lengthy publications in 1922 and 1923, based on all the previous data and on early tidal studies, he calculated the non-tidal residual currents of the outer Bay of Fundy and promised a third paper in which the results of 'hydrodynamic methods,' presumably based on Sandström's techniques in the report of the Canadian Fisheries Expedition, would be presented.[9] This paper never appeared, and Mavor disappeared, at least as far as oceanography is concerned, into teaching in a liberal arts college in the United States. Although Huntsman led a last major expedition with physical oceanographic as well as biological objectives in 1923 to the Strait of Belle Isle,[10] very little was published on its results, and it appeared that there, too, an opportunity had been lost to bring the sophistication of dynamic physical oceanography into conjunction with his over riding interest in the links between environment and fisheries.

Although Huntsman's interest in physical oceanography remained high during the mid- years of the 1920s,[11] his attention was truly concentrated on it when serious proposals were put forward in the late 1920s to use the tides of the Passamaquoddy Bay region to generate power.[12] Closing off the Bay that was the home sea of the St Andrews laboratory would have great effects on the climate of the region, on the circulation of its waters, and on its fisheries, but little information was available that would allow the effects to be predicted with certainty.[13] At the same time, Huntsman's close colleague and friend in the North American Council on Fishery Investigations, Henry Bigelow, had just published a definitive monograph on the descriptive physical oceanography of the adjacent Gulf of Maine.[14] Huntsman was asked to testify in early 1927 before a Royal Commission under Mr Justice A.K. Maclean that was charged with recommending improvements in all aspects of the fishing industry of Eastern Canada. With the Passamaquoddy power project in mind, Bigelow's example at hand, and the need for information on physical oceanography becoming more important day by day, it is no surprise that he recommended to the Commission that the Biological Board be asked to appoint a 'biological hydrographer,' that is, a physical oceanographer, to allow a full investigation of East Coast fisheries.[15] This recommendation was included in the report of the Commission (see the epigraph to this chapter), setting the stage for the reintroduction of dynamic oceanography to Canadian marine science.

'Bio-hydrographical Investigations' of the Strait of Georgia

Only a year after A.G. Huntsman's appointment at St Andrews, in 1912 Charles McLean Fraser (1872–1946), later a zoologist at the University of British Columbia, was appointed director of the Pacific Biological Station in Nanaimo.[16] Despite – or perhaps because of – his professional interest in hydroids, Fraser began 'hydrographic' studies of the Strait of Georgia near and at Nanaimo, sometimes farther afield, in the summer of 1914. Collaborating with A.T. Cameron (1882–1947), a young physiological chemist from the University of Manitoba,[17] Fraser's were the first physical studies of British Columbia waters, giving an initial series of temperature and density observations that he expanded into a daily record of the water properties of Departure Bay (just off the biological station) into the 1920s. He stated early on that he expected a relationship between temperature and salinity and the distribution of marine plants and animals, and that they would have to be known if successful transplantations of oysters and lobsters were to take place.[18]

A.T. Cameron expanded Fraser's work beginning in the summer of 1921, collaborating with the botanist Irene Mounce (1894–1987) of the University of British Columbia (UBC), attempting to account for the distribution of plants and animals, also the level of plant photosynthesis, by variations in temperature, specific gravity, pH and water chemistry.[19] But the most extensive and ambitious studies of British Columbia waters were those directed by Andrew Hutchinson (1888–1975), then head of the Department of Botany at UBC, and a series of young assistants, including Mounce, Colin Lucas (1903–1981), and Murchie McPhail (1907–1989), between 1926 and 1929.[20] These studies took place just before the work of H.W. Harvey and W.R.G. Atkins in England brought together, between 1929 and 1934, all the factors governing plankton production in one influential model.[21] Initially Hutchinson and his colleagues' work was aimed at surveying the oceanographic properties of the southern Strait of Georgia in an attempt to account for diatom production, which was at the base of the marine food chain.[22] In this, salmon bulked large. Hutchinson and Lucas explained the rationale for the last summer's work:

> The original problem which gave the primary impetus to the present investigation was the feasibility of diverting the migrations of salmon by closing Canoe Pass, the most southerly outlet of the Fraser River. This necessitated a general survey of the area and an estimate of the relative effect of this

river in comparison with the other rivers entering the Strait of Georgia; a study of the effect of tides and other influences upon the salinity, temperature, pH, phosphate, silicate, nitrogen content, and plankton, including zooplankton and phytoplankton of the Strait of Georgia; and in turn the effect of physical and chemical conditions and the distribution of fish food upon marine organisms including fish and shellfish.[23]

The disadvantages of having synoptic research programs centred at the Pacific Biological Station (PBS) carried out only in summer somewhat informally by university researchers was evident to W.A. Clemens (1887–1963), who took over as director of PBS from Fraser in 1924:[24]

At the time of my appointment, I was given no written 'terms of reference' but Dr. Knight, Chairman of the [Biological] Board, had indicated a number of matters to which I should give consideration:

(1) For some years prior to 1924, Dr. Knight had been greatly concerned about the efficiencies of fish hatcheries and so he instructed me to consider the efficiency of the salmon hatcheries in British Columbia and to present a program of investigation to the Board at an early date.

(2) There had occurred a reorganization of the Board involving representatives of the fishing industry and a special grant had been made available for the establishment of technological stations at Halifax and Prince Rupert. I was directed to give attention to locating the station at Prince Rupert and organizing a staff.

(3) With the steady expansion of the fisheries, an increasing number of questions had arisen concerning the stocks of fishes and adequate conservation measures for them and my appointment was expected to lead to the development of a program of fishery research and the acquisition of basic information on the fish and fisheries.

(4) Practically all the investigations of the Station had been carried on by volunteer summer investigators from the Universities. I was given to understand that I was to make this phase an effective part of the Station's program.

Quietly, he began to look for a full-time employee who could carry on an oceanographic program at PBS, for, as he said, '... a real knowledge of the physicochemical environment is essential for an understanding of the fisheries.' But clearly he did not consider that the 'volunteer' oceanographic work had been satisfactory, for as he wrote, 'being trained as a limnologist, I was rather surprised on coming to the Station to find out how little was known of the physicochemical conditions of the sea along the British Columbia coast.'[25]

After an abortive attempt to hire T.G. Thompson of the Friday Harbor Laboratory of the University of Washington to begin an oceanographic program, Clemens was directed to a young chemist, Neal Carter (1902–1978), who cut short a postdoctoral fellowship in Germany to return to Canada and take up a position as a chemist and oceanographer at PBS.[26] He arrived in Nanaimo in September 1930, and was soon at work:

> My first job was to put on a suit of old clothing and go aboard the station vessel *A.P. Knight,* and start taking water samples out in the Strait of Georgia. After getting enough samples to make it worthwhile, I brought them back to the laboratory and taught myself how to analyze them for what was needed.[27]

Along with the routine duties, Carter began work on the oceanography of British Columbia fjords, expanding PBS's program into coastal oceanography, until he left Nanaimo in 1933 to direct Clemens's new technological station in Prince Rupert. At Nanaimo he was the first mentor of the young researcher J.P. Tully, who devoted himself to developing the first full program in physical oceanography on the West Coast of Canada, following the introduction of that discipline on the East Coast.

Harry Hachey and Jack Tully Enter Oceanography

Within four years, 1927 to 1931, physical oceanography became part of the research program of the Biological Board on both coasts of Canada. The precipitating factors were different and yet the same. In St Andrews, Huntsman was faced with the problem of what would happen if Passamaquoddy Bay was dammed and, even without that problem, believed that there were firm links between oceanic mixing (as in the outer Bay of Fundy) and levels of fish production. W.A. Clemens in Nanaimo found himself expanding a program in fisheries centred mainly around salmon, its capture, and processing, along with a few lesser resources. He too believed that there must be a close link between environment and fisheries. There is no evidence that Huntsman and Clemens ever talked of these things (although they may have, at the annual meetings of the Biological Board), but for reasons unique to each program and individual, they came to the conclusion that oceanography was essential to fishery research. Trained physical oceanographers were hard to come by, and each found himself an ambitious young scientist ready to re-educate himself from another science into oceanography.

Henry Benedict Hachey (1901–1985) (fig. 8.1) arrived at the Atlantic Biological Station in St Andrews on 1 June 1927. At the time, he was on the faculty of the University of New Brunswick in Fredericton teaching physics. Of New Brunswick Acadian origin, he attended St Francis Xavier University in Nova Scotia, graduating in 1922, and went from there to McGill, where he received an M.Sc. in physics in 1925. He then began a Ph.D. program, also in physics, in Toronto, but, probably short of money, found a position teaching physics at the University of New Brunswick (UNB).[28]

There seems to be no documentary evidence of how Hachey found his way to St Andrews in 1927,[29] but it seems likely that it was through his colleague at UNB, Philip Cox (?–1940), an ichthyologist and professor of natural history, who had worked at St Andrews in summers since at least 1913, and whose daughter Katherine was in one of Hachey's classes.[30] Once in St Andrews, he met Huntsman's friend Henry Bigelow, who was on one of his periodic visits to the biological station. Bigelow was persuasive about the importance of oceanography, and Huntsman needed physical oceanographic information.[31] By the time Hachey returned to Fredericton in early September, he had carried out a study of the circulation of Passamaquoddy Bay for Huntsman. A year later, he left university teaching and joined the staff of the biological station as assistant hydrographer – timing that coincided with Huntsman's testimony before the Maclean Commission and the urgent need for more information on Passamaquoddy Bay.[32]

Within a year of his appointment, Hachey had expanded his work on Passamaquoddy Bay, picked up Mavor's work on the circulation of the Bay of Fundy, taken over drift-bottle studies from Huntsman, begun the analysis of thermograph data from steamers travelling between Montreal, Halifax, Bermuda, and the West Indies, and begun a program of temperature data collection at coastal stations throughout the Atlantic Provinces.[33] He also began to toy with the idea of using what he called a 'mathematical treatment of temperature and salinity data to determine current direction and velocity.'[34] This became a reality in 1930 when the attention of the Biological Board turned to Hudson Bay.

The west coast of Hudson Bay had been the subject of hydrographic (i.e., charting) surveys since 1884, when a suitable port was being sought to serve as the terminus of a railway from Winnipeg that could carry Prai-

8.1 Canada's first physical oceanographer, H.B. Hachey (left), in St Andrews, New Brunswick, probably during the summer of 1928. From Photograph Archives, St Andrews Biological Station, Fisheries and Oceans Canada. With permission of Department of Fisheries and Oceans.

rie Provinces grain to Hudson Bay, shortening the distance to Europe and thus reducing the cost of transport to market. Churchill, Manitoba, was the eventual choice.[35] But in 1930 the Biological Board became involved in Hudson Bay when business interests in Manitoba put pressure on the federal Department of Marine and Fisheries to open a fishery there. Organized by Huntsman, the Hudson's Bay Fisheries Expedition, in a small steamship, *Loubyrne,* under Hachey's direction, spent the late summer of 1930 trawling for fish and collecting oceanographic information throughout the Bay.[36] The biological results were striking: as Hachey wrote, 'as a result of the total work covering the whole of Hudson Bay, not a single commercial fish was taken.'[37] Hudson Bay proved to be an arctic, rather than subarctic, environment, permanently stratified, with a slow, cyclonic circulation unfavourable for the growth of commercially useful fish.[38]

But if the fishery investigation was a failure, the oceanographic one certainly was not, at least from Hachey's point of view. Despite the inadequacies of the data (there were only seventeen oceanographic stations, mainly in two southwest to northeast sections north of the latitude of Churchill, to represent properties in an enormous inland sea), with some panache he taught himself the techniques of dynamic oceanography and published the first calculations of currents for Hudson Bay, using techniques learned from Sandström's monograph of 1919 and E.H. Smith's 'A Practical Method for Determining Ocean Currents.'[39] His discussion, curiously, does not mention the lack of adequate data, but he was aware of the difficulty of finding a level of no net motion so that absolute rather than relative currents could be calculated, and he introduced a graphical method of representing the stability of the water column that later was used in an influential way in biological oceanography.[40] Dynamic oceanography had returned to Canada.

Hachey's work in Hudson Bay in 1930 coincided with an increased emphasis at St Andrews on the fishery resources of the Bay of Fundy.[41] Two years later, he became involved in an extended study of the Scotian Shelf, based in Halifax, involving the collection of oceanographic data several times a year along sections between Canso and Shelburne, Nova Scotia. This also allowed him to do a preliminary dynamic analysis of that area, using the techniques he had learned for the Hudson Bay work, in 1932 and succeeding years.[42] By 1934, only four years after his first use of the dynamic method, Hachey felt sufficiently secure with it to publish an independent derivation of Bjerknes's circulation theorem, intended to show that it had a 'solid foundation' in hydrodynamics.[43]

During the 1930s, the Scotian Shelf work provided a steady supply of new data on a major fishing area, including the complex vertical structure of the water column (notably the cold intermediate layer, derived from Labrador Current and Gulf of St Lawrence waters) and frequent short-term fluctuations of temperature. These Hachey linked initially to the passage of tropical storms along the North American east coast, later to the effect of tidally induced internal waves, and eventually to Ekman transport of surface waters causing nearshore upwelling.[44] He was also beginning to visit Woods Hole Oceanographic Institution to arrange cooperative cruises between WHOI and the Biological Board, which were intended to delimit the slope water and give expanded information on seasonal and irregular variations in temperature and other hydrographic properties along the east coast, including variations in the position of the Gulf Stream.[45]

Hachey's contacts with oceanographers in the eastern United States and a good deal of self-education gave him an eclectic tool kit of techniques to apply to the Scotian Shelf by the late 1930s. He found, for example, that T-S diagrams could be used to show the degree of mixing and the arrival of water from outside the region.[46] He used George McEwen's early application of V.W. Ekman's work on the effects of the wind and the Earth's rotation to show how coastal upwelling would occur along the Nova Scotian coast when southwest winds blew for lengthy periods, and expanded it to calculate the temperature and volume of the upwelling water and show that it originated in the cold intermediate layer.[47] After meeting R.B. Montgomery at WHOI in 1937, Hachey used his technique of isentropic (isopycnal) analysis to confirm current patterns on the Scotian Shelf.[48] His later dynamic calculations of transport used modifications of Helland-Hansen and Sandström's basic method, modified by Helland-Hansen for use in shallow water in 1934 and later by A.E. Parr.[49] When he summarized a decade's worth of thermograph records across the Gulf Stream in 1939, Hachey used C.-G. Rossby's theoretical work on the distribution of mass in currents, along with Montgomery's studies of the effect of changes in sea level between Bermuda and Charleston, South Carolina, to suggest mechanisms for changes in the position of the northern edge of the Gulf Stream.[50]

When Harry Hachey left St Andrews to go overseas with the Canadian Army in August 1940, he had come a long way since his first application of the dynamic method in 1931. His use of dynamic oceanography was only the first step in the self-education of a physical oceanographer. His initial purpose, to provide links between the physical environment

244   The Fluid Envelope of Our Planet

and fisheries, was not forgotten totally – Atlantic cod inhabited and were fished on the Scotian Shelf after all[51] – but by 1940 he had provided a descriptive physical oceanographic account of one of Canada's premier fishing areas and had done a respectable dynamic analysis as well.[52] These summary papers, written after little more than a decade at St Andrews, make no reference to fish. Without qualification, Harry Hachey was not a servant of fisheries biologists but a physical oceanographer.

A similar process unfolded in British Columbia at the same time, even though surveys of the literature show that the contribution of Canadian scientists to Pacific oceanography was small until the mid-1950s. T.W. Vaughan's *International Aspects of Oceanography,* surveying international ocean science in 1937 for the U.S. National Academy of Sciences, contains no oceanographic information from Canadian sources, and Grier's bibliography of North Pacific oceanography in 1941 lists only a handful of chemical and hydrographic works, most of them very short, by Canadian authors.[53] Although the whole Pacific was poorly known until the Second World War, the Canadian contribution seems disproportionately small. And yet a change of focus to a different scale shows that this was not the case, for in Nanaimo, at W.A. Clemens's Pacific Biological Station, a thriving program in coastal oceanography began to unfold in the early 1930s.

Neal Carter, who had been hired by Clemens in 1930 as chemist and oceanographer, took over Hutchinson's survey of the Strait of Georgia, and began to study the hydrography of mainland fjords as habitats for fish.[54] But Carter was badly overworked and needed an assistant for chemical analyses. A young chemist from Manitoba, John P. Tully (1906– 1987) (fig. 8.2), was recruited in 1931.[55] Even though he had no experience in the marine environment and had lost a leg in an automobile accident in his youth, Tully proved to be highly motivated, extroverted, and expansionist in personality. When Carter left PBS in 1933 to direct the Board's experimental station in Prince Rupert, BC, Tully stepped into his place. Unencumbered by any preconceptions about marine science, he set about learning oceanography and applying it in the context established by Clemens and the aims of the work at the station.

Records of sea temperatures had been kept since 1914 in Departure Bay, and since 1917 at William Head, near Victoria. In 1932, Tully arranged to have lighthouse keepers at five locations record daily sea temperatures and meteorological variables. As he said later,

... the direct object of this program is the preparation of continuous charts of the hydrospheric and atmospheric variations to show their geographical

8.2 J.P. Tully (back row centre) and colleagues at the Pacific Biological Station, Nanaimo, British Columbia, in 1935. From J.P. Tully retirement scrapbook, Pacific Biological Station, Nanaimo, BC, Scientific Archives, Pacific Biological Station, Fisheries and Oceans Canada. Photo courtesy of Department of Fisheries and Oceans.

daily, seasonal and annual variations and their inter-relations, with a view to determining the meteorological factors that affect the movements of the various commercial fishes. It is well known that the sea and the air affect each other and that the behavior of one cannot be entirely separated from the other.[56]

Furthermore, the lighthouse observation program, which expanded to more than twenty stations later, was a proxy for work at sea, for, although the Strait of Georgia could be studied from PBS's small boats,[57] the open ocean could not. Its characteristics and variations off the British Columbia coast were totally unknown in the early 1930s. Tully's solution, along with the lighthouse observations, was to use ships of convenience. Viewing red tape as a way to pull oneself along, not as an impediment,[58] he began hitching rides on hydrographic surveying vessels to study the west coast of Vancouver Island and the open Pacific. Aboard the Canadian Hydrographic Service's vessel *William J. Stewart,* he spent a few days studying the currents of Nootka Sound in 1933,[59] several weeks between Cape

Flattery and Esperanza Inlet in 1934, and three months in the area of the Queen Charlotte Islands in 1935. These experiences were valuable but frustrating. He found that oceanographic work took second place to charting and to the whims of the chief hydrographer, and was hindered by the unsuitability of the hydrographic vessels for oceanography.[60] The solution was to find a better vessel that could be devoted, at least part time, to oceanography. Such a vessel was HMCS *Armentières*, loaned for a few months each year by the Royal Canadian Navy for oceanographic surveys from 1936 to 1938.

Between February and September of the first year of his offshore surveys, Tully and colleagues occupied one hundred oceanographic stations between the entrance to the Strait of Juan de Fuca and Queen Charlotte Sound. In the next two years, he concentrated on Swiftsure and Lapérouse Banks at the entrance to the Strait of Juan de Fuca, showing the variability of water properties, the presence of rapidly changing eddies, and the existence of areas of cold water offshore, bounded by even cooler coastal and offshore waters (a situation quite contrary to expectations that a warm 'Japanese Current' extended across the Pacific to British Columbia).[61] Tully summarized the rationale for his work in 1937, stating that

> ... the hydrographical investigations ... have been primarily directed towards the discovery and measurement of the factors affecting the physical environment of the food fishes. Primary consideration has been given to the discovery of the elements of the problem so that it would be possible to reduce the phenomena to their primary causative forces and to determine their cyclic nature with a view to forming a firm basis for the predictions of the physical conditions in the fishing areas which might be used in fisheries prediction.[62]

The years of the *Armentières* cruises, ending just before the Second World War, were ones in which Tully's capabilities and orientation as an oceanographer changed rapidly. A B.Sc. in chemistry had not prepared him for the complexity of current analysis, and certainly not for dynamic oceanography. Initially, Tully used conventional methods of analysis presented by the American tidal expert H.A. Marmer, in which hydrodynamic forces were responsible for the difference between total currents measured using current meters and those resolved as tidal by mathematical analysis.[63] Rapidly, however, Tully began to teach himself the dynamic method, based on Bjerknes's theorem, but in the form used by so many others, the adaptation by Sandström and Helland-Hansen, which by 1936 was easily available in E.H. Smith's monograph. His first text was Sandström's classic monograph (1919). A year later, in 1937, he referred to a variety of works by Bjerknes, V.W. Ekman, George McEwen, and E.H. Smith.[64] That year he applied what he called the calculation of 'gradient currents' to the area off the west coast of Vancouver Island and the Strait of Juan de Fuca, publishing the results in short papers that also showed his knowledge of V.W. Ekman's work on wind-driven ocean circulation.[65]

Dynamic oceanography became part of an important proposal to his superiors, primarily the director of PBS, W.A. Clemens. Under the title 'Oceanographic Program. Pacific Biological Station, Nanaimo, B.C. 1937,' Tully recommended a new program

> ... designed to study the currents in the ocean off the Pacific coast of North America within range of operation, to determine their cause, and tidal, seasonal and annual variations and cycles; and to correlate the factors of tide, physical properties of the water, and meteorological variations; so that the resultant movements and changes in the sea may be reduced to a calculable basis.[66]

Its value to the aims of the Fisheries Research Board[67] was patent:

> The value of accurate knowledge of the physical factors constituting a good, poor, or average year in the various fisheries, lies in the possibility of correlation to biological factors and consequent forecast of abundance and availability. As a result, regulations may be adjusted to suit the conditions, and industrial organization to handle the catch in the most efficient manner.
> ...
> Besides these commercial applications the study offers at least a partial solution of one of the major scientific problems, namely oceanic circulation. Consequently this research is classic in its field and scope.[68]

In such a program, the lighthouse observations, meteorological observations (many of them being made at lighthouses), and the study of currents were inseparable:

> The meteorological aspect of the study is so important both from the standpoint of the factors themselves and for their bearing on the direct measurement of the currents, and the whole value of these observations lies in their

continuity, that this phase of the study must be assured in its entirety before any other phase is attempted.[69]

However, although meteorological observations were of great importance, Tully claimed that 'the problem in its final analysis is a study of the currents in the sea,'[70] and outlined in some detail what was known of wind, tidal, and 'gradient' currents, citing the classic authorities such as Bjerknes, Ekman, Smith, and McEwen. He also gave details of a plan of attack, citing the locations and spacings of stations designed to give data satisfactory for mathematical analysis. His conclusion was that 'a program is required that will observe all the significant factors affecting the sea in this area, and reduce them by correlation in[to] the fewest possible factors affecting the fisheries directly.'[71]

Tully's proposal in 1937 was an attempt simultaneously to provide wide-ranging scientific support to fisheries research in Nanaimo and to enlarge his own research group. Although it appears that Tully never regarded these as separate, competing, aims, there was one practical difficulty, his lack of ability in mathematical oceanography. Never a very skilful mathematical oceanographer, he knew his limitations and in 1936–7 corresponded with Bjørn Helland-Hansen about the possibility of doing a Ph.D. in the Geophysical Institute in Bergen.[72] This came to nothing; instead, he began to work toward a Ph.D. in T.G. Thompson's Oceanographical Laboratories at the University of Washington, where, probably for the first time, he encountered the full range of oceanographic literature.[73]

The war interrupted Tully's doctoral research on the oceanography of Alberni Inlet, where in 1939 he had begun work to assess the impact of a sulfite pulp mill being planned for Port Alberni.[74] Research at PBS was frozen for a time by the onset of the war. Work on the open ocean became impossible. Redirecting himself to Alberni Inlet, on the west coast of Vancouver Island, Tully completed his surveys in 1942. In 1940 he had begun to construct a small hydraulic model of the inlet to simulate and simplify estimations of the effect of the pulp mill. He quickly developed an interest in experimental approaches to oceanic circulation that complemented his surveys of coastal waters. Oceanographic surveys, long time-series of measurements (for example, the lighthouse observations), and hydraulic modelling began to loom larger in Tully's view of oceanography. But it was the war that allowed them to coalesce, and that brought his work together with that of Harry Hachey on the East Coast.

'Oceanographers at War

In Canada, as elsewhere, the threat posed by submarines was responsible
for bringing oceanographers and a variety of other physical scientists
into the Second World War. Physics was brought to bear on improvements to Asdic, the early form of sonar, used in submarine detection, also
on problems of how to overcome the threats of magnetic and acoustic
mines and acoustic torpedoes.[75] But because the Royal Canadian Navy
(RCN) had no research capabilities, it had to find help where it could.
First, using the expertise of physicists at Dalhousie University in Halifax, countermeasures against magnetic mines were undertaken, including research on degaussing ships and the establishment of a degaussing
range in Halifax in 1940. This work expanded to include a degaussing
and sound range in the main channel of Halifax harbour in 1942, and
led late in the war to the foundation of Canada's first defence-related
scientific establishment, the Naval Research Establishment on Halifax
harbour.[76]

Acoustic mines were first laid by German vessels in September 1940,
posing a new threat to shipping. At the same time, attacks by submarines on convoys were increasing. Both made more work on underwater
sound essential. The Acoustics Section of the Division of Physics and
Electrical Engineering at the National Research Council (NRC) in Ottawa was asked by the British Admiralty in January 1941 to begin work on
defences against acoustic mines. George S. Field (1905–?) of the Acoustics Section visited defence laboratories in the United States and Britain
during the early months of 1941, learning all he could of countermeasures against mines and submarines. By June 1941 he was organizing a
team in Ottawa to work on defences against acoustic mines in addition
to the early work by the group on Asdic. Late in 1941, as the importance
of its work on these and other fronts grew, the NRC was appointed the
Scientific Research and Development Establishment of the RCN.[77]

Late in 1941 the Acoustic Section's work was enlarged when the British
Admiralty asked U.S. and Canadian laboratories to take over the majority of anti-submarine acoustic work. After another visit by George Field
to the United States, the Canadian group agreed to put its effort into the
properties of sound in the sea. In the following year, after six bathyther-
mographs (BTs) were brought from Woods Hole to Canada, the Section
began to survey the influence of variations in physical properties of the
water on the effectiveness of Asdic in detecting submarines.[78] In 1943,
collaborating with the RCN, Acoustic Section employees and naval of-

ficers made an extensive survey of acoustic conditions using BTs off the Canadian east coast.[79]

At the beginning of the war there were only two physical oceanographers in Canada, Harry Hachey in St Andrews and Jack Tully in Nanaimo. Tully could not join the military because of his disability, but Hachey, who had been a member of the Canadian Army Reserve before the war, went on active service, leaving Canada for England in 1940. He remained there until 1942, when he returned to work in a weapons laboratory in Suffield, Alberta.[80] Tully had been forced to end his open ocean research when the war began, redirecting it to inshore studies of Alberni Inlet, and hydraulic modelling. His attention began to turn to defence-related work, and in 1943 Tully was assigned to duty with the RCN in Nanaimo and to acoustics-related research. In April 1944, Lieutenant-Colonel Hachey was formally seconded from the Canadian Army to the RCN and returned full-time to St Andrews to work on anti-submarine research. Hachey, working with the RCN, had been involved in the first BT surveys of temperature on the East Coast,[81] and in 1944 this kind of work on both coasts was formalized with the establishment of the Pacific Oceanographic Research Group in Nanaimo under Tully and the Atlantic Oceanographic Research Group in St Andrews under Hachey (the names were soon shortened to Pacific Oceanographic Group – POG – and Atlantic Oceanographic Group – AOG), both integrated with the NRC's Acoustic Section in Ottawa under the direction of George Field and reporting to him. Tully's and Hachey's work for the rest of the war involved the oceanography of antisubmarine warfare. Nanaimo, in particular, which was assigned the use of the CNAV *Ehkoli*, an 84-foot converted seiner, for use in surveys of physical conditions, became the main centre for Canadian acoustic work in the sea, especially after submarine attacks nearshore practically halted activity on the East Coast early in 1945.[82]

A New Profession, Forged in War

Immediately after he left wartime service with the RCN in 1946, Jack Tully completed a Ph.D. at the University of Washington. Even before that, he had begun to take steps to continue military-related research at Nanaimo. When the American oceanographer Waldo Lyon (1914–1998), an acoustician at the U.S. Navy Radio and Sound Laboratory (later called USNEL) in San Diego, was visiting the Seattle area in 1945, Tully contacted him with the proposal that they begin a program of collaborative research on acoustic oceanography in the protected inshore waters of British Columbia.[83] The result was a lengthy period of work together and the involvement of Tully's POG with defence-related research in the North Pacific and Arctic that had no counterpart on the East Coast.

When Tully returned to Nanaimo in 1946, Ph.D. in hand, he had not only *Ehkoli* but soon also the 165-foot HMCS *Cedarwood*, capable of offshore work, available for new programs.[84] These began in 1948 with a physical study by Tully, W.M. Cameron (1914–2008), and G.L. Pickard (1913–2007) of the effect of the Skeena and Nass Rivers on Chatham Sound, near Prince Rupert, a waterway traversed by sockeye salmon on their way to and from the rivers.[85] Nearby was Nodales Channel, well mixed and isothermal, and thus ideal for studying the sonar signatures of submarine-like objects (and submarines themselves). In what was described as 'probably the largest joint oceanographic research operation undertaken in Canadian waters,' POG and USNEL personnel, including Tully, Cameron, and Waldo Lyon, in four ships and two smaller craft, studied the acoustic signatures of iron spheres, a triplane target, and a submarine (fig. 8.3).[86]

Tully and Cameron's collaboration with USNEL expanded far beyond Nodales Channel in 1949, when the POG in *Cedarwood* and USNEL scientists began oceanographic work, much of it security classified, in the Bering and Chukchi Seas. This was only a first step. Yearly Arctic cruises between 1950 and 1954, involving POG, the Canadian Defence Research Board (DRB), and USNEL scientists, culminated in 1954, when the new Canadian icebreaker HMCS *Labrador* joined a group of U.S. vessels in the Canadian Western Arctic after negotiating the Northwest Passage.[87]

Closer to home, the availability of *Cedarwood* made studies in the open Pacific – begun in HMCS *Armentières* between 1936 and 1938 – possible once again. Tully's colleague L.A.E. Doe (1916–2002) was put in charge of an extensive dynamic survey of offshore waters between Cape Flattery to the south and Dixon Entrance to the north, extending to 140°W. 'Project Offshore' under Doe amplified and extended Tully's early conclusions about the current regime west of Vancouver Island, verifying that the warm water not far offshore was of local, seasonal origin, not the result of the North Pacific ('Japan') Current. When J.L. Reid (b. 1923) of the Scripps Institution of Oceanography visited Nanaimo in 1953, he suggested amalgamating data from Project Offshore with that taken by the California-based Marine Life Research Group off the U.S. west coast. As a result, Doe's publication of the results was the first synoptic account of currents off the North American west coast, showing the divergence

8.3 CNAV *Ehkoli* towing an acoustic target in Nodales Channel, British Columbia, in 1949. From J.P. Tully retirement scrapbook, Pacific Biological Station, Nanaimo, BC, Scientific Archives, Pacific Biological Station, Fisheries and Oceans Canada. Photo courtesy of Department of Fisheries and Oceans.

of the North Pacific Current at the latitude of British Columbia, its variations, and the source of the Alaska and California Currents.[88] As a logical extension of this work, Tully and his POG, using HMCS *Ste Thérèse,* became involved in an even more ambitious survey, NORPAC, the study of the remaining unknown central regions of the North Pacific, north of 20°N into the Bering Sea, during the summer of 1955. This joint project of Japan, Canada, and the United States provided baseline studies of the subtropical and subarctic North Pacific upon which all subsequent work has been based.[89] The Canadian contribution was modest – only one ship among more than twenty involved in the project – but it indicated the ability of Tully and his POG oceanographers to make significant contributions to international oceanography with limited resources but unbounded ambition only a decade after the war.

Harry Hachey's program of research on the East Coast after the war was less ambitious, partly because he had not developed collaboration with oceanographers in the United States to the same extent. He concentrated instead on extending his pre-war studies of the Bay of Fundy and the Scotian Shelf, turning much of the work over to new colleagues in the AOG like Louis Lauzier (1917–1981) and Hugh McLellan (b. 1921) and beginning new, intensive seasonal monitoring of oceanographic conditions in those areas and in the Gulf of St Lawrence. Operating off-shore much more that before, AOG turned its attention increasingly to the slope water lying inshore of the Gulf Stream[90] Like Tully, Hachey had negotiated the use of small to medium-sized RCN ships to allow oceanographic programs to be extended to more distant waters and throughout the year.[91] And in 1950, AOG, along with DRB scientists, took part in a U.S.-organized study of the Gulf Stream named Operation Cabot.[92]

Hachey had been reluctant to involve himself and the AOG in the Arctic; there was too much else to do with the Group's limited resources. But the pressure to move north was inexorable due to the operations of the RCN's new icebreaker HMCS *Labrador,* the presence of U.S. ships in the Arctic (which had given Tully and the POG their opportunity to expand their horizons), and the completion or planning of lines of early-warning radar stations across the North American Arctic by 1954. In that year, the AOG agreed to take the main responsibility for physical oceanographic research in the Arctic.[93] There, over the next eight years, the first concerted Canadian physical oceanographic research in the Eastern Arctic took place in Hudson Bay, Foxe Basin, and Baffin Bay.[94]

Taking Oceanography from the Coast to the Classroom

In the immediate post-war years, the autonomy and effectiveness of Tully's POG and Hachey's AOG, similar research groups but always rivals for resources, were increased by their inclusion in the Canadian Joint Committee on Oceanography (JCO) when it was established in April 1946. Intended to coordinate and promote Canadian oceanographic research, the JCO was made up of senior members of the Fisheries Research Board of Canada, the RCN, the NRC, and, a little later, the Canadian Hydrographic Service, the Meteorological Service, and the Defence Research Board. The JCO's members were in close contact with research, had influence with their chiefs or directors, and did not hesitate to find resources, ranging from money to ships, for oceanographic work.[95] But one problem that the JCO and its successor the Canadian Committee on Oceanography (CCO) could not solve was the supply of new oceanographers, much in demand during the early days of the Cold War, and as government programs on the coasts and in the Arctic expanded. The scope and scale of the problem went well beyond the kind of self-education that Hachey and Tully had accomplished in the 1930s.

A memo from J.P. Tully, probably written in 1948, diagnosed the problem. Although, as Tully said, oceanography was important to fisheries, shipping, sewage disposal, ocean mineral exploitation, and submarine warfare, the field had expanded too rapidly for the supply of personnel from the pure sciences to keep pace. Speaking from experience, he found that too much time was spent on training new personnel, at the expense of research. Even then, training was, as he said,

> ... largely in techniques and procedures when it should embrace familiarity with the sciences and their applications to the sea. Obviously this [the current situation] is inefficient, but it appears to be the only course unless adequate fundamental training is provided in Universities.[96]

But even if the universities rose to the challenge, they lacked faculty members qualified to teach oceanography. Thus there was a need for newly trained staff and for oceanographic research in Canadian universities. He predicted an ample supply of jobs for graduates in oceanography, because experience at the Scripps Institution of Oceanography indicated that new jobs in the ocean sciences were appearing at least as fast as its students graduated.[97]

By the middle of 1949, more than a year of machinations by the JCO (especially George Field, by then deputy director general of the Defence Research Board), President N.A.M. MacKenzie of the University of British Columbia, and J.P. Tully had taken place. According to President MacKenzie's notes, the JCO, concerned about increasing personnel in oceanography, had instigated a discussion of the problem at a meeting of the Royal Society of Canada held in Vancouver in June 1948.[98] The president of Section V (Biological Sciences), Andrew Hutchinson of UBC (the pioneer of oceanographic studies in the Strait of Georgia during the 1920s), seconded a resolution that:

> Whereas oceanographical investigations conducted by the Fisheries Research Board of Canada over a period of thirty years were of particular value to national defence during World War II, as well as of importance to fisheries research and
> Whereas there has been formed a National Committee on Oceanography [the JCO] under the joint auspices of the National Research Council, the

Royal Canadian Navy and the Fisheries Research Board for the carrying out
of oceanographic investigations applicable to defence and fisheries problems and
Whereas there is an urgent need for training of young men for participa-
tion in the investigational programmes being developed and for the con-
duct of fundamental research
Be it resolved that the Royal Society recommend that there be established
in Canada an Institute of Oceanography, and that copies of this resolution
be sent to the Department of National Defence, to the National Research
Council, to the Royal Canadian Navy, to the Fisheries Research Board and
to the Universities of Canada.[99]

Then, according to the records of the meeting, 'following the presentation of the report ... Dr A.G. Huntsman stressed the importance of the
establishment of an Institute of Oceanography in Canada and suggested
that the recommendations put forth ... be carried out at once.'

But events were already in motion outside the sanctum of the Royal
Society. Sometime in 1948, George Field, newly established on the Ottawa headquarters staff of the Defence Research Board, visited President
MacKenzie of UBC in Vancouver to suggest that UBC start 'instruction
and fundamental research in oceanography,' promising that DRB would
provide 'substantial assistance' toward the establishment of an institute
of oceanography.[100] By August 1948 an institute at UBC was being discussed as a certainty by Field and Tully. Field suggested that UBC attempt
to recruit the eminent British tidal expert Joseph Proudman as leader
of the new UBC group and said that DRB would seriously consider supporting the research of the UBC physicist George L. Pickard since he was
showing signs of interest in oceanography.[101]

Early in 1949, President MacKenzie met with George Field, H.B.
Hachey (by then chief oceanographer of Canada), W.A. Clemens (formerly director of PBS, then head of the UBC Zoology Department), J.S.
Johnson (director of scientific services of the RCN), and Gordon Shrum
(head of the UBC Physics Department) to consider the location of an
institute of oceanography, which clearly had to be on one coast or the
other – that is, at UBC in the west, or at Dalhousie University in the
east.[102] They agreed to support UBC first because of its location and
facilities and because of strong support from Tully at PBS. Then, or a bit
later, Field agreed to provide salary for one position in the new institute,
to pay for sending George Pickard to the Scripps Institution of Oceanography for a year's training in oceanography, and to make available as

a teacher W.M. Cameron, by then a DRB employee at the Pacific Naval Laboratory in Esquimalt, who had worked with Tully in Nanaimo, for the first academic year, 1949–50. And by agreement of the JCO, Tully would also be free to teach in Vancouver when the new institute opened.

On 26 August 1949 the UBC Senate approved President MacKenzie's proposal for an Institute of Oceanography. Three days later, the Board of Governors of the university approved the new Institute and granted it interim funding for the remainder of the year. Students were waiting in the wings. Within days, the Institute of Oceanography at UBC was in action, alone in the field of oceanographic education for the next decade.[103] Its first students, frequently government scientists looking for qualifications in oceanography, were taught by Tully, who commuted to Vancouver to teach chemical oceanography, by other POG staff from Nanaimo, and by Cameron from his base in the Pacific Naval Laboratory across the Strait of Georgia in the Victoria area.[104] The UBC graduate students in oceanography got their feet wet quickly, frequently working with the POG in Nanaimo. Only later did Harry Hachey begin to provide working space and projects for students in the AOG in St Andrews, although his role grew in importance for a time when graduate training began on the East Coast in 1959, and when the federal government's first oceanographic laboratory, the Bedford Institute of Oceanography, opened in the Halifax area in 1962 and required a well-trained staff.

It might appear a form of tunnel vision to concentrate on the importance of only two scientists, Harry Hachey and Jack Tully, in the development of quantitative physical oceanography in the Canadian national setting or any other. But their careers, and their contributions as scientists, administrators, and eventually teachers of physical oceanography, show yet another way that the marine sciences changed from the early years of the twentieth century into the 1950s. Each was hired to provide expertise to fisheries research – and each did just that. But each was caught up too in the international expansion of oceanography that Hjort represented first in Canada in 1915 and that Helland-Hansen promoted from Bergen. Individual ambition, the examples set first by Hjort, the catholic scientific tastes of A.G. Huntsman and W.A. Clemens, and the standards and techniques that came to define the practice of physical oceanography, were incorporated by Hachey and Tully into new settings and organizations unique to their scientific environments. But what distinguishes these paths into a new science more than anything else is the imperative to *teach* oceanography, not just in practice but in a formal academic setting. For Jack Tully, in particular, Harry Hachey to a lesser extent, physical oceanography had changed during their lifetimes from a minor service to others to a full science that could – and indeed had to be – professed to others. Dynamic oceanography had become part of an expanding scientific discipline throughout North America.

9

Studying *The Oceans* and the Oceans

'The Oceans' will appeal to many seafarers; many ships would be glad to have a copy. There is no doubt that it will be very effective in promoting research, and increasing our knowledge of the oceans.

　　　　　　　　George Deacon, 1945, reviewing *The Oceans, Nature* 155, p. 654

On the whole we have done pretty well. A new field of geophysical fluid dynamics ... has grown, with deep connections to meteorology and astrophysics. Leaf through the 1942 treatise by Sverdrup, Johnson and Fleming and you are struck by the absence of any dynamical theory beyond Ekman's 1902 spiral and the elder Bjerknes's practical method of doing dynamical current calculations. The most elementary problems had not been posed, nor the most primitive models constructed. Since then some of the vacuum has been filled ...

　　　　　　　Henry Stommel, 1989, quoted in Luyten and Hogg 1992, pp. 52–3

A New Science of the Sea?

In the last chapter, I carried the story of the careers of J.P. Tully and H.B. Hachey closer to our own times than for any other protagonists in this book. There were two reasons for this, beginning with the need to provide narratives that had some logical outcome, that moved to a satisfying conclusion. But of greater importance was my wish to show that in at least a couple of careers there was evidence that something new had come out of the varied fortunes of dynamic oceanography since the first attempts of Henrik Mohn and Vilhelm Bjerknes to provide a satisfying mathematical framework – and satisfying conclusions – to the study of oceanic circulation. Especially in Tully's case, the logical outcome and the satisfying conclusion came through education – the passing on of new approaches and a canon of scientific practice to aspiring ocean scientists. A new profession had appeared and was being expanded academically during the 1940s and 1950s. By the 1960s, physical oceanographic practice and theory, too, had been transformed by a change of approach at least as radical as that introduced by Vilhelm Bjerknes and his Scandinavian disciples sixty years before.

Compared with the 1930s, it was clear that ocean scientists after the Second World War were working in a new realm of scientific practice. This was evident first of all in what they read – the first argument of this concluding chapter, to be expanded upon shortly – and also in the virtual explosion, after decades of stasis, of new teaching and research institutions, especially in North America. In this, Scripps Institution of Oceanography (SIO) led the way. W.E. Ritter's marine laboratory in San Diego, formally established in 1903, became the rootstock of the newly named Scripps Institution of Oceanography (the name was formally applied in 1925), the only laboratory dedicated to oceanography (and teaching it) until the foundation of Woods Hole Oceanographic Institution (WHOI) and the funding of the Oceanographical Laboratories of the University of Washington and the Bingham Oceanographic Laboratory at Yale in 1930. But Europe had been there early. The Institut für Meereskunde of the University of Berlin was founded in 1900 (although a case can certainly be made – see chapter 5 – that it was not truly oceanographic until the regime of Albrecht Penck and Alfred Merz in the 1920s). In France, the Prince of Monaco's Institut océanographique in Paris, although it got a building and professorships later, was formally established in 1906[1]. In Norway, the Geophysical Institute in Bergen (the progenitor of the post–Second World War University of Bergen) was established in 1917, with Bjørn Helland-Hansen and Vilhelm Bjerknes occupying its principal chairs. In England, William Herdman, powerful and influential professor of zoology at the University of Liverpool, established a chair of oceanography in that university in 1919 – and finding no one more suitable initially, occupied it himself until succeeded by James Johnstone in 1920. After the Second World War, expansion came rapidly, especially in North America; for example, the formation of Texas A & M's Department of Oceanography and Meteorology and the University of British Columbia's Institute of Oceanography in 1949, the University of Washington's Department of Oceanography in 1951, the Department of Oceanography of Oregon State University in 1959, Dalhousie University's Institute of Oceanography the same year, and the Bedford Institute of Oceanography in 1962, to list only the most prominent.[2] Oceanography – including physical oceanography – was a very different science by the 1950s and after in content, practice, and execution than it had been in the immediate pre-war years. What accounts for this remarkable transformation?

The Oceans – the Origin of a Textbook

It is a nearly unexamined truism among historians of oceanography that the importance of the oceans during the Second World War led to the expansion of oceanography. Oceanographers played important parts in research on anti-submarine warfare, safe sea-states for amphibious landings, and other aspects of the war, but despite the common-sense relation between the expansion of oceanography and the war, the evidence is mainly scattered and anecdotal.[3] There is better evidence for a relationship a bit later, when oceanographers found various ways of capitalizing on the Cold War, but those years fall well outside the period that I have been analysing.[4]

The consolidation of oceanography as a separate science came earlier, undoubtedly over a period of decades, and the beginning of its new status was a seemingly modest event, the appointment of the Norwegian geophysicist Harald Sverdrup as director of the Scripps Institution of Oceanography in 1936 (see chapter 7). Sverdrup had studied under Vilhelm Bjerknes and followed Bjerknes to Leipzig in 1917, where his doctoral thesis on the North Atlantic Trade Winds was written. He had honed his expertise in field science and in completing scientific projects under adverse conditions during seven years on Amundsen's *Maud* in the Arctic Ocean (1918–25), and during a break in the expedition, had worked up some of the results in the United States at the Carnegie Institution of Washington. His New World connections, along with his distinction as a scientist and the recommendation of Sverdrup's Bergen colleague Bjørn Helland-Hansen, brought Sverdrup to the attention of the SIO director T.W. Vaughan, and, after some soul-searching, Sverdrup left a professorship in Bergen and began the job of director of SIO in the late summer of 1936.[5]

In La Jolla, Sverdrup faced a daunting prospect. The Scripps Institution was small, isolated geographically and intellectually, lacked a suitable ship, and was regarded as academically suspect by the University of California scientists in Berkeley and Los Angeles who oversaw its programs.[6] But, as Sverdrup noted of the Scripps Institution, 'The group here is more than helpful and competent. It is largely a collection of unusually pleasant people, who have done everything possible to make us feel at home.' As a result, he wrote, '... it would be especially satisfying to do something for them; get them better housing, increase contact with the university, bring in bigger and more interesting problems for discussion.'[7]

The teaching program at SIO was particularly troubling. It was scattered and lacked focus. Rapidly, Sverdrup instigated reform, beginning a general class in oceanography given by all the Institution's faculty members (Sverdrup and the young chemical oceanographer Richard Fleming taught the physical oceanography section, not George McEwen), and a series of specialized classes at an advanced level, including one on dynamic oceanography shared by Sverdrup and McEwen.[8] But he was troubled my the lack of a textbook, and, after nearly turning down the opportunity, Sverdrup signed a contract to write one in September 1938, collaborating with his SIO colleagues Martin Johnson, a zoologist, and Richard Fleming, a chemist.[9] The project, originally scheduled for one year and a few hundred pages, took until late in 1941 to complete and was a monograph of nearly 1,100 pages. The first printed copies reached La Jolla in December 1942. *The Oceans* had been born.

The Oceans – a Philosophy of Oceanography

During the writing of *The Oceans*, Sverdrup wrote to his Swedish colleague Hans Ahlmann that '... it will not resemble any prior existing textbook.'[10] Recently, Munk and Day, focusing primarily on its significance in La Jolla, pointed out that '*The Oceans*, Sverdrup's vision of oceanography, represents what he was trying to accomplish at the institution and what he was trying to build into the curriculum.'[11] Other commentators have seen it as more than a contribution to the teaching program of the Scripps Institution. In a celebration of the fiftieth anniversary of the book, D. James Baker wrote that '*The Oceans* pulled together the field of oceanography by being both a textbook and a documentation of the state of research ... *The Oceans* defined oceanography in a way that would be difficult to do today.'[12] Both are correct, although both require some amplification if the significance of Sverdrup's accomplishment is to be appreciated historically and scientifically.

In retrospect, it is not surprising that Sverdrup's contribution to a new textbook took the form it did. In an early talk, prepared for the first U.S. National Academy of Sciences Committee on Oceanography in 1928, he put physical oceanography in an unexpectedly broad setting:

In order to arrive at correct results regarding the general circulation it is necessary to know the configuration of the bottom of the sea, especially the extent of submarine ridges which are of fundamental importance to the movement of the deeper and deepest strata. Knowledge of the sediments covering the bottom may be helpful. The same also applies to the knowledge of the gaseous content of the sea water; e.g., the amount of oxygen. In order to make proper use of these data the effect of biological processes must be taken into account.

...

Studies of the variations of the currents and the physical and chemical properties of the sea water have a bearing on a number of practical questions ... [t]he reaction on the meteorological conditions and thus indirectly on the plant growth, the time of beginning of seasonal fisheries and the time of greatest output, the occurrence of certain organisms in certain localities, the southern limit of drift ice and icebergs, etc.

...

Physical oceanography is interested in several other subjects which have not been mentioned above: penetration of light, conduction of heat, melting and freezing of ice, etc. The oceanographer must also be familiar with the main chemical problems, the composition of the sea water and the determination of the chemical properties which are of special importance to the organic life.[13]

Later, early in his directorship at SIO, Sverdrup found it useful, pragmatically and scientifically, to bring together all the sub-fields of oceanography described above, involving all his colleagues in oceanography, and thus uniting the disunited Institution that he found in La Jolla in 1936. A year later, he responded positively when the California Department of Fish and Game asked for oceanographic assistance in tackling the problem of the crash of California sardine populations.[14] Part of the payoff came in the ability to do physical oceanographic work in coastal waters on State vessels,[15] but Sverdrup readily appreciated that keeping oceanography broadly based in the biological, chemical, and geological sciences was of intellectual benefit as well as general advantage. As he expressed it in 1940 for non-scientific readers:

It is evident that an enterprise of this nature requires the complete cooperation of a number of specialists. The physical oceanographer has to account for the character of the currents and has to examine why the observed patterns exist. The chemist has to determine the salinity and oxygen content of the water and to pay particular attention to the small amounts of nutrient salts which are of the utmost importance to the development of the tiny floating plants. The marine botanist has to determine the number and character of plants present, the zoologist has to study the populations of animals which graze on the microscopic plants, and together the two latter have to enter upon the question of interrelationships between the different groups of organisms. The bacteriologist has to provide the answers to a number of questions dealing with the decomposition of organic matter, and in the course of this work the specialists in each line have to turn to all the others in order to obtain as complete information as possible as to the events which take place.[16]

Later, after the publication of *The Oceans,* Sverdrup expressed his views of science and scientists in a seminar at SIO in January 1948:

... let me say that I think it is a pity that the designation 'naturalist' has obtained a specific connotation and has come to mean a biologist primarily interested in the descriptive aspects of biology. I think that the word 'naturalist' should apply to every scientist who is principally interested in the study of our surrounding nature, regardless of whether dead or alive. All of us engaged in the larger parts of the biological sciences, geology, geophysics, have in common our curiosity as to the world which surrounds us, our love of nature, and I think we all of us ought to be called 'naturalists.'[17]

Clearly, Sverdrup's eclectic viewpoint in 1928, long before he faced the problem of keeping the Scripps Institution of Oceanography alive – perhaps even making it thrive – was a precondition of his success in La Jolla. It was not an expedient but a long-held and unchanging belief about how oceanography should be conducted and was part of his scientific fibre.

In 1928, reviewing physical oceanography in Europe, Sverdrup commented that the International Council for the Exploration of the Sea had been envisioned as a means of coordinating large-scale work on the oceans that was beyond the means of single nations. But this had not happened, and, as he said, 'the interest in physical oceanography within this council has been placed in the background,' subordinated to fisheries problems, leaving the physical problems to individuals without the means to tackle them on a large geographic or scientific scale.[18] However, there were still opportunities for small-scale research, contributing to larger scales of knowledge, through physical oceanographic programs carried out systematically and in several areas at the same time with modest-sized vessels (like the Norwegian *Armauer Hansen*), as his colleagues Albert Defant and Bjørn Helland-Hansen were attempting to do in European waters.[19] Eight years later, in La Jolla, he had the opportunity to begin just such a program, contributing both to science and the future of the Scripps Institution. Writing in 1940, he said that 'it has been my hope, since I became acquainted with the Scripps Institution, to place emphasis on what all the special fields have in common – the ocean itself – in order to develop a program for the Institution as such, in which as many as possible of our specialists can participate. In this manner results in one field may find application in another and more rapid advances may be expected.'[20] New scientific programs soon united much of the effort in La Jolla, and Sverdrup's vision of a unified curriculum for the students at SIO was realized quickly. But a textbook incorporating Sverdrup's surprisingly catholic approach to oceanography did not exist.

The appendix in this book (page 287) lists the advanced-level books available to students of oceanography up to the time of publication of *The Oceans*. They range from general accounts of the sciences of the sea, such as Murray and Hjort's classic *The Depths of the Ocean* (1912), to highly specialized monographs on specific subjects like Defant's *Dynamische Ozeanographie* (1929). Nothing was suitable as a text around which an integrated graduate program in oceanography could be built. Sverdrup, Johnson, and Fleming apparently surveyed what was available, for in his memoirs Martin Johnson commented that 'lack of an integrated textbook was a serious handicap. This led Dr. Sverdrup, Dr. Fleming, and me to write such a text after having abandoned an earlier idea of preparing only a syllabus for class work.'[21] After three years' hard labour, Sverdrup and his colleagues had what they wanted.

The Content of *The Oceans*

As the writing of the manuscript began, Sverdrup, Johnson, and Fleming found their aim changing. As they stated,

> Four years ago when we started the preparation of this book, we hoped to give a survey of well-established oceanographic knowledge, but it soon became apparent that the book could not be brought up to date without summarizing and synthesizing the wealth of information that has been acquired within the past dozen years, as well as the many new ideas that have been advanced. Consequently, the book has grown far beyond its originally

planned scope, and the presentation has become colored by the personal concepts of the authors.

Furthermore,

> At the risk or premature generalizations, we have ... preferred definite statements to mere enumerations of uncorrelated observations and conflicting interpretations, believing that the treatment selected would be more stimulating.[22]

As a result, *The Oceans* became a magisterial and definitive treatise in all the fields it surveyed, biological, chemical, geological, and physical oceanography, creating syntheses of knowledge available nowhere else.[23] This was especially true of Sverdrup's treatment of physical oceanography in eight chapters, more than a third of the total pages, reflecting in their organization and content a prescription of the way oceanography should be approached as well as a delineation of accepted knowledge.

The first 'physical' chapter, the third, sets out in an orderly and thorough way the physical properties of seawater, including the properties of pure water, salinity and chlorinity, physical units, density, thermal properties, viscosity, conductivity, sea ice, plus sound and light transmission. It is clear early that the aim was not just general information, but especially the information needed for dynamic oceanography: sigma-t, the dependence of density on salinity, pressure, and temperature, specific volume and specific volume anomaly, and potential temperature are introduced seriatim, although with no explanation at this stage of their greater significance, which comes later. For the first time in an easily accessible form, Sverdrup introduced the concepts of eddy coefficients of thermal conductivity, diffusion, and viscosity, indicating that laboratory coefficients did not apply in the sea because of the constant state of motion – '... where turbulent motion prevails, it is necessary to introduce an "eddy" coefficient that is many times larger and that is mainly dependent on the state of motion.'[24] In a four-page discussion of eddy coefficients, he makes it clear that due to the complex internal motions of the sea, 'no basis exists for the application to processes of the sea of the coefficients of thermal conductivity, diffusion, and viscosity that have been determined in the laboratory.'[25] The sea is a physical entity unto itself, fully justifying unique treatment.

A lengthy section on the heat budget of the earth and oceans in the fourth chapter is followed by a detailed discussion of density and thermohaline processes. Oceanic deep water forms through cooling and sinking, and although direct evidence was lacking, Sverdrup speculated that North Atlantic deep waters were formed in the Labrador Sea, the Irminger Sea, and possibly in the Norwegian Sea, from where they flowed southward throughout the Atlantic basins, as W.B. Carpenter had suggested several decades before.[26] To the south, conditions for deep-water formation existed mainly on the Antarctic continental shelf, although only in very localized areas south of the Atlantic Ocean (as summarized only a few years before in George Deacon's work on the Southern Ocean).[27] He introduced a distinction between water masses, which had distinctive temperature and salinity curves, as shown in the T-S curves introduced in 1916 by Helland-Hansen, and water types, which were far more localized and characterized by single T-S values.[28] Such close characterization of water masses and water types lay behind the newly introduced 'core method' of Georg Wüst, by which water could be traced from the places where it attained distinctive values until they were lost by mixing (an example being the tracing of warm, saline Mediterranean water throughout the central North Atlantic). As Sverdrup described the method,

> The spreading of the water can ... be described by means of a T-S curve, one end point of which represents the temperature and salinity at the source region and the other end point of which represents the temperature and salinity in the region where the last trace of this particular water disappears. Having defined such a T-S curve, one can directly read off from the curve the percentage amount of the original water type that is found in any locality. The core method has proved very successful in the Atlantic Ocean and is particularly applicable in cases in which a well-defined water type spreads out from a source region.[29]

Once again, the emphasis was on the geophysical basics – such as the generalities of heat budgets – and on the direct application of the basics to ocean circulation using the most up-to-date examples. New concepts and methods were inseparable from their uses, linking the theoretical power of the quantitative approach to its application. Nowhere was this more true than in Sverdrup's treatment of the distribution of variables in the sea (chapter 5 of *The Oceans*), considered as scalar and vector fields, both of which could be represented mathematically in differential equations. Any scalar quantity (he considered temperature, salinity, pressure, and oxygen content) could be modified by local effects (not involving motion) and by motion of the water, changing it in the two horizontal directions and in depth (represented by x, y and z axes). Moreover, those properties could be of two kinds, conservative (that is, changed only by diffusion and current motion, such as temperature and salinity) and non-conservative (changed by biological activity, such as oxygen content), for which he developed differential equations specifying the changes with and without biological activity and making use of the relevant eddy coefficients to give the changes of properties in time in a quantitative framework. Once this was established, Sverdrup took pains to emphasize that the overall distribution of properties in the sea was not static but the result of dynamic processes of supply and removal (he used the distribution of oxygen as a non-conservative example, the resultant of biological production and utilization) so that constancy of properties did not necessarily mean lack of change. As a result,

> ... it is evident that in the discussion of the distribution of concentrations in the sea it is as yet impossible to apply a method of deduction based on knowledge of all processes involved in maintaining the distribution. Instead, one has to follow a winding course, discuss processes and their effects whenever possible, discuss actual distributions if such have been determined, and either interpret these distributions by means of knowledge gained from other sources as to acting processes or draw conclusions as to these processes from the distribution.[30]

But although the process of unravelling the causes of processes in the sea might be a 'winding course,' he left no doubt, without editorializing, that only a systematic quantitative treatment would be an acceptable way to reach that end.

Ocean currents were at the heart of Sverdrup's physical oceanography. Once again, in chapter 11, his approach was an orderly systematic one, a non-mathematical treatment of the principles of dynamic oceanography that owed its origins to Vilhelm Bjerknes and his Scandinavian followers. This Sverdrup called 'the method that has ... become standard procedure in physical oceanography.'[31] As he summarized the value of the dynamic method,

> Currents related to the distribution of density can be computed from the more easily observed temperatures and salinities, following the procedure which has been outlined here ... Herein lies the value of the application of hydrodynamics to oceanography and the necessity of familiarity with this application if all possible conclusions shall be drawn from the observed distributions.[32]

Before 1942, the only extensive presentations of dynamic oceanography in English after Sandström (1919) had been by Edward H. Smith (in 1926) and George McEwen (in 1932).[33] Although both monographs were important and influential, neither had placed the technique in a thorough-going physical setting, starting from first principles. Sverdrup did just that by establishing a physical setting, then distinguishing between currents resulting from the distribution of pressure in the oceans and those resulting from wind-stress on the water.[34] Within the water column, the distribution of pressure was not uniform, and as a result of the slope of isobaric surfaces (resulting from differences from place to place in pressure gradients) and the effect of the Earth's rotation (the Coriolis effect), water moved along the contours of isobaric surfaces:

> From a chart of the topography of an isobaric surface the corresponding currents are readily obtained on the assumptions that have been made, namely that the component of gravity acting along the isobaric surface is balanced by Coriolis force. On this assumption the current is directed along the contours. If these are drawn at equal intervals, the slope of the surface is inversely proportional to the horizontal distance between the contours, and the velocity is therefore inversely proportional to the distance, the factor of proportionality depending on the latitude ...[35]

But there were two problems, the first that the currents calculated using the dynamic method were relative rather than absolute because they were calculated from a surface that itself might be in motion. Sverdrup discussed at some length the solutions to this problem, none of them unequivocal.[36] But of greater significance, he made the first clear statement of the uncertainties about the direction of causation in working with the dynamic method:

> ... no mention has been made of cause and effect. The reason is that any given distribution of density can remain unaltered in the course of time only in the presence of currents ... Therefore, all that can be stated is that a mutual relation exists between the distribution of density and the corresponding currents, but it is impossible to tell *whether the distribution of density causes the currents or the currents cause the distribution of density.*

And a few lines later,

> The effect of the wind on the ocean currents is twofold. In the first place,

the stress that the wind exerts on the sea surface leads directly to the development of a shallow wind drift; in the second place, the transport of the water by the wind drift leads to an altered distribution of density and the development of corresponding currents.[37]

Here, in uncharacteristic italics, and in comments on the effect of the wind, he diagnosed and warned against the difficulties that had led George McEwen to question the proper physical foundations of the dynamic method.[38]

The heart of Sverdrup's presentation of dynamic oceanography lies in a single chapter titled 'Dynamics of Ocean Currents' (chapter 13). Bruce Warren has commented (from a 1990s perspective) that 'the dynamical material in *The Oceans* ... is of mainly antiquarian interest now,'[39] and so it is from a scientific point of view. But it is important historically because it allows us to see how Sverdrup believed that dynamic oceanography should be presented.

Typically, Sverdrup introduced the dynamic approach to the oceans by starting with the principles of statics and kinematics (in chapter 12), first units and dimensions in physics, then the fields of gravity, mass, and pressure. Then he moved on to kinematics and the equations of fluids in motion, including the field of motion and a quantitative treatment of continuity. From this quantitative treatment of the basics, his chapter on the dynamics of ocean currents expanded on much that had gone before, based on his opening statement that

> our knowledge of the currents of the ocean would be very scanty if it were based entirely on direct observations, but, fortunately, conclusions as to currents can be drawn from the distribution of readily observed properties such as temperature and salinity. When drawing such conclusions, one must take into account the acting forces and apply the laws of hydrodynamics and thermodynamics as derived from theoretical considerations or from generalizations of experimental results.[40]

Not explicit, but clear upon historical analysis, is the importance of the innovations in approaches to ocean circulation that had been begun by Vilhelm Bjerknes and expanded by his Scandinavian followers, among them Sverdrup himself.

After thirty pages of preliminaries, based largely on Bjerknes et al.'s *Physikalische Hydrodynamik* (1933), including an extensive discussion of the modification of the equations of motion to account for rotation, the relation between currents and the pressure field, and the problem of determining absolute current velocities from dynamic calculations, Sverdrup returned, this time in a mathematical framework, to Bjerknes's circulation theorem and its modification by Sandström and Helland-Hansen (see chapter 3 for background), along with discussions of transport and turbulence, eddy coefficients, and friction.

At the outset, Sverdrup emphasized that there were only two fully 'dynamical' processes involved in ocean currents, the creation of motion due to the pressure field (derived from the distribution of temperature and salinity, and thus density – the basic data exploited by Bjerknes and his followers) and the effect of the wind.[41] In a few pages, he developed the mathematical basis of the effect of the wind on surface waters as first formulated by V.W. Ekman in 1902, bringing it up to date by the addition of new approaches to wind stress, the roughness of the sea surface, density varying with depth,[42] the dissipation of energy in the production of wind-driven currents, and the production of upwelling (using examples from the California Current first examined by George McEwen and later by Sverdrup and Richard Fleming).[43] And in concluding his discussion of dynamics, he devoted a few pages to the topic that had fitted so awkwardly into schemes of ocean circulation from the time of Rumford and Humboldt through the Carpenter – Croll controversy of the 1860s and 1870s, ocean thermodynamics and thermohaline circulation. Referring to work by Bjerknes and Sandström[44] showing on thermodynamic principles that for a thermal circulation to exist, heating would have to occur at a greater depth than cooling, he concluded that

> ... it is ... evident that in the oceans conditions are very unfavorable for the development of thermal circulations. Heating and cooling take place mainly at the same level – namely, at the sea surface, where heat is received by radiation from the sun during the day when the sun is high in the sky, or lost by long-wave radiation into space at night or when the sun is so low that the loss is greater than the gain and heat is received or lost by contact with air.[45]

A thermally driven circulation will certainly be established if the sea is heated at the surface in one region (the tropics, for example) and cooled elsewhere (at high latitudes, typically), but this circulation 'cannot become very intensive, particularly because the heating within the return flow must take place by the slow processes of conduction'[46] (just the kind of circulation defended by Carpenter and attacked relentlessly by Croll, as discussed earlier). Sverdrup's view was that a slow thermohaline circulation from equator to high latitudes was certainly a possibility, but that it *was* dominated by the wind:

> It is probable ... that the existing current system bears no similarity to the one that would result from ... a thermohaline circulation, but is mainly dependent upon the character of the prevailing winds and the extent to which the circulation maintained by the wind is checked by the thermal conditions. In other words, the wind system tends to bring about a distribution of density that is inconsistent with the effect of heating and cooling and the actual distribution approaches a balance between the two factors.[47]

Nonetheless, despite these reservations, he saw a role for vertical convection in deep-water formation, and referred his readers to earlier pages in which he described in general terms how dense bottom water, formed at high latitudes during intense cooling, should spread slowly throughout the ocean basins: 'In general, the water of the greatest density is formed in high latitudes, and because this water sinks and fills all ocean basins, the deep and bottom water of all oceans is cold.'[48] The treatment is uncharacteristically tentative, mainly because of Sverdrup's disinclination to speculate without at least a modicum of data about the extent of vertical convection and of thermohaline processes in general. The Carpenter-Croll debate was not ended by *The Oceans*.[49]

Almost a book in itself, the final chapter on physical oceanography, titled 'The Water Masses and Currents of the Oceans,' has been described as 'Sverdrup's greatest contribution to oceanography.'[50] Their great geographic sweep and outstanding scientific content have made these 120 pages probably the most influential in all of physical oceanography. In 1992, Bruce Warren summarized the lasting value of this achievement by asking,

> Is there any benefit ... in reading ['The Water Masses and Currents of the Oceans'] today? I certainly think so: not so much for exact numerical values or other details, but for the overview. The water-property fields in the world ocean are so multifarious that we cannot do without an organizing framework within which to comprehend them, and to discuss our observations of them. Sverdrup's judicious water-mass scheme, with its implied linkages to sources of water types and its possibilities for refinement, still works, still meets this need better than anything else yet devised.[51]

Several generations of oceanographers, including my own, learned the basics of worldwide ocean circulation from Sverdrup's chapter on water masses and currents, and parts of it, such as the twenty pages he devoted to the Southern Ocean,[52] along with a series of diagrams mainly derived from the work of George Deacon in the 1930s on the Discovery Investigations, have become iconic.[53]

Characteristically of Sverdrup, the treatment of each ocean (beginning with the Southern Ocean because of its connection with all the others, its geographical extent, and relative simplicity) is a systematic one, in which geographical features, the ensemble of water properties, and then the current regime are treated in sequence. The world's oceans become part of a detailed taxonomy based largely upon Sverdrup's signature properties, their water masses and currents, beginning, after the Antarctic (Southern) Ocean, with the Atlantic, Indian, and Pacific Oceans, considered, as Sverdrup wrote, 'as deep bays that are in open communication with the Antarctic Ocean to the south but are closed at their northern ends,'[54] then moving on to the various 'adjacent' seas such as the Caribbean, Mediterranean, Baltic, Red, and so on. The approach is spare, highly focused, and largely non-historical, using quantitative assessments whenever adequate data were available.

Sverdrup's treatment of water masses and currents concludes with a summary in figures, charts and T-S diagrams of the water masses of the world oceans, and a discussion of their deep water circulation. The figures and diagrams were intended, as he wrote, to 'together illustrate the concepts that water masses are formed at the sea surface and sink and spread in a manner that depends on their density in relation to the general distribution of density in the oceans.'[55] Deep waters still remained something of an enigma, and Sverdrup's treatment is tangibly restrained and diffident because of the absence of solid data. Nonetheless, he made it clear that

> in order to understand the deep-water circulation, one has to bear in mind that the deep and bottom waters represent water the density of which became greatly increased when the water was in contact with the atmosphere, and that this water, by subsequent sinking and spreading, fills all deeper portions of the oceans. The most conspicuous formation of water of high density takes place in the subarctic and in the Antarctic regions of the Atlantic Ocean. The deep and bottom water in all oceans is derived mainly from these two sources ...[56]

With that, the discussion of ocean circulation in *The Oceans* was complete.

One of the first reviewers of *The Oceans,* J.N. Carruthers, called *it* 'a wonderful book' and claimed that 'all things considered, "The Oceans" may be considered as epoch-making.'[57] So it was. George Deacon was more measured in his praise, but he made it clear, as he said, that 'the success of the book is unmistakable; it will be invaluable to both specialist and beginner,' and later was more fulsome: 'It is difficult to convey a true impression of the great wealth of information contained in the book It is never likely to fail as a reference book, and it will be the starting point of many inquiries ...'[58] Modern assessments of this great achievement, notably that by Bruce Warren, have been complimentary, but justifiably have emphasized the ways Sverdrup's synthesis is differently oriented from, or deficient in comparison to, modern knowledge. Among these is Sverdrup's emphasis on the overall meridional orientation of deep-water circulation, derived from the influential work of Merz and Wüst in the early 1920s (see chapter 4), the lack of emphasis on the asymmetric, gyral circulation of the oceans (to which Sverdrup and his student Walter Munk contributed a decade or so later), the lack of deep boundary currents, the absence of deep western boundary currents, mistakes in the locations where North Atlantic Deep Water forms, and the absence of some major currents that became known especially in the 1950s.[59] We cannot blame Sverdrup for what he did not know, especially considering the brilliance of his insights into ocean circulation, and his bringing together for the first time all the available information, placing it into a theoretical framework of water types and water masses, along with their modes of formation, and thus giving physical oceanography its first fully formal, lasting, structure.

Oceanography after *The Oceans* – the Wind-Driven Circulation

With our perfect hindsight, the absences from *The Oceans* and the mistakes can help us build an outline of physical oceanography's development during the past half century. Harald Sverdrup himself, continuing his original contributions to ocean dynamics after *The Oceans,* was the first to open a new chapter. In 1938, speaking to a group of mechanical engineers about fluid mechanical problems of ocean circulation, he emphasized the importance of the wind, particularly its 'tangential stress' in causing surface currents, relating it to upwelling along the California coast.[60] This approach, delayed by the war and the preparation of *The Oceans,* became the basis of a far more ambitious approach to wind-driven circulation in 1947, when Sverdrup's paper 'Wind-Driven Currents in a Baroclinic Ocean ...' appeared, presenting his examination of the problem as applied to the eastern equatorial Pacific (part of his adopted home-ocean), examining, as he wrote, 'whether the equatorial currents, including the counter currents, can be accounted for on the basis of our knowledge of the wind stress only.'[61] The basis of his calculations was a portion of an ocean in which there was an eastern boundary but no western boundary, and in which, to the north, the wind blew from the west, while closer to the equator, it blew from the east, involving in both cases transport to the right of the downwind direction (the case in the Northern Hemisphere). The result was a set of equations[62] relating the pressure gradient and vertically integrated mass transport in two horizontal directions to the stress of the wind and in which, realistically, the Coriolis effect varied with latitude. These were tested against oceanographic observations from the eastern equatorial Pacific, allowing Sverdrup to arrive at the conclusion that 'the distribution of density and the mass transport by the accompanying currents of the eastern equatorial Pacific depend entirely upon the average stress exerted on the sea surface by the prevailing winds.'[63] In short, not just the pattern of the currents in the eastern part of the North Pacific gyre could be explained mathematically, but the volume of the meridional (north-south) transport depended solely on the curl of the wind stress applied to the ocean surface, not the absolute velocity of the wind.[64] Wind-driven currents, of the kind described qualitatively a century earlier by James Rennell, were now being approached in a fully geophysical, quantitative way.[65]

Not unexpectedly, it was Walter Munk (b. 1917)[66] (fig. 9.1) of SIO, Sverdrup's former student, who in 1950 addressed the modest geographic scope of Sverdrup's model. His approach was to model the situation in an ocean basin extending to higher latitudes, and which had a western boundary along which an intense flow occurred, like the Gulf Stream in the North Atlantic and the Kuroshio in the North Pacific. Using a rectangular framework representing an ocean from high latitudes to the equator and from an western boundary to an eastern one, Munk incorporated, realistically, the Coriolis effect varying with latitude,[67] the friction of the water within the ocean due to varying velocities and with both coastal boundaries, and the most up-to-date information available on the wind stress in the North Pacific and the Central and North Atlantic. From modified equations of motion, he then calculated the circulation and volume transport of the two ocean areas, as shown in figure 9.2A. There was a further striking outcome. As Munk wrote, 'The zonal wind system divides the ocean circulation into a number of gyres, each bounded by latitudes ... The major dividing lines between gyres correspond

9.1 Walter Munk of the Scripps Institution of Oceanography about 1956. With permission of Scripps Institution of Oceanography Archives, UC San Diego Libraries.

to the latitudes of maximum west winds, of the northerly and southerly trades, and of the doldrums ...'[68]

The pattern of gyres (fig. 9.2B) was so striking that it led Munk to propose a classification applicable to all oceans, based on gyres, yielding a consistent pattern:

> In high (north or south) latitudes the cyclonic subpolar gyre corresponds to the region of the cyclonic storms, the anticyclonic subtropical gyre to the anticyclones. The equatorial currents and countercurrent enclose two additional gyres ... corresponding to the trades and the doldrums, but these are so narrow that they are best not regarded as gyres. A typical gyre is composed of zonal currents to the north and south, a strong persistent current and boundary vortex on its western side, and a compensating drift in the central and eastern portion, upon which a variable eastern current and wind-spun vortex are superimposed.[69]

Thus a consistent nomenclature of the gyral circulation of the oceans, noted many times in the past in a qualitative fashion, took on a quantitative form in which the Earth's rotation and the effect of the wind blowing across the ocean surface acted dynamically and could be calculated. The results were fully consistent with observation – but based on the quantitative assessment of a few crucial causal factors, such as the wind stress and the effect of the Earth's rotation.

Munk had been influenced significantly in the development of his model not only by Sverdrup, whom he acknowledged as a contributor of 'many helpful suggestions,'[70] but by a short and initially little-known paper by the Wood Hole theoretical oceanographer Henry Stommel titled 'The Westward Intensification of Wind-Driven Ocean Currents,' published two years before. In Stommel's remarkable insight, for the first time, were the elements of a quantitative theory of both wind-driven and abyssal circulation that finally would put the Carpenter-Croll controversy to rest. It also provided the first elements of a framework for the development of physical oceanography through the 1960s.

A Global Theory of Circulation – Linking the Mixed Layer and the Abyss

Henry Stommel (1920–1992) (fig. 9.3) came to Woods Hole Oceanographic Institution in 1944 after an undergraduate degree at Yale in astronomy.[71] When he died in 1992, the *New York Times'* eminent science

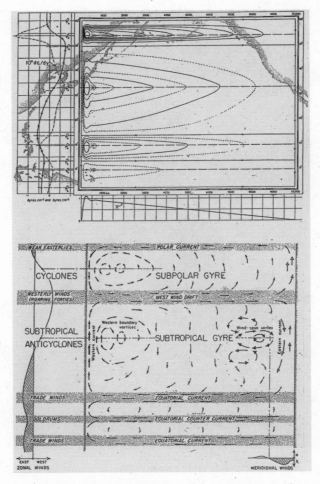

9.2 Walter Munk's diagrams of wind stress, mass transport, and the gyral circulation typical of all oceans, taken from his paper 'On the Wind-Driven Ocean Circulation' (1950).

A. Top, mean annual zonal wind stress (solid line) and curl (torque) of the wind stress (dashed line); on the right, mass transport streamlines from 60°N to just south of the equator, extending from a western boundary to an eastern one.

B. Bottom, the representation of winds and the resulting gyres in a model rectangular ocean representing the features of the North Pacific and North Atlantic Oceans. This was the first systematic application of the concept of gyres and of the names of the associated features to the oceans.

9.3 Henry Stommel of Woods Hole Oceanographic Institution, probably in the late 1950s. With permission of Woods Hole Oceanographic Institution Archives.

writer Walter Sullivan described him as 'widely acclaimed as an authority on ocean currents.' Less austerely, in an obituary for the National Academy of Sciences, his former Harvard student and colleague Carl Wunsch described him as 'probably the most original and important physical oceanographer of all time.'[72] The origins of Wunsch's assessment lie in the late 1940s, when Stommel took on the problem of the asymmetry of oceanic circulation – the gyres described and given dynamic formulations by Sverdrup and later Munk – particularly the puzzling fact that the western boundaries of those gyres were exceptionally strong currents like the Gulf Stream and Kuroshio. Stommel soon provided a dynamic explanation, one with enormous implications for the overall circulation of the oceans, whether wind-driven or thermohaline. He oversaw, and contributed remarkably to, a recasting of the kind of general oceanic circulation described in *The Oceans*. Stommel's young colleague at Woods Hole, Bruce Warren, summarized the revolution in knowledge of ocean circulation he began, saying that

> what has changed greatly [since *The Oceans*] is the conception of their [oceanic distributions of properties] basis: that the underlying mean deep circulation, largely masked by lateral mixing of the water characteristics, is a system of western boundary currents and poleward interior flows, as in the dynamically-consistent models of Stommel and Arons ... rather than the Merzian system of basin-wide slow meridional flows from source regions.[73]

Stommel's short paper in 1948 was the first step in this transformation.

It was the influence of his colleagues Ray Montgomery and Columbus Iselin at Woods Hole, rather than Sverdrup's 1947 paper, that directed Stommel to the problem of the causes of the Gulf Stream and similar western boundary currents.[74] He set out the problem succinctly:

> Perhaps the most striking feature of the general oceanic wind-driven circulation is the intense crowding of streamlines near the western borders of the oceans. The Gulf Stream, the Kuroshio, and the Agulhas Current are examples of this phenomenon. The physical reason for the westward crowding of streamlines has always been obscure. The purpose of this paper is to study the dynamics of wind-driven ocean circulation using analytically simple systems in an attempt to discover a physical parameter capable of producing the crowding of streamlines.[75]

His approach was to investigate the solutions of an equation (based on the equations of motion) representing the pattern of circulation in a rectangular ocean with western and eastern boundaries, over which the winds were symmetrical (westerlies to the north, trades to the south). There were three cases: a non-rotating ocean, an ocean with constant Coriolis parameter, and one in which (as applied approximately to the real Earth) the Coriolis parameter varied linearly with latitude. Only in the last case did the streamlines of current flow crowd together on the western side of the ocean – leading to the concluding sentence of his paper: 'The similarity that the velocity field of this simple case bears to that of the actual Gulf Stream suggests that the westward concentration of streamlines in the wind-driven oceanic circulation is a real result of the variation of the Coriolis parameter with latitude.'[76] A little later, he restated this conclusion in terms of vorticity, the form in which it is now best recognized.[77] And within two years, Stommel was attempting to make sense in similar terms of the abyssal circulation.[78]

When he reviewed the results of work on the abyssal circulation at an international meeting held in Toronto in September 1957, Stommel had arrived at a consistent picture of westward intensification of ocean currents applying to the abyss as well as to the surface waters. The northward-directed Gulf Stream should be underlaid by a deep southward-directed western boundary current, which he equated with North Atlantic Deep Water, along the bottom close to the continental margin.[79] Stommel had been mentioning the idea since late in 1954, and proof was already at hand. Using neutrally-buoyant floats developed by the British oceanographer John Swallow, Swallow and the Woods Hole oceanographer L.V. Worthington had showed during a cruise in March-April 1957 that the predicted return flow could be measured off the eastern United States,[80] lending enormous credibility to Stommel's developing theory of oceanic circulation.[81]

By late in 1957, it is clear that Stommel had arrived at a comprehensive theory of oceanic circulation involving both the surface and abyssal circulation. Its essence was the origin of abyssal waters of the Atlantic in two source regions, the northern North Atlantic and the Weddell Sea, most of the North Atlantic Deep Water flowing into the Southern Hemisphere (and eventually with the Antarctic Bottom Water into the other oceans) as an intense deep western boundary current. Return flows from the western boundary currents were more diffuse, but eastward, poleward, and upward, creating and maintaining the thermocline of the oceanic regions and countering surface warming (fig. 9.4).[82] Moreover, as Stommel, Arnold Arons (1916–2001),[83] and Alan Faller soon showed, the elements of this simple model of oceanic circulation, in which the main components were intense western boundary currents and a steady, broad return flow poleward and upward, could be verified experimentally using a fluid-filled sector of a rotating cylindrical tank.[84] Examining in detail the formation of the thermocline, Allan Robinson and Stommel calculated vertical velocities and eddy conductivities at the base of the thermocline, and provided the first quantitative model of its formation

9.4 The abyssal circulation as envisioned by Henry Stommel in 1958, showing the origin of abyssal waters at high latitudes, intense deep western boundary currents, and general eastward and poleward return flows that maintained the thermocline (from Stommel 1958a).

and variation with depth as a component of a complete theory of ocean circulation. They commented in closing that

> ... this model of the thermohaline circulation implies many definite things about the circulation of the deep waters of the ocean. For example, we should be able to derive the total amplitude of the deep circulation of the ocean by integrating the deep upwelling velocity over the whole water-covered globe ... In this way we should be able to predict the average age of the deep waters, a topic of current interest to geochemists, and a subject of much present day experimental enquiry, and sea-going activity.[85]

This remarkable tour de force of innovation in the theory of oceanic circulation culminated in a series of papers between 1960 and 1972 on many aspects of the problems that Stommel had first broached in 1948. Their results have come to be known as the Stommel-Arons circulation model. The first two were the most wide-ranging and profound, permanently changing the way oceanographers envisioned global ocean circulation.

Stommel, Arons, and Faller's original experimental test of Stommel's western boundary current and global circulation model had used a sector of a cylindrical tank to investigate the reality of the postulated circulation. In 1960, Stommel and Arons examined what happened when the analysis of sinking water forming a deep western boundary current and returning through generalized upwelling to form the thermocline was done for a rotating sphere – that is, made more representative of the Earth.[86] Then, in the next paper they moved on, as they stated it, 'to construct a highly idealized model of the general abyssal circulation of the world ocean' in quantitative terms, based on observational information on deep-water formation,[87] turning the old simplistic idea of a thermohaline circulation driven by cooling and sinking at high latitudes on its head:

> We envisage the ultimate cause of the abyssal circulation to be the mechanism of the main thermocline which in effect, as a result of surface heating and applied wind stress, demands a certain upward flux at mid-depth.
>
> ...
>
> This way of looking at the abyssal circulation is quite different from the traditional view which implies that the magnitude of the abyssal circulation is determined essentially by the amount of winter cooling at the polar regions and hence that a warming of the polar regions of one or two degrees would

largely stop the abyssal flow. From our point of view a warming of polar regions of one or two degrees would not affect deep transports except possibly to shift the location of the sources, and to reshuffle the western boundary currents ... The overall transports and the depth of the thermocline etc. would remain unchanged, providing the climatological factors which determine the thermocline remain unchanged.[88]

On this basis, incorporating the Robinson and Stommel model of the thermocline, they then calculated the amplitude and velocity of abyssal upward circulation, first providing a box-model and budget of the transports into and out of the various basins of the world oceans,[89] then estimating vertical velocities and the replacement time worldwide of abyssal waters.

In later papers, Stommel and others used the basic model formulation to estimate the flow rates of various water masses, to estimate the rate of distribution of dissolved oxygen, and the newly useful oceanographic tracer carbon-14, and to model the effect of bottom slope on the width of deep boundary currents.[90] But these were elaborations of the basic model of global circulation, not a recasting of any of its foundations.

Among working oceanographers in the 1960s, it is unlikely that anyone but Henry Stommel knew of the bitter dispute between James Croll and W.B. Carpenter described in chapter 2. He and Allan Robinson were dismissive of that sparring match over abyssal circulation, commenting that 'the exchange of letters in NATURE during the 1870s between W.B. Carpenter and James Croll shows how futile simple verbal arguments can be in discussing such issues.'[91] Indeed, if there is any single factor that separates the oceanography of the late nineteenth century from that of the twentieth century after the Second World War, it is the increasing use of mathematics, pioneered by Bjerknes and his followers (as set forth in most of this book), but made mandatory by the need to understand the processes governing oceanic circulation recognized by Harald Sverdrup, Walter Munk, Henry Stommel, and new generations of mathematically able physical oceanographers who followed their example between the 1940s and the 1960s. The Stommel-Arons model of 1960 united the elements over which Croll and Carpenter fought so bitterly – wind stress and its production of surface ocean currents, and a deep thermohaline circulation, which, unexpectedly, was linked intimately to the surface circulation through the production of the thermocline and could be most dramatically observed in intense western boundary currents. In so doing, they ended the long-forgotten controversy and verified that there was a new way of doing oceanography centring on the search, using mathematics, specifically mathematical modelling, for the causal factors of ocean circulation.

Beginning a New Tale

This book ends with events, now more that forty years old, that resolved nearly two centuries of speculation and controversy about ocean circulation, and that were the outcome of an increasing attention, particularly from about 1900 onward, to ways of expressing ocean circulation in numbers rather than in words and qualitative mappings. There is a satisfying finality to the story – a problem has been solved, right has triumphed, and a tale has been told. But this is illusory. By concentrating on the search for solutions to the problem of how to represent and measure oceanic circulation, it is possible to tell a coherent story, as shown in the preceding chapters. But a new tale is unfolding, one far harder to formulate and even (to use a physical oceanographer's term) parameterize. What are the dominant themes in the search for solutions about oceanic circulation since the Stommel-Arons model, with its close links to the clever, profound contributions of Sverdrup and Munk, not to mention the oceanographers whose work was so changed by the insights of Vilhelm Bjerknes and his followers?

This is akin to asking what should be in the next book about the development of physical oceanography. My view of this is sketchy and constrained by all the problems of dealing with contemporary history, that is, lack of perspective, too much detail, actors still alive or too near in time to allow objective assessment, and the difficulty of separating real scientific progress from evanescent fashions. One way to tackle the problem would be to take the subjects regarded as canonical in a good recent textbook as research topics for a history of the development of physical oceanography – particularly ocean circulation – since the early 1960s and work backward – perhaps a historical use of what geophysicists call the inverse method.[92] But why not let Henry Stommel have the last word?

The Oceanography Society was founded in 1988 in the United States to unify what many felt was a fragmenting field, to provide a forum for discussions in all branches of oceanography, to overcome dissatisfaction with existing institutions, and to produce a magazine that would also fulfil these functions. At the society's first meeting in 1989, Henry Stommel was one of the invited lecturers. He used the opportunity to talk about his own very personal approach to oceanographic research, but also to review some of the achievements of the field since the 1960s.[93] He was fully aware of the debt owed by oceanography to other disciplines, a debt that had increased in the post-war years:

> Many new sophisticated mathematical techniques that have increased our power to study the oceans have come from other disciplines. Fortunately there have been oceanographers with broad enough skills to translate these techniques into useful tools for oceanography. Singular perturbation theory and linear programming have come from applied mathematics. Most of the fundamental dynamical ideas, objective analysis, and numerical modeling with data assimilation have come from meteorology.[94] Modern theories of time-series analysis and techniques for detecting signals have come from electrical engineering. Inverse theory has come via geophysics.[95]

He, perhaps more than anyone else, appreciated the contribution of new instruments to oceanography – John Swallow's neutrally buoyant floats, for example, had indicated the presence of the deep western countercurrent below the Gulf Stream, and also the at first dimly perceived prevalence of eddies in the Gulf Stream circulation and then everywhere in the oceans.[96] Instruments were the key to new observational data and to new theory:

> The chief source of new ideas in oceanography comes, I think, from new observations ... There was a time when eddies and meanders were only dimly perceived (1948); a time when we didn't know of the existence of the equatorial undercurrent (1952), or that the slope of the isotherms in the Gulf Stream extends to the bottom (1954), or that there was a deep recirculation, a time when the deep western boundary currents of Greenland Sea water had not been discovered flowing along the slope of Greenland around into the Labrador Sea (1952). There was a time when we didn't have reliable estimates of the flow through the Florida Straits (1960), when the ubiquity of inertial motions in the deep sea was not suspected (1957), when it could be thought that the velocity in deep water was too small to measure by current meter, and when we had no clear observational description of a deep, winter-time, bottom-water-formation event (1969).[97]

Stommel's contribution to many of these (though not all) will be easy to detect after reading the earlier parts of this chapter. But there was more:

More recently, we were surprised by multiple jets at the equator of the Indian Ocean (1976), by hot vents and the great helium plume in the Pacific,
and by the red spectrum of the nine-year drift of SOFAR [Sound Fixing and
Ranging] floats in the Atlantic. The geochemists persist in unsettling our
mental equilibrium with new current patterns revealed by exotic tracers
like Freons. Who would have foreseen 'meddies' (1981),[98] and who knows
their role in deep-sea mixing? There is the wonderful unfolding development of our knowledge about El Niño ... And there is that amazing large-
scale horizontal coherence of persistent double diffusive layers revealed in
the Caribbean Sheets and Layers Transects (1987) expedition that is preg-
nant with implications concerning deep-ocean mixing processes.[99]

Small-scale oceanographic research was Stommel's forte and his deep
personal preference. He referred to the increase of collaborative research
and multi-investigator, multi-ship, and sometimes multi-disciplinary programs as 'the Genie of Big Planning'[100] and late in his career avoided
being involved in them at all costs. And yet he, with Allan Robinson,
co-directed the MODE (Mid-Ocean Dynamics Experiment) of 1971–3 to
study small-scale variation in the Western North Atlantic, a project that
was typical of large-scale investigations persisting into our own time. He
recognized, with regret and ambivalence, it appears, that Big Science was
becoming a way of life in oceanography, and that as with particle physics,
it was a necessary part of the development of science if the insights of
the individual were to be tested on the proper physical scale. His life was
at the interface between science that could be carried out by individuals
on the deck of a ship or at a personal computer keyboard and science
requiring (or is it hooked on?) masses of data from the use of satellites,
whole fleets of ships, large-scale deployments of telemetering and profil-
ing floats, massive computing power, and megabucks.

Putting the nature of oceanography after Stommel in these terms may
exaggerate the changes that have occurred in oceanography since the
time of Vilhelm Bjerknes and those following who exploited the development of mathematical approaches to the study of ocean circulation – but
not by much. There is a whole new story to be told. Those first seek-
ers after knowledge of ocean circulation, Count Rumford, Alexander
von Humboldt, Emil von Lenz, W.B. Carpenter, James Croll, Vilhelm
Bjerknes, Björn Helland-Hansen, Johan Sandström, Johan Hjort, Alfred
Merz, Georg Wüst, George McEwen, Edward H. Smith, John Tully and
Harry Hachey – even Julien Thoulet – would today scarcely recognize
the field to which they contributed. The ocean is largely unchanged, but
the science that studies its circulation has been transformed.

APPENDIX: TEXTBOOKS OF PHYSICAL OCEANOGRAPHY

The following is a list of textbooks or treatises of oceanography arranged chronologically from the middle of the nineteenth century until the publication of *The Oceans* in 1942. Asterisks indicate works influential in their own time; double asterisks indicate especially influential works.

**Maury, M.F. 1855. *The physical geography of the sea.* New York: Harper & Sons.

Jilek, A., ed. 1857. *Lehrbuch der Oceanographie zum Gebrauche der k.k. Marine-Akademie.* Wien: Kaiserlich-Königlichen Hof- und Staatsdruckerei. x + 298pp.

Kayser, J. 1873. *Physik des Meeres: für gebildete Leser.* Paderborn: Ferdinand Schöningh. x + 359pp.

Reclus, E. 1873. *The ocean, atmosphere and life. Section I.* Translated by B.B. Woodward and edited by Henry Woodward. London: Chapman and Hall. viii + 304pp.

Attlmayr, F., ed. 1883. *Handbuch der Oceanographie und maritimen Meteorologie. Im Auftrage des K.K. Reichs-Kriegs-Ministeriums (Marine-Section).* Wien: Kaiserlich-Königlichen Hof- und Staatsdruckerei. 2 vols. xx + 1–598pp; xi + 599–990.

*Boguslawski, G. von. 1884. *Handbuch der Ozeanographie. Band I. Räumliche, physikalische und chemische Beschaffenheit der Ozeane.* Stuttgart: J. Engelhorn. xviii + 400pp.

Krümmel, O. 1886. *Der Ozean: eine Einführung in die allgemeine Meereskunde.* Das Wissen der Gegenwart 52. Leipzig and Prague: G. Fremtag und G. Tempsky. viii + 242pp. (Second edition in 1902.)

**Boguslawski, G. von, and O. Krümmel. 1887. *Handbuch der Ozeanographie. Band II. Die Bewegungsformen des Meeres,* von Dr Otto Krümmel.

Mit einem Beitrage von Prof. Dr K. Zöppritz. Stuttgart: J. Engelhorn. xvi + 592pp.

*Thoulet, J. 1890. *Océanographie (statique)*. Paris: Librarie militaire de L. Baudoin et Cie. x + 492pp.

Mill, H.R. 1892. *The realm of nature*. London: John Murray. vi + 369pp.

Walther, J. 1893. *Allgemeine Meereskunde*. Leipzig: J.J. Weber. xvi + 296pp.

*Thoulet, J. 1895. *Guide d'océanographie pratique*. Encyclopédie scientifique des aide-mémoire. Paris: Gauthier-Villars. 224pp.

*Thoulet, J. 1896. *Océanographie (dynamique), première partie*. Paris: Librairie militaire de L. Baudoin et Cie. 131pp.

Schott, G. 1903. *Physische Meereskunde*. Sammmlung Göschen. Leipzig: G.J. Göschen'sche Verlagshandlung. 162pp. (Later editions 1910, 1924.)

Thoulet, J. 1904. *L'océan, ses lois et ses problèmes*. Paris: Hachette. vii + 382pp.

**Krümmel, O. 1907. *Handbuch der Ozeanographie. Band I. Die räumlich, chemischen und physikalischen Verhältnisse des Meeres*. Stuttgart: J. Engelhorn. 526pp. (The second part, dealing with dynamic aspects of oceanography, was published in 1911.)

*Richard, J. 1907. *L'océanographie*. Paris: Vuibert et Nony. vi + 398pp.

*Thoulet, J. 1908. *Instruments et opérations d'océanographie pratique*. Paris: Libraire militaire R. Chapelot et Cie. vi + 186pp.

Murray, J. 1910. *The ocean*. New York: Holt. 248pp. (Published also in London in 1913.)

**Krümmel, O. 1911. *Handbuch der Ozeanographie. Band II. Die Bewegungsformen des Meeres (Wellen, Gezeiten, Strömungen)*. Stuttgart: J. Engelhorn. xvi + 766pp.

Fowler, G.H. 1912. *Science of the sea: An elementary handbook of practical oceanography for travellers, sailors, and yachtsmen*. London: John Murray. xviii + 452pp. (See also Fowler and Allen 1928.)

*Helland-Hansen, B. 1912. The ocean waters: An introduction to physical oceanography. I. Teil. *Internationale Revue der gesamten Hydrobiologie und Hydrographie* 3, Supplement 1: 1–84. (A second part appeared in 1923, but the third, intended to discuss dynamic oceanography, was never published.)

**Murray, J., and J. Hjort. 1912. *The depths of the ocean: A general account of the modern science of oceanography based largely on the scientific researches of the Norwegian steamer 'Michael Sars' in the North Atlantic*. London: Macmillan. xxi + 821pp.

Jenkins, J.T. 1921. *A textbook of oceanography*. London: Constable and Co. x + 206pp.

Rouch, J. 1922. *Manuel d'océanographie physique*. Paris: Masson. 229pp.

Thoulet, J. 1922. *L'océanographie*. Paris: Gauthier-Villars. ix + 287pp.

*Helland-Hansen, B. 1923. The ocean waters: An introduction to physical oceanography. Second part. *Internationale Revue der gesamten Hydrobiologie und Hydrographie* 11(5–6): 393–488. (The second part of an unfinished treatise. See also Helland-Hansen 1912.)

*Herdman, W.A. 1923. *Founders of oceanography and their work: An introduction to the science of the sea*. London: Edward Arnold. 340pp.

Johnstone, J. 1923. *An introduction to oceanography with special reference to geography and geophysics*. London: Hodder and Stoughton. xii + 351pp.

Gellert, G. (ed. W.B. Sachs). 1926. *Die Wunder des Meeres: Allgeimeinverständliche Darstellung des Lebens und Treibens im Meere, der Their- und Pflanzenwelt, der maritimen Einrichtungen und der Eroberung und Nutzbarmachung des Meeres durch den Menschen*. Berlin-Schöneberg: Peter J. Oestergaard. xii + 399pp. (A popular work, but with authoritative sections by Victor Hensen, Otto Krümmel, etc.)

Johnstone, J. 1926. *A study of the oceans*. London: Edward Arnold. vi + 215pp.

Defant, A. 1928. Physik des Meeres. In *Handbuch der Experimentalphysik* 25(2), pp. 569–686. Leipzig.

Fowler, G.H., and E.J. Allen. 1928. *Science of the sea: An elementary handbook of practical oceanography for travellers, sailors, and yachtsmen*. 2nd edn. Oxford: Clarendon Press. xxiii + 502pp.

*Defant, A. 1929. *Dynamische Ozeanographie* (Band III der *Einführung in der Geophysik*). Berlin: Julius Springer. x + 222pp.

Berget, A. 1930. Leçons d'océanographie physique professées à l'Institut océanographique de Paris. 1ère partie. Généralités. Océanographie physique. *Paris. Annales de l'Institut océanographique, nouvelle série*, 9, 352pp.

Marmer, H.A. 1930. *The sea*. New York: D. Appleton and Co. x + 312pp.

Berget, A. 1931. Leçons d'océanographie physique professées à l'Institut océanographique de Paris. 2e partie. L'océan et l'atmosphère. *Paris. Annales de l'Institut océanographique, nouvelle série*, 11, 396pp.

Rouch, J. 1938. Leçon d'ouverture du cours d'océanographie physique. *Monaco. Bulletin de l'Institut océanographique* No. 749, 22pp.

**Sverdrup, H.U., M.W. Johnson, and R.H. Fleming. 1942. *The oceans: Their physics, chemistry and general biology*. Englewood Cliffs, NJ: Prentice-Hall. x + 1087pp.

NOTES

Introduction: The Fluid Envelope of Our Planet

1 Both air and water are fluids, of course. Humboldt clearly recognized this – see the epigraph to this Introduction – but his interest in the oceans is my concern.

2 There is ambiguity about what Aristotle meant. In Book I, ch. 3, of his *Meteorologica*, he mentions a high stratum (above the mountain tops) in which clouds cannot form because of the motion imparted by the celestial sphere; later, in Book II, Ch. 4, he clearly mentions surface winds being given a horizontal motion by the motion of the heavens. See chapter 1.

3 An exception was the great western boundary current of the Pacific, the Kuroshio, which was known in part to residents of some of the Japanese Islands from ancient times and to some eighteenth-century navigators, but which was not appreciated as a major current system until after the middle of the nineteenth century (see Wüst 1936, and especially Kawai 1998, and Jones and Jones 2002).

4 Humboldt 1997, p. 23.

5 A change akin to that in physics in the eighteenth century, described by Gingras (2001) as 'mathematics ... replacing verbal formulations as the final arbiter and true explanation of phenomena.'

6 Interest in their history is rare, but not unknown, among physical oceanographers; for an example dealing with the past five decades, see Jochum and Murtugudde 2006.

7 This viewpoint has old roots; see the discussion in Gingras 2001 about the reception of mathematics by physicists between the Scientific Revolution and the early nineteenth century.

8 Charnock and Deacon 2001, pp. 156–7; Wells, Gould, and Kemp 1996, pp. 51–6.

9　Deacon and Summerhayes 2001, p. 21.

10　See, for example, the changes in technology described by Gould 2001.

11　Deacon 1996, 9–16; Deacon 1997, pp. ix-xiv; also Deacon and Summerhayes 2001, pp. 3–11.

1. The Way of the Sea: Knowledge of Oceanic Circulation before the Nineteenth Century

1　On Berghaus, see Engelmann 1976.

2　Camerini 1993b, pp. 489–98, 506–10; Engelmann 1964. On the life of A.K. Johnston, see J.M.R. 1892. Camerini (1993a, pp. 701–4) discusses the role of the visual, including maps, in science.

3　In an unpublished manuscript on ocean currents; see Kortum 1990. Other works on Humboldt's oceanography include Engelmann 1969; Defant 1960; Dietrich 1970, and Wüst 1959.

4　A.K. Johnston 1848, Hydrology, p. 1. Here the Argument from Design is applied to the oceans in a particularly British way. The list of possible causes of currents is derived from Humboldt, as discussed later in this chapter.

5　Parry 1963, 1966, 1971, and especially 1974. See also Tracy 1990, especially the chapters by Phillips and van der Wee.

6　For extensive discussion, see Parry 1974, chs 1, 2, 4, 5, and 8.

7　Allen 1992, pp. 12–19; Stevenson 1932. This map, produced in several editions in the late fifteenth century, was obsolete before it was printed – but its beauty and comprehensiveness, showing the extent of the known world, was very influential.

8　Parry 1974, pp. 97–100; see also Newitt 1986 and Russell 1984, 2000. Hall (1996, pp. 155–83) describes the Portuguese accomplishments in the Indian Ocean.

9　Parry 1974, pp. 283–7; Peterson et al. 1996, pp. 16–17; Whitfield 1996, pp. 30–3.

10　With the exception that it incorporated Gerhard Mercator's novel projection (used first in his map of 1569), useful to navigators because a straight line on its surface gave constant bearing. On this and the Ortelius map, see Whitfield 1996, pp. 68, 69–71; Allen 1992, pp. 36–9; Nordenskiöld 1973, pp. 116, 124, xlvi; and Whitfield 1994, pp. 66–9.

11　Peterson et al. 1996, pp. 16–20.

12　Bourne 1578; Deacon 1997, pp. 43–5; Peterson et al. 1996, pp. 21–2.

13　Peterson et al. 1996, p. 22.

14　Peterson et al. 1996, p. 11.

15　For an account of the monsoonal circulation of the Indian Ocean, see Tomczak and Godfrey 1994, pp. 196–208.

16 Aleem 1967, p. 459; Aleem 1968, 1980. For a full discussion of early knowledge of the Indian Ocean, see especially Tibbetts 1971; also Warren 1966, 1987, and Peterson et al. 1996, pp. 10–11, along with Aleem's publications.

17 The title was an example of journalistic hype, intended to attract readers. For discussion of this and later knowledge of the Indian Ocean, see Warren 1987, pp. 149–53, and a brief mention in Warren 2006, p. 2.

18 Hall 1996, pp. 155–83.On the identification of their pilot, see Tibbetts 1971, pp. 9–12.

19 Peterson et al. 1996, pp. 13–14 (who may have been mistaken about what and when Columbus reported); Pearce 1980; Lutjeharms et al. 1992; Kohl 1868, pp. 28–30, and Warren 2006, p. 2.

20 Martire d'Anghiera 1516, 1555; Peterson et al. 1996, pp. 14–15.

21 Martire 1555, pp. 118–19; Burstyn 1971c, pp. 10–11; Peterson et al. 1996, pp. 14–15. This argument was the origin of the persistent belief in a Northwest Passage to the Indies through or around North America.

22 Kohl 1868, pp. 34–6; Peterson et al. 1996, p. 15.

23 Apparently the Labrador Current, also observed by Gilbert in 1583. See Peterson et al. 1996, pp. 20–2.

24 Peterson et al. 1996, p. 22.

25 Burstyn 1971c, pp. 23–5; Peterson et al. 1996, pp. 23–5 (despite their claims, Varen actually said nothing about a current identifiable with the Kuroshio – see Jones and Jones 2002, pp. 87–8). On Varen, see Bullough 1976.

26 Burstyn 1971c, pp. 26–7; Deacon 1993a (reprint of Voss 1677).

27 Pickard and Emery 1982, pp. 50–1.

28 Voss 1677, pp. 1–2 (in Deacon 1993a).

29 Voss 1677, p. 15 (in Deacon 1993a).

30 Voss 1677, p. 39 (in Deacon 1993a); see also Burstyn 1971c, pp. 26–7.

31 Kangro 1973; for brief accounts of Kircher's geology, see Oldroyd 1996, pp. 35–8, and Rudwick 1972, p. 56. Findlen 2004 gives a remarkable account of the man and his knowledge.

32 For an account of the recovery of Greek writings from the Arab world, see Grant 1996, pp. 18–32.

33 Aristotle 1937, 1952.

34 A good outline, emphasizing later interpretations, is given by Grant 1996, pp. 54–69. See also Lindberg 1992, pp. 46–62, 245–61, especially on medieval–early Renaissance Aristotelian cosmology.

35 Aristotle 1952, *Meteorologica,* Book I, chs 9, 13, and Book II, chs 1–3, are the relevant parts.

36 The salt of the oceans cannot originate on land and be carried to the sea in rivers and streams (the modern explanation) because, according to Aristotle, river water does not taste of salt (*Meteorologica* II, ch. 3). This explana-

tion also accounted for the slightly salty taste of rainwater blowing in off the Mediterranean.

37 Aristotle, *Meteorologica* II, ch. 4; for discussion, see Biswas 1970b, pp. 61–9; Burstyn 1966b, pp. 169–70, and 1971c, pp. 8–9.

38 *Meteorologica* I, pp. 69–71.

39 This and the following quotation from Fulke's *Book of Meteors* (1563), in Hornberger 1979, pp. 95, 107–8. I have reproduced Fulke's account largely in twentieth-century English. I am indebted to Andrew Cranmer, Department of History, Dalhousie University, for information from his unpublished essay 'On the Subject of the Sea: The History of Aristotle's *Meteorologica* and Marine Science' (1998).

40 Both quoted by Burstyn 1971c, p. 11, whose comprehensive discussion of winds and ocean currents through the seventeenth century has never been superseded.

41 Voss 1677, pp. 1–2 (in Deacon 1993a).

42 Burstyn 1971c, pp. 23–5; Peterson et al. 1996, p. 24.

43 Burstyn 1971c, p. 26; Peterson et al. 1996, p. 26.

44 Voss 1677, p. 1 (in Deacon 1993a). For a longer discussion of Voss's ideas, see Deacon 1993a, pp. 9–21.

45 Voss 1677, pp. 4–5 (in Deacon 1993a).

46 Burstyn 1966b, p. 171. For an outline of Halley's life and work, see Cook 1998; and, centred on astronomy, Ronan 1972.

47 Burstyn 1966b, p. 171.

48 Ronan 1968; Thrower 1969, especially pp. 17–33; Deacon 1986a. These disciplinary names, of course, were applied long after Halley's time.

49 Well analysed by Burstyn 1966b, pp. 171–80, upon which part of my account is based.

50 Mariotte 1686, quoted by Burstyn 1966b, p. 173. On Mariotte, see Mahoney 1974; and on the *Traité* and its discussion of winds, Mahoney 1974, p. 119. Wolf 1950, pp. 316–24, gives a summary of Halley's and Mariotte's work.

51 Mariotte 1686, quoted by Burstyn 1966b, p. 175.

52 Halley 1686, pp. 153–63.

53 Naylor 2007.

54 Halley 1686, p. 164.

55 Halley 1686, p. 165.

56 Halley 1686, pp. 165–6.

57 Halley 1686, p. 167.

58 Halley 1686, pp. 167–8. His explanation can scarcely be bettered in our own times; see Tomczak and Godfrey 1994, pp. 196–7.

59 Burstyn 1966b, pp. 178–80.

60 Halley 1691, p. 471. See discussion of Halley's work on the water cycle in
   Wolf 1950, pp. 320–3; Ronan 1968, pp. 246–7; Tuan 1968, pp. 100–4; Biswas
   1970b, pp. 221–8; and Biswas 1970a.
61 Deacon 1997, p. 170.
62 On Dampier's life, see A. Gill 1997; Jones and Jones 1992, pp. 16–29; and
   Preston and Preston 2004. Thrower 1969, pp. 14–15, suggests the link between Halley's and Dampier's charts of the winds.
63 Perhaps for the first time. See Deacon 1993a, p. 171; Burstyn 1966b, p. 182;
   also Jones and Jones 1992, pp. 16–29.
64 R. Porter 1980, p. 292.
65 Deacon 1997, chs 9, 10.
66 Cuvier 1810, pp. 161–2.
67 Frängsmyr et al. 1990.
68 Heilbron 1990, pp. 4–10.
69 Austin and McConnell 1980; McConnell 1982, pp. 22–5.
70 McConnell 1982, p. 24. On the voyage, see Brosse 1983, pp. 114–23.
71 Parry 1971; see also Williams 1966, 1997 (esp. chs 6–10)
72 Parry 1971, ch. 11. See especially Williams 1997, ch. 9, and Williams 1999,
   on Anson's disastrous circumnavigation of 1740–4 and the effects of scurvy.
73 On the oft-told tale of John Harrison (1693–1776), marine chronometers,
   and the background of longitude determination, see Parry 1971, ch. 12;
   Forbes 1974; Howse 1980; Sobel 1995; Andrewes 1996; and Peterson et al.
   1996, pp. 51–7.
74 As Peterson et al. 1996, p. 53, suggest, based upon the statements of James
   Rennell.
75 Williams 1997, chs 4, 5.
76 Gurney 1977, pp. 72–85.
77 Williams 1966, pp. 56–67; 1997, ch. 9; 1999.
78 Sorrenson 1996.
79 Stoye 1994.
80 Marsigli 1681; Deacon 1978a, pp. 25–6, 33–4; 1986b; 1997, pp. 147–9; A.E.
   Gill 1982, pp. 96–7; McConnell 1982, pp. 11–12; Stoye 1994, pp. 15–27;
   McConnell 1999.
81 Maffioli 1994, pp. 297–303.
82 Deacon 1978a, pp. 175–80; Marsigli 1999.
83 Stoye 1994, p. 309.
84 Mills 2001b.
85 Levere 1993, pp. 143–5; on its origin, see Knight 1974.
86 Deacon 1978a, pp. 185–9, 191–2; Rubin 1982; Jones and Jones 1992,
   pp. 39–44. On Wales and Bayly, see Deacon and Deacon 1969, pp. 37–9.

87 Rubin 1982, p. 35. All these measurements were useful in determining latitude, at a time when purely astronomical and chronometric methods of latitude determination were in competition.

88 Deacon 1997, p. 186; Rubin 1982, pp. 37–8. It is likely that the thermometers had extra thick glass, because they did not appear to show the effect of pressure on the subsurface readings, according to information given by Rubin, p. 38.

89 R.E.A. 1890, p. 434; Burstyn 1966b, p. 183.

90 Hadley 1735, p. 58.

91 Hadley 1735, p. 60.

92 Hadley 1735, p. 62.

93 Burstyn 1966b, pp. 186–7. On Maclaurin, see C.P. 1893; Scott 1973.

94 Neither recognized that *any* motion, not just meridional, will be affected by the Earth's rotation. For discussion, including the contributions of Coriolis, Poisson, and Foucault in the nineteenth century, see Burstyn 1966a; 1966b, p. 187. On earlier work, see Burstyn 1965.

95 Peterson et al. 1996, p. 64.

96 Biographical information is in Eyles 1975, Markham 1895, and Pollard and Griffiths 1993.

97 Bravo 1993, p. 43; Markham 1895, pp. 41–59.

98 Bravo 1993, pp. 44–5.

99 Bravo 1993, p. 45; Griffiths 1993.

100 Rennell 1793, pp. 198–9. The addition in square brackets is mine. The Royal Navy did not use marine chronometers routinely until 1810, more than thirty years after their utility and reliability had been proved.

101 Peterson et al. 1996, p. 60.

102 For the complexities of interpretation, see Deacon 1997, pp. 222–4; Peterson et al. 1996, pp. 62–3; and Gould 1993.

103 The reference here is to another navigational work, W.G. De Brahm's *The Atlantic Pilot*, published in London in 1772.

104 Peterson et al. 1996, pp. 60–1.

105 Deacon 1997, p. 222; Gould 1993, pp. 31–2.

106 Rennell 1793, p. 186.

107 Rennell 1832, p. 141. His complete argument is on pp. 146–51.

108 Rennell 1832, pp. 149–50.

109 Bravo 1993, pp. 47–8.

110 Kortum 1990, p. 127; Kortum and Lehmann 1997, p. 50; Peterson et al. 1996, p. 64. On Humboldt, see Biermann 1972, Jahn 2001, Rupke 1997, and Stuardo 2004.

111 Engelmann 1969, pp. 100–1; Kortum 1990, p. 125, and 1994a.

112 Humboldt 1997, p. 7 (in an English edition of Humboldt 1845); Bowen 1970.

113 See especially Humboldt 1975, pp. 380–9; here he debunks vital forces.

114 Cannon 1978, ch. 3.

115 On measurements of the ocean, see Defant 1960, p. 92. Cannon 1978, pp. 75–6, gives a full list and discussion.

116 R. Porter 1980, p. 312.

117 Humboldt 1997, p. 23.

118 For detailed outlines, see Engelmann 1969, Wüst 1959, Defant 1960, and Dietrich 1970.

119 Humboldt 1837; Stuardo 2004; also Engelmann 1969, pp. 102–4. Quotations from Humboldt 1837 about the cool waters off Peru are in Wüst 1959, pp. 98–100; Engelmann 1969, p. 105; Peterson et al. 1996, p. 64; and Kortum 1993. See also Humboldt 1975, pp. 7, 97, dating from the first edition of the work in 1808.

120 Engelmann 1969, pp. 104–5; Wüst 1959, pp. 94–8; Peterson et al. 1996, p. 64. The original account is Humboldt 1814, pp. 66–73.

121 Kortum 1994b; Kortum and Lehmann 1997.

122 Engelmann 1969, pp. 106–7; Kortum 1990. It is located in the Cotta-Archiv, Marbach, Germany,

123 Kortum 1990, p. 129.

124 Humboldt 1814, especially pp. 63–77.

125 The North Equatorial Current had not yet been distinguished, and the Equatorial Countercurrent was not noted – or at least not reported – until the 1820s.

126 Humboldt 1997, p. 307.

127 Humboldt 1814, pp. 72–3. 'The discovery of a group of inhabited islands is of less interest than knowledge of laws that bring together a number of isolated facts.'

128 The effect is actually the reverse – but, of course, none of the relevant magnitudes was known, and Humboldt was writing two decades before the effect of the Earth's rotation was demonstrated.

129 Deacon 1997, p. 183.

130 Benjamin Thompson, Count Rumford (see Thompson 1798, Volume 2, pp. 199–313), had already speculated on the sinking of polar waters to form deep, cold water in the oceans, a view that Humboldt certainly knew and that helped to form his opinion.

131 Humboldt 1814, pp. 73–4.

132 Humboldt 1997, p. 303.

133 Humboldt 1997, p. 306.

134 Camerini 1993b, p. 491; Rupke 1997, pp. xv–xviii.
135 Defant 1960, p. 85. 'A physical description of the world in the Humboldtian sense can no longer be maintained.'
136 R. Porter 1980, pp. 322–3.
137 T. Porter 1994, p. 404. The reference is to Latour 1987, ch. 6.

2. Groping through the Darkness: The Problem of Deep Ocean Circulation

1 Maury 1844, p. 181. For a more complete exposition of his teleological approach, see Maury 1861, pp. 81, 190–2, 217; 240–1, and 247.
2 See Leighly 1963, 1968.
3 Beechey 1851, pp. 52–3, in Knight 1974.
4 Hopkins 1852, p. 10.
5 Herschel 1861, pp. 46–7.
6 Warren 2006, p. 3.
7 Findlay 1853, p. 236.
8 Deacon 1997, pp. 134–5.
9 Discussed in detail by Deacon 1997, ch. 7; ch. 10, pp. 207–8; ch. 11, p. 239. Also Deacon1968, 1978b, 1985a, 1998.
10 Deacon 1997, pp. 134–42.
11 A kite or sail-like device used for determining the direction and velocity of ocean currents. See Deacon 1996, p. 12.
12 Halley 1687; Deacon 1997, pp. 146–7. In fact, Halley never discussed the possibility of an undercurrent; his concern was the atmospheric water cycle.
13 Waitz 1755, quoted in Deacon 1985a, p. 96. On Waitz, see also Deacon 1996, p. 14, and 1998.
14 Marsigli 1681. See especially Soffientino and Pilson 2005; also Deacon 1997, pp. 147–9; Gill 1982, pp. 96–8; McConnell 1982, pp. 11–12; Peterson et al. 1996, pp. 32–3; and Stoye 1994, pp. 26–7.
15 Marsigli's study was published in Italian in 1681, republished in German a few year later, and then was not resurrected until 1935, once again in Italian (see Marsigli 1681). But Buffon knew of it during the writing of his great *Histoire naturelle* in the mid-eighteenth century and debunked it (Deacon 1998, pp. 84, 91). Recently Soffientino and Pilson (2005) have redirected attention to Marsigli and the Bosphorus.
16 Deacon 1978b, pp. 28–9.
17 Spratt 1871. It is not known what he thought of the direct demonstration of two-layered circulation in both the Bosphorus and the Dardanelles by his colleague W.J.L. Wharton in H.M.S. *Shearwater* in 1872 (see Deacon 1978b, p. 51; Wharton 1873, 1886).

18 For an informed nineteenth-century summary, see Prestwich 1875, pp. 623–4.

19 Brown 1967, 1979, on Thompson. Krümmel (1911) attributed the idea of deep convection to the physical geographer J.F.W. Otto (1800), though Otto almost certainly got it from Thompson (see Warren and Wunsch 1981, p. 40).

20 Thompson 1798, p. 305. See the discussion in Deacon 1997, pp. 208–9; and Peterson et al. 1996, pp. 48–51.

21 Rubin 1982. For longer discussions of the investigation of deep-sea temperatures, see Deacon 1997, chs 9–11, 13, 14; Jones and Jones 1992, chs 1–3; and especially the wonderful pioneering analysis by Prestwich (1875). My summary is based largely on their more detailed accounts.

22 Prestwich 1875, p. 592; Jones and Jones 1992, pp. 56–78; Deacon 1997, pp. 192–3. Brosse 1983, pp. 95–107, describes the expedition.

23 Known as 'Hales' bucket.' Temperature was sometimes determined using a thermometer sent down in the bucket or by measuring the temperature of the sample when the bucket was back on deck. On Hales' bucket, see McConnell 1982, pp. 16–17. On the Krusenstern expedition, see Brosse 1983, pp. 117–23; Deacon 1997, pp. 226–7, 232; Jones and Jones 1992, pp. 99–102; Krusenstern 1813; and Prestwich 1875, pp. 592–3.

24 Dumont d'Urville 1834–5; Brosse 1983, pp. 152–9; Rosenman 1987.

25 Jones and Jones 1992, p. 108.

26 Dumont d'Urville 1830–3; 1833.

27 Deacon 1997, pp. 277–9; Jones and Jones 1992, pp. 112–15; Prestwich 1875, p. 600. Dumont also noted, as had Péron, that cool water was closer to the surface at the equator than farther north and south.

28 Deacon 1997, pp. 208–9: see also pp. 206, 230, and 238–41, and Prestwich 1875, pp. 613–16, on the history of this controversy.

29 Marcet 1819; Erman 1828. Despretz (1837) provided further evidence.

30 Deacon 1997, p. 278; Jones and Jones 1992, pp. 113–14.

31 Which forced the indicating fluid higher than it should have been in the arm of the thermometer indicating minimum temperature.

32 On the Wilkes and Ross expeditions, see Brosse 1983, pp. 195–204; Deacon 1997, pp. 280–1; Jones and Jones 1992, pp. 158–67, 179–203; and Prestwich 1875, pp. 614–16.

33 Lezhneva 1973. Lenz was born in Dorpat, Russia, now Tartu, Estonia, and had a long career in St Petersburg in association with the university and the Academy of Sciences.

34 Brosse 1983, p. 131. On his first voyage around the world (1815–18), Kotzebue and his officers took 116 subsurface temperature measurements as

deep as 746 metres using Six thermometers. These were used later by Lenz
in his summaries of deep oceanic temperatures and circulation (Jones and
Jones 1992, pp. 116–17).

35 Lenz 1830, 1831, 1832; Prestwich 1875, p. 599.
36 Dortet de Tessan 1844. See also Deacon 1997 p. 279; Jones and Jones 1992,
    pp. 126–9; and Prestwich 1875, pp. 601–2. On the voyage, see Petit-Thouars
    1840–55.
37 Including experimental work by Parrot (1833) on the effect of pressure on
    thermometers (Lenz 1845, 1847).
38 A conclusion arrived at also by Arago (1838).
39 Lenz 1848, pp. 623–4; Deacon 1997, p. 293; Jones and Jones 1992, pp.
    117–20; Prestwich 1875, pp. 625–30. Lenz noted in passing that the course
    of the upper and lower water masses would be affected by the Earth's rota-
    tion.
40 For an extended account of the work of Thomson and Carpenter, see Mills
    1983, pp. 10–18. Carpenter's life and career are outlined by J.E. Carpenter
    1888; Hall 1979, pp. 132–6; Lankester 1885; and Thomas 1971.
41 Sivertsen 1968.
42 For an account of British dredging through the mid-1860s, see Mills 1978.
43 See Mills 1978, and 1983, pp. 6–15, for an account of the origin and out-
    comes of this belief.
44 Thomson 1873, pp. 50ff. .
45 W.B. Carpenter to A.M. Norman, 10 August 1868, Alder-Norman Letters
    #325, General Library, Natural History Museum, London.
46 Carpenter 1868, pp. 185–9.
47 Both ships are described by Rice 1986, pp. 97–100, 118–24. On the cruise of
    Porcupine, see Carpenter, Jeffreys, and Thomson 1870.
48 Deacon 1997, p. 311; Mills 1983, pp. 15–16.
49 A graphic account is given by Thomson 1870.
50 Carpenter, Jeffreys, and Thomson 1870, p. 398.
51 The phrase was Thomson's (Deacon 1997, p. 314). Thomson and Jeffreys
    both disavowed Carpenter's developing hypothesis of oceanic circulation.
52 Carpenter 1868, p. 187, including the footnote. The text was Buff 1850,
    1851.
53 Carpenter 1868, footnote, pp. 187–8; 1869, p. 4.
54 Carpenter 1869, p. 4.
55 Carpenter 1868, pp. 193–5; 1869, pp. 5–8. Carpenter's argument is phrased
    in terms of similar 'representative species,' based on some pre-Darwinian
    problems of species formation – but by 1868, Carpenter was a Darwinist,
    though not convinced of the efficacy of natural selection.

56 Carpenter, Jeffreys, and Thomson 1870, pp. 472–4 (Carpenter wrote the sections on temperatures and their implications); Carpenter 1870a; 1870b, pp. 67–71; Deacon 1997, pp. 313–14.

57 Carpenter and Jeffreys 1870, 1871.

58 Carpenter and Jeffreys 1870, p. 181. Carpenter wrote the sections on ocean circulation; Jeffreys, the biology. See Tomczak and Godfrey 1994, pp. 300–2, and especially Tsimplis and Bryden 2000, for a modern review of the oceanography of the area. When Sir John Herschel (1871) learned of Carpenter's results, he wrote, 'So, after all, there *is* an under-current setting outwards in the Straits of Gibraltar.'

59 Carpenter 1872d, pp. 565–78. Further evidence for the Gibraltar undercurrent and of those in the Dardanelles and Bosphorus was provided by W.J.L. Wharton (1873) using *Shearwater* in summer 1872.

60 Carpenter and Jeffreys 1870, pp. 210–19.

61 Carpenter and Jeffreys 1870, p. 211.

62 Carpenter and Jeffreys 1870, p. 214.

63 Late in 1870, when Carpenter first formalized his ideas, he did not know of Lenz's nearly identical model of large-scale ocean circulation.

64 Carpenter and Jeffreys 1870, p. 215.

65 Carpenter 1868, footnote, pp. 193–4.

66 The definitive account of his life is Irons 1896. Shorter accounts are Burstyn 1971a, and Imbrie and Imbrie 1979, pp. 77–81. Smith and Wise (1989, pp. 593–7) indicate the significance of Croll's geology.

67 Imbrie and Imbrie 1979, pp. 69–75; Hamlin 1982; pp. 573–7.

68 Imbrie and Imbrie 1979, pp. 81–5.

69 Croll 1864, p. 129.

70 Croll 1864, pp. 134–5; 1867, pp. 121, 124–7.

71 Croll 1864, pp. 136–7; 1867, pp. 121–3, 126–30.

72 Croll 1870a, pp. 97–106.

73 Croll 1870b, p. 181.

74 Maury 1861; Croll 1870c, p. 235, where he refers to an 1870 edition. In the 1861 edition of Maury's book (probably the final revision), his schemes of the dynamics of ocean circulation are given in Chapters 2 and 8–10.

75 Croll 1870c, pp. 235–51 (quotations from pp. 235, 239); see also Croll 1875b, pp. 104–105, 109–14. Maury was no 'scientist of the sea,' despite the claims of one of his biographers (Williams 1963).

76 Croll 1870c, pp. 254–8.

77 Croll 1870c, p. 233.

78 Croll 1870c, pp. 249–50. In his calculation, Croll assumed that the temperature of the equatorial column decreased uniformly with depth, which

he knew to be incorrect. He pointed out that a more realistic temperature structure, in which there was only a shallow warm layer, would result in a much smaller slope.

79 Croll 1870c, pp. 250–1, where he cites Du Buat 1816, vol. 1, p. 64. On Du Buat and his work in hydraulics, see Rouse 1971, and Rouse and Ince 1957, especially pp. 129–34.

80 Croll 1870c, p. 251.

81 Croll 1870c, p. 234.

82 Croll 1875b, p. 17.

83 Deacon 1993b, pp. 57–8. See also Deacon 1997, chs 14 and 15, for a detailed account and an extensive bibliography of the controversy.

84 Anonymous 1871, p. 97.

85 The most significant of which are Croll 1871a-c, 1872a-f, 1874a-c, 1875a-f, and 1876.

86 Croll 1871c, pp. 278–9.

87 Croll 1874b, p. 95.

88 Summarized in his later work in Croll 1874b, pp. 184–90, and *in extenso* in 1875b.

89 Croll 1871c, pp. 272–8; 1874b, pp. 168–77.

90 Carpenter 1872a, p. 469, footnote; 1874a, pp. 393–407.

91 Carpenter 1872a, pp. 435–6; 1872d, pp. 553–4, 557–8; 1875b, p. 533; 1875d, p. 403; 1876a, pp. 666–7.

92 Croll 1874b, pp. 100–1. Here he argues that increased salinity due to evaporation at the equator will counteract increased density due to cold in high latitudes, preventing any difference of level. On Croll's use of Du Buat, see 1874b, pp. 116–18.

93 See Croll 1871c, pp. 247–69; 1874b, pp. 111–22.

94 Croll 1874b, p. 114.

95 Croll 1874b, p. 177. See also Croll 1875b, ch. 13, and 1875f.

96 Croll 1874b, p. 99; see also 1872c, pp. 502–3, and 1874b, p. 179.

97 For nearly fifty years after Croll's time, the 'Humboldt' or Peru Current was believed to flow strongly from the south; the British *Discovery* investigations of the 1930s showed that the 'current' also involved an extensive system of upwelling along the South American coast (Gunther 1936).

98 Croll 1874b, pp. 168–9,179–83; and 1874a,c. See also Croll 1874b, pp. 99, 101–4; 1875b, pp. 130–3, and ch. 13; and 1875f, pp. 289–90.

99 Croll 1875f, p. 289.

100 In his last publication on the subject, Croll (1879) enlisted the aid of the German geophysicist Karl Zöppritz, who purported to show that wind-

driven currents could extend to great depths, although over very long periods of time. This was in response to the Dutch hydrographer Maris Jansen, who, in an exchange of letters with Croll in 1876, pointed out that Croll's claim that winds caused ocean currents was no more than an assumption (Irons 1896, pp. 291–305).

101 On the cruise, see especially Deacon 1997, ch. 15.

102 Carpenter 1874a, p. 301; 1875a, p. 454; Croll 1875a, p. 242; 1876, p. 191.

103 Carpenter 1875e.

104 Carpenter 1874a, p. 365.

105 Croll 1875a, 1876.

106 Croll 1875a, pp. 242–6; 1875c,e; 1876, pp. 191–2. Using standard tables of the volume of seawater related to temperature, Croll calculated the height of the whole water column from bottom to surface at three locations along a N-S section of the Atlantic at 38°N, 23°N, and the equator. Although based upon very little data, Croll's calculations are borne out by the dynamic topography of the western North Atlantic and by satellite altimetry (Open University Course Team 2001, pp. 10, 61–3).

107 Carpenter 1875e, pp. 507–9. Croll (1875a, p. 249) argued in opposition that the North Pacific surface temperatures could not be accounted for by any scheme (like Carpenter's) deriving circulation from the difference between equatorial and polar temperatures.

108 Croll 1875a, p. 243.

109 Carpenter 1875a.

110 Croll 1875d, p. 494.

111 For Croll's own explanation of his change of interests, see Irons 1896, pp. 346–7.

112 F.R.S. 1874. This attack was stimulated by the exchange in Croll 1874b and Carpenter 1874d.

113 To such an extent that Carpenter was accused (though not by name) of having vetoed a Civil List pension to Croll after his retirement from the Geological Survey of Scotland (Irons 1896, p. 374)

114 Buchanan 1874, p. 127.

115 Carpenter 1872d, p. 644.

116 Croll 1874b, p. 99

117 Carpenter 1872a, p. 465.

118 Petermann 1871. Carpenter 1872a, pp. 430–52, 467–9, gives a lengthy, historically based discussion. See also Carpenter 1872d, pp. 592–638, and 1875c, which include a thorough-going attack on Croll's estimates of the velocity, volume-transport, and heat content of the Gulf Stream.

119 Carpenter 1872d, pp. 565–86.
120 Seasonal vertical convective formation of Mediterranean deep water was detected much later, in the mid-twentieth century.
121 Carpenter 1872d, p. 550. Presented in another form in Carpenter 1874a, pp. 353–4.
122 Carpenter 1872d, pp. 552–65; 1875b; 1875d, p. 403; 1876a, p. 666. Carpenter admitted that little was known of the viscosity of water in nature. Croll's and Carpenter's arguments were framed entirely in terms of molecular viscosity; the relevant, and much higher, eddy viscosities of water were formulated and applied after the turn of the twentieth century by V.W. Ekman and J.P. Jacobsen.
123 Prestwich 1871, pp. xlii–lxii; 1874; 1875. Prestwich did not take sides overtly, but it is clear that he favoured Carpenter's position. On Prestwich, see Bonney 1901, Challinor 1975, Evans 1897, G.A. Prestwich 1899, and Woodward 1893.
124 Carpenter 1874b, p. 171.
125 Carpenter 1874d, p. 423; see the longer discussion integrated into Carpenter 1874a.
126 Carpenter 1875b; 1875d, pp. 402–3.
127 Carpenter 1880, in J.E. Carpenter 1888, pp. 340–1.
128 Carpenter 1875a, p. 454, a suggestion borne out by the revelation in the 1960s of deep convection of Weddell Sea water (on the history of this work, see Mills 2005, 2007).
129 On the expedition, see Deacon 1985b; Deacon and Savours 1976, pp. 132–7; Hattersley-Smith 1976; and Levere 1993, pp. 261–306. Rice 1986, pp. 13–15, 147–9 describes the ships and the role of *Valorous*.
130 Carpenter 1876b, pp. 232–3.
131 Carpenter believed that the higher temperature of North Atlantic deep water compared to the Pacific was due to the small inflow of Antarctic deep water in the south and to ridges in the north blocking the flow of abyssal water from the Arctic. Evidence accumulating from the 1930s through the 1960s has shown that, contrary to Carpenter's expectations that the deep water of the North Atlantic was formed at very high latitudes, most North Atlantic Deep Water (NADW) forms during irregular episodes of deep convection in winter in the Norwegian and Greenland Seas, spilling over the Greenland-Iceland ridge. Some deep water also forms in the Labrador Sea (Bacon 2002). Actual evidence of a ridge between the Shetland and Faroe Islands (it was named for Wyville Thomson) did not come until 1880, after Carpenter had given up active work on ocean circulation (Deacon 1977).
132 In Croll's case, the study of evolution in a theological context. In a letter

to Osmund Fisher in 1879, he wrote, 'I have left physics altogether ... and commence the study of my old subject Theism, or rather Evolution in its philosophical aspects' (Irons 1896, pp. 346–7).

133 See especially Hamlin 1982, and Imbrie and Imbrie 1979, ch. 7. Geikie's *The Great Ice Age* (1874) was greatly influenced by Croll. Ball (1891) gave general support to the astronomical basis of Croll's theory but recalculated the amount of heat received by the Earth in summer and winter.

134 Imbrie and Imbrie 1979, ch. 8.

135 Croll 1879, p. 202.

136 Irons 1896, pp. 261–2.

137 See especially the discussion of Carpenter's philosophy of nature by Hall (1979) and in Carpenter 1850.

138 Hall 1979, pp. 148–51.

139 W.B. Carpenter 1873; reprinted in J.E. Carpenter 1888, pp. 185–211.

140 Carpenter 1873, p. lxxii.

141 Carpenter 1873, p. lxxiv. There are clear links here to John Herschel's *Preliminary Discourse ...* (1830); Carpenter knew Herschel and, like Herschel, appears to have been greatly influenced by Edinburgh philosophy (see Olson 1975, ch. 10).

142 Carpenter 1873, p. lxxxii.

143 Carpenter 1872c.

144 Croll 1876.

145 George Carey Foster (1835–1919), professor of physics at Anderson's College, Glasgow, where Croll met him, then professor of physics at University College, London (1865–98).

146 G.C. Foster to James Croll, 1 October 1875, in Irons 1896, pp. 283–5.

147 Osmond Fisher to James Croll, 24 September 1875, in Irons 1896, pp. 285–7.

148 Lengthy correspondence in January and February 1876 (Irons 1896, pp. 291–305). See Anonymous 1895 on Jansen.

149 An excellent summary of theoretical and mathematical physics in the nineteenth century is given by Harman 1982. On hydrodynamics, see Darrigol 1998, 2005.

150 Maffioli 1994 gives an excellent account of the mathematics of fluids in the seventeenth and eighteenth centuries. Standard chronological accounts, centring on hydraulics and without historical analysis, are in Ince 1987, Rouse and Ince 1957, and Rouse 1987. The most comprehensive modern account is that of Darrigol 2005.

151 In later editions titled *Hydrodynamics*. On Lamb, see Rouse and Ince 1957, pp. 211, 213, 217, and Bullen 1973.

152 Lamb 1879, p. 1.

153 Ferrel 1872a, p. 385; 1872b. A more important point of difference was the magnitude of the moon's effect on the tides compared to the wind (Croll 1872b,d). Ferrel's distinguished work on atmosphere and oceans is summarized by Burstyn 1971b.

154 Beginning in 1874 (see Carpenter 1874e, pp. 361–2) when Mohn was at work on a mathematical formulation of atmospheric motions (Kutzbach 1979, pp. 76–83, 101–10).

3. Boundaries Built with Numbers: Making the Ocean Mathematical

1 Thomson 1878, p. 452.

2 See especially Jörberg 1973 and Mills 1989, pp. 75–81, on Scandinavian economies and fisheries in the nineteenth century. Nornvall 1999 gives the Swedish context.

3 Pettersson and Ekman 1899, pp. 13–15.

4 Sars is best known as one of the great carcinologists. Biographical information is given by Broch 1954, Calman 1927, Christiansen 1993, Frey 1982, Hjort 1927 and Nordgaard 1918.

5 See V.W. Ekman 1930a,b; Pettersson 1930; and Hubendick 1950.

6 Perhaps the most influential marine scientist of his generation. See Anonymous 1923, 1941; V.W. Ekman 1941; Euler 1941; Maurice 1923; H. Pettersson 1923; and Thompson 1941, 1947, and Svansson 2006. Many of Pettersson's papers, now in the Oceanographical Laboratory of the Swedish Meteorological and Hydrological Institute, Gothenburg, will soon be transferred to the county archive in the same city. Letters to Pettersson, 1887–1905, are archived in the University Library, Gothenburg (information from Dr Artur Svansson).

7 Mills 1989, ch. 3. For an influential example, see Pettersson 1896. Nornvall (1999) discusses changing Swedish motivations for this work.

8 On F.L. Ekman, see H. Pettersson 1950. He was a distant relative of Gustaf Ekman and father of the distinguished theoretical oceanographer V.W. Ekman.

9 F.L. Ekman 1870; 1875a,b; 1881.

10 F.L. Ekman 1875a,b; 1876.

11 Krümmel 1911, pp. 470, 476–7.

12 F.L. Ekman and O. Pettersson 1893; Pettersson 1894, pp. 281–2, 617–24; Nornvall 1999, p. 36.

13 Jörberg 1973, pp. 437–49.

14 Jörberg 1973, p. 430.

15 On Mohn's life and career, see especially Halvorsen 1896, pp. 74–83 (including a list of publications to that date), Spiess 1935, Thorade 1935, Hesselberg 1940, Pedersen 1974, and Kutzbach 1979, pp. 240–1; and on the Norwegian Meteorological Institute, Barlaup 1966.

16 Mills 1989, p. 77.

17 Wille 1882b, pp. 1–6.

18 Wille 1882a,b describes the work year by year, the ship, and the equipment used. See also Blindheim 1995, pp. 4–5, and Mills 2004.

19 Pedersen 1974, p. 443.

20 Kutzbach 1979, pp. 3–10, 64–71.

21 For a detailed history of the thermal theory of cyclones, see Kutzbach 1979.

22 Kutzbach 1979, pp. 71–5.

23 Kutzbach 1979, pp. 76–83. Mohn's text, which went into several editions, was particularly influential in swinging German opinion toward the thermal theory of cyclones. Georg von Neumayer, the director of the Deutsche Seewarte in Hamburg, was responsible for the publication of Mohn's work in German in an attempt to modernize meteorological thought in Germany (see Kutzbach 1979, pp. 89–90).

24 Guldberg, born in Christiania (Oslo), studied in Christiania, Paris, and Zürich before becoming professor of applied mathematics in the University of Christiania in 1869 (see Oettingen 1904, pp. 551–2).

25 Kutzbach 1979, pp. 99–100.

26 Colding 1870, 1871; see especially Colding 1877 for a summary in English. Colding examined the properties, especially the distribution of pressure, associated with a current within a rotating system, but without explicitly taking into account the Earth's rotation.

27 One of the greatest geophysical theoreticians of the nineteenth century, especially in meteorology. See Leighly 1968, Burstyn 1971b, 1984, and Kutzbach 1979, pp. 232–3.

28 Ferrel 1856, 1859, 1859–60, 1861, 1874. See Gill 1982, pp. 208–14, for the historical setting and modern developments related to Ferrel's contributions.

29 Guldberg and Mohn 1876, extended and expanded in Guldberg and Mohn 1877, 1880.

30 The meteorological implications and the contributions of contemporaries like Peslin and Ferrel are discussed by Kutzbach 1979, pp. 84–117.

31 Peterson et al. 1996, p. 100.

32 Kutzbach 1979, p. 18, n. 18. This portion of the Guldberg and Mohn equations was a formalization of Buys Ballot's Law, according to which the wind blows perpendicularly to the pressure gradient and with velocity propor-

tional to the steepness of the gradient (in fact, there is a slight inward component near the Earth's surface due to friction).

33　Notably in Guldberg and Mohn 1880.

34　Mohn 1887, p. 195.

35　Mohn 1885, p. 1.

36　Mohn 1887, p. 109.

37　Mohn 1885, 1887; Mills 2004, pp. 49–53.

38　Mohn 1880; 1883; 1887, pp. 9–60; Schmelk 1882 (including a historical survey of seawater analysis); Tornøe 1880.

39　Mohn 1887, p. 165.

40　Mohn 1885, pp. 7–11; 1887, pp. 108–30; Peterson et al. 1996, p. 101.

41　Recent estimates indicate that this was an overestimate; see Peterson et al. 1996, p. 101.

42　Mohn 1885, and 1887, pp. 108–95, describe the method in detail. A very clear summary and analysis is given by Peterson et al. 1996, pp. 101–5. A reviewer has pointed out flaws in Mohn's approach, including the relation between wind direction and the direction of the resulting current, and especially his error in using the horizontal pressure gradient to balance the wind-induced flow at the surface. V.W. Ekman (1902, 1905) produced the first widely accepted approach to the problem of wind-driven currents long after Mohn had moved on to other studies.

43　Mainly Miller-Casella maximum-minimum instruments, although in the last year of the Expedition, new Negretti and Zambra reversing thermometers were used whenever possible. Mohn often used a mercury piezometer devised by the *Challenger* physicist John Young Buchanan to correct the thermometers, and as a thermometer in its own right. On the thermometers and piezometer, see Mohn 1883 and especially 1887, pp. 9–60.

44　Mohn 1887, p. 195.

45　Mohn 1887, p. 196.

46　Peterson et al. 1996, p. 105.

47　Kutzbach 1979, p. 110, based on a comment by Vilhelm Bjerknes in 1917. In addition, Mohn was aware that the deficiencies of his thermometers and other instruments and techniques could have affected his results – a caveat later borne out by the work of Helland-Hansen and Nansen in the Norwegian Sea (see Nansen 1901, pp. 149, 161, and Helland-Hansen and Nansen 1909b, pp. 7–8, 28).

48　On Krümmel, see Eckert 1913, Kortum 1993a, Matthaus 1967, Schott 1987, Ulrich 1986, Ulrich and Kortum 1997, and Wegemann 1915.

49　Boguslawski and Krümmel 1887, Krümmel 1907, Krümmel 1911.

50　Krümmel 1910, p. 90. The Darmstadt physicist Carl Forch (1870–?) arrived at the same view, partly in opposition to the conclusions of Zöppritz and perhaps partly under Krümmel's influence (see Forch 1906; 1911, pp. 361–2).

51 Engelhardt 1899a,b; Wegemann 1899, 1900; Castens 1905a,b; Wissemann 1906; Merz and Wüst 1922 (on this, see Mills 1997, pp. 58–9). Castens's work incorporated newer Scandinavian approaches – the modification of Bjerknes's theorem by Sandström and Helland-Hansen described later in this chapter – along with Mohn's technique.

52 Deficiencies and errors in Mohn's analysis certainly would have been manifest to Vilhelm Bjerknes, whose work is discussed later, but I have found no evidence that he referred to them.

53 See Helland-Hansen and Nansen 1909b, p. 7. In Mohn's study, bottom temperatures were too low and there were errors in specific gravity and salinity determinations, leading to incorrect calculations of circulation based on those factors alone.

54 Zöppritz 1878a,b; 1879a,b; also Boguslawski and Krümmel 1887, pp. 342–52 (Zöppritz had begun to write the second volume of Boguslawski's monograph, but when he died in 1885 it was taken over by Krümmel). Zöppritz's equations of motion totally ignored the Earth's rotation, and his analysis used a molecular coefficient of the viscosity of water (shown to be incorrect by J.P. Jacobsen and V.W. Ekman nearly a quarter of a century later). On his career, see Anonymous 1885, Günther 1900, Kerz 1979, and Wagner 1885.

55 V.W. Ekman 1939, p. 6. Ekman's conclusions were incorporated by H.U. Sverdrup in the great text *The Oceans* (Sverdrup et al. 1942, p. 460).

56 Thorade 1935, p. 182. In chapter 5, I examine the transition from physical geography to mathematical physics in German oceanography.

57 Thorade 1935, p. 182.

58 Friedman 1989, which outlines in detail, based on many Scandinavian sources, Vilhelm Bjerknes's life and scientific work, emphasizing especially his contributions to meteorology.

59 For an outline, see Purrington 1997.

60 Especially Harman 1982, 1998; Purrington 1997; Schneer 1984; Smith 1998; and Smith and Wise 1989. Merz's (1903) account is still stimulating and useful; and Jungnickel and McCormmach (1986) provide a valuable account for Germany alone.

61 For example, in 1901 the German theoretical physicist Arnold Sommerfeld instructed H.A. Lorentz that the contribution he was preparing for the new *Encyklopädie der mathematischen Wissenschaft* was to concentrate on electromagnetics, ignoring the mechanical hydrodynamic approach. This was a serious blow to Vilhelm Bjerknes. See Friedman 1989, p. 25.

62 McCormmach 1982.

63 The history of ether and field theories is thoroughly discussed in essays in Cantor and Hodge 1981; Harman 1982 (ch. 4), 1998; Hesse 1961; and Schneer 1984 (ch. 13, 14, and 16).

64 Hesse 1961, p. 200.

65 On Faraday's views, see Siegel 1981, pp. 239–40, and Harman 1982, pp. 73–9. The meaning of 'field' in all its intricacy is explored by Hesse 1961, ch. 8, and Cantor and Hodge 1981, pp. 37–44. More recent interpretations are in Smith and Wise 1989, Purrington 1997, and Harman 1998.

66 Siegel 1981, p. 239, Harman 1998, pp. 4–6.

67 Kargon and Achinstein 1987, pp. 3, 206.

68 Harman 1987, pp. 269–71; 1998, ch, 2; Purrington 1997, pp. 63–71; Smith 1998, ch. 11; Harman 1998, Ch. II.

69 Harman 1982, pp. 103–7.

70 Action at a distance played no part in Maxwell's electrodynamics, which dealt with the transmission of electromagnetism by fields of force and the ether within space. See Harman 1982, pp. 106–7, and Buchwald 1990 on Helmholtz.

71 Eddington 1942, p. 197.

72 On Hertz and his work, see Aitken 1985, pp. 22–5, and ch. 3; Baird et al. 1998; Buchwald 1985, pp. 190–3; 1990; 1992; Friedman 1989, pp. 17–19; Harman 1982, pp. 107–11; Hesse 1961, pp. 212–15; Jungnickel and McCormmach 1986, pp. 29–30, 85–97, 141–3; Klein 1972, pp. 73–5; McCormmach 1972; and Schneer 1984, pp. 283–4.

73 Jungnickel and McCormmach 1986, p. 96.

74 Jungnickel and McCormmach 1986, pp. 141–3; Klein 1972, pp. 73–5. Hertz's book of 1894 was published in English in 1900; a reprint edition is still available (on its background, see Buchwald 1990).

75 Klein 1972, p. 75. See also Friedman 1989, pp. 17–18.

76 Friedman 1989, pp. 17–19.

77 Kröber 1965.

78 Bergeron et al. 1962, pp. 7–8; Friedman 1989, pp. 11–13; Jewell 1984, pp. 785–7; Pihl 1972a, pp. 166–7; Kragh 2002, p. 60. The fullest account of C.A. Bjerknes is V.F.K. Bjerknes 1925, republished in German in 1933.

79 On Vilhelm Bjerknes, see Anonymous 1942; Anonymous 1951; Bergeron et al. 1962; Eliassen 1982; Hesselberg 1923; Jewell 1984; and Pihl 1972b. A definitive account, especially of his meteorological work, from Scandinavian sources is Friedman 1989, on which much of the detail in my account is based.

80 At this time, Hertz was interested in meteorology as well as electrodynamics

and theoretical mechanics, a fact likely to have been significant in Bjerknes's later career. See Mulligan 1987, p. 716.

81 Bjerknes 1900b, 1902.

82 The terms *barotropic* and *baroclinic* were first defined by Bjerknes 1921, pp. 562–3 (with historical notes on pp. 595–6), and later by Bjerknes et al. 1933, p. 3. By definition, barotropic conditions exist when density is a function of pressure alone so that lines of equal pressure (isobars) and of equal density (isopycnals) are parallel. Under baroclinic conditions, variation in properties (in the ocean) such as temperature or salinity causes isobars and isopycnals to be inclined to each other, resulting in dynamic instability and the possibility of motion, that is, current flow. Baroclinicity in the atmosphere is caused by variations in temperature and moisture content. For a discussion of oceanic conditions, see Open University Course Team 2001, pp. 49–53, and for an advanced treatment, Gill 1992, pp. 117–28.

83 Bjerknes 1898a, pp. 28–9. See note 91 for the definition of circulation.

84 Bjerknes 1898b, pp. 4–5, 26–35. For discussion of the background of this and the preceding paper, see Friedman 1989, pp. 19–21, 33–4.

85 On Ekholm, see Kutzbach 1979, pp. 159–62, 183–4, 231.

86 Friedman 1989, pp. 36–7. Kutzbach 1979, pp. 159–71, deals with Bjerknes, Ekholm, and the application of Bjerknes's theorem to meteorology. On the ill-fated balloon expedition, see Neatby 1973, ch. 13.

87 Pettersson 1900, p. 342, is an indication of his early interest. Six years later, Bjerknes (1906) outlined his methods and their relevance to the oceanographic programs of the ICES nations, and Pettersson (see International Council for the Exploration of the Sea 1906, p. 36) recommended that the physical results of ICES cruises should be calculated using Bjerknes's method.

88 Friedman 1989, pp. 39–42. Bjerknes and Sandström 1901, 1903, is apparently a printed version of this presentation.

89 Bjerknes 1901.

90 For his explanation of the term, see Bjerknes 1901, footnote, pp. 745–6, in which he makes it clear that the term not only had meaning in electrical terms but had a history of use in vector analysis and theoretical hydrodynamics as well. Solenoids have been defined as 'parallelograms delineated by the intersection of isosteres [isopycnals] and isobars' (Kutzbach 1979, p. 163).

91 A good non-mathematical description of Bjerknes's theorem is by Thorade 1935, pp. 184–5, but see also Kutzbach 1979, pp. 159–65.

92 Defined as 'the integral along a closed curve of the components of the velocity field tangential to the curve' (Kutzbach 1979, p. 162).

93 V.W. Ekman became one of the great theoretical oceanographers of the twentieth century and was also a distinguished developer of instruments (a highly unusual combination). On his career, see Hansen 1954, Kullenberg 1954, Svansson 1996, and Welander 1971. Walker 1991a,b deals with the background of Ekman's accomplishment in developing the 'Ekman spiral.'

94 See V.W. Ekman 1902, 1904, 1905, 1906.

95 Nansen 1902, pp. 285, 353–6 (in which Ekman is credited with the calculations). This report was published in sections between November 1899 and February 1902, making dating uncertain, but the work seems to have been done by 1901 at the latest. Also included were Ekman's calculations of the effect of the wind on surface circulation (pp. 369–77, 386–93) and of an eddy coefficient of friction involved in wind-driven circulation (the factor that had vitiated Zöppritz's calculation of the effect of the wind on ocean circulation).

96 Described in detail in Friedman 1989 and Monmonier 1999, ch. 4.

97 Dickson 1901, p. 64. On the Scottish surveys, which owed much to – and, indeed, were for a time part of – the extensive surveys done under the Copenhagen Program of Otto Pettersson and Gustaf Ekman, see Turrell and Angel 1995 and Adams 1995. On Dickson, see Adams 1995, p. 15.

98 It was, in fact, the first of the quarterly cruises conducted by the Fishery Board for Scotland for the newly founded International Council for the Exploration of the Sea.

99 Devik 1956, 1958; Dietrich 1958; Hylleraas 1957; Mamayev 1987; Mosby 1958, 1959; Solemdal 1997; Sverdrup 1934; and Tait 1958, 1959.

100 Nansen 1901, pp. 129–30. Nansen's oceanographic work deserves further attention. For an introduction, see Helland-Hansen and Worm-Müller 1940 and Mosby 1961.

101 Biographical information on Sandström is sparse: Eliassen 1982, pp. 4–5; Friedman 1989, pp. 38, 47, 56–7, 62–3, 82; Gold 1947; Kutzbach 1979, pp. 242–3; and Liljequist 1993, pp. 459–60.

102 Sandström and Helland-Hansen 1903, 1905 (the 1905 paper is part of Thompson 1905). The original German-language version was applied in studies of the western Baltic in 1905 by Rudolf Kohlmann, a doctoral student of the ever-alert Otto Krümmel. This was the first application of the method after its use by Sandström and Helland-Hansen.

103 Helland-Hansen to D'Arcy Thompson, 15 May 1903, St Andrews University Library, D'Arcy Wentworth Thompson Papers, MS 14307.

104 Helland-Hansen 1905, p. 5.

105 Note that this is not an absolute current velocity, as seen, for example, by an observer at a distance, but a velocity at the upper level relative to

the one below, which itself might be in motion. Many later developments of the technique concentrated on the problem of determining absolute current velocities, often by determining, or assuming, that there was no motion at the greater depth (the so-called 'depth of no net motion').
106 Helland-Hansen 1905, p. 6. The reformulation was presented in a short section of a paper on the Faeroe-Shetland Channel in 1902.
107 Robertson 1907, pp. 28–34.
108 Proudman 1953, p. 72.
109 See almost any textbook of dynamic oceanography; for example, Sverdrup et al. 1942, pp. 448, 460–2, and Open University Course Team 2001, pp. 53–9.

4. Evangelizing in the Wilderness: Dynamic Oceanography Comes to Canada

1 Letter by Fred Cook to G.J. Desbarats, deputy minister of the Naval Service, in Library and Archives Canada (LAC), RG 23, vol. 1204, file 726-2-4 [3].
2 Department of the Naval Service 1919.
3 For useful surveys of Canada up the the First World War, see Cook 1987; Brown and Cook 1974, chs 1, 4, 5; and Granatstein et al. 1983, ch. 2.
4 Cook 1987, pp. 377–9.
5 Nova Scotia, New Brunswick, and Prince Edward Island. Newfoundland became the fourth Atlantic Province of Canada in 1949 after a long history as British colony and self-governing Dominion.
6 Thompson and Seager 1985, ch. 1, provide a summary of Canada based on the census of 1921.
7 This was 9 per cent of goods production in the Maritime Provinces in 1910, although only 2 per cent for the whole nation. The distribution of value by major species was cod $3.9 million, herring $3.3 million, lobster $3.6 million (all from the Atlantic coast), and Pacific salmon $8.2 million. These figures appear to be landed value (see Gough 1991, pp. 76–7).
8 Gough 1988, 1991, 1993.
9 The first steam trawlers appeared in Canso, Nova Scotia, in 1910. For smaller vessels, the transition from sail to engines was rapid: in 1910/11 there were 5,000 gas-engined fishing boats on the Atlantic Coast; in 1915 more than 50,000 (Gough 1991, pp. 35–6; see also Cowie 1912, pp. 107–9). For another contemporary view, see Prince 1897, 1924. Hubbard 1993, pp. 229–56, synoptically reviews Canadian fisheries from Confederation to 1930.
10 Jarrell 1985.
11 The impetus came from the technological needs of the war. For background and the effect on physics, see Gingras 1991.

12 They were, in fact, the nucleus of the Royal Canadian Navy (the 'Naval Service') created by the Laurier government in 1910; see Gough 1991, pp. 21–2; Milner 1999, ch. 1, 2.

13 Prince 1924, pp. 270–1, outlines the fisheries-related responsibilities of the department. See also Hubbard 1993, pp. 175–85, and 2006, ch. 4.

14 See especially Zaslow 1975, and Zeller 1987, ch. 1–5.

15 Johnstone 1977, pp. 22–4.

16 On Prince, see Anonymous 1912; Hubbard 1993, pp. 12–13, 50–4, and 2006, pp. 17–19; Huntsman 1945; Johnstone 1977, pp. 43–4; and Needler 1985. As fish embryologist, Prince would bring scientific expertise to the department's fish hatcheries and other programs (Hubbard 1993, pp. 13–14; Prince 1913, pp. 89–90).

17 Promoted especially by J.P. McMurrich, then of the Ontario Agricultural College, in an influential article in 1884. For Prince's view, see his summary, 1913, pp. 89–90.

18 See the discussion and further references in Hubbard 1993, pp. 40–51, and 2006, pp. 22–6, 33–7.

19 Developments treated *in extenso* by Hubbard 1993, pp. 11–32, 81–6, and 2006, pp. 33–4; and Johnstone 1977, pp. 24–41.

20 For summaries, see Hubbard 1993, pp. 105–8, and 2006, ch. 2; and Johnstone 1977, pp. 52–76. The Biological Board was renamed the Fisheries Research Board of Canada in 1937 and remained such until its demise in the 1970s (Anderson 1984; Johnstone 1977, ch. 21).

21 Anonymous 1914, 1943; Mackenzie 1943.

22 Cowie 1912, pp. 98–9.

23 Cowie 1912, p. 94. Prince agreed with his mentor McIntosh in this and brought the opinion to Canada (see Prince 1913, pp. 101–2, 103–4; Hubbard 1993, pp. 292–304, and 2006, pp. 17–18, 159–160).

24 Cowie 1912, p. 99.

25 Cowie 1912, pp. 102–4; Prince 1916b, pp. 48–50, 1916a, pp. 37–9.

26 Prince 1916a, p. 37.

27 Byrne 1916, pp. 16–17, 21–4. The quotation is from E.E. Prince to J.D. Hazen: 'MEMO re HERRING FISHERY DEVELOPMENT INVESTIGATION,' LAC, RG 23, vol. 1204, file 726-2-4 [1].

28 Prince 1916a, p. 39.

29 Among many biographical notices of Hjort, see especially Andersson 1949; Hardy 1950; Hubbard 1993, pp. 114–26; Hubbard 2006, pp. 67–75; Maurice 1948; Merriman 1972; Nordgaard and Kielhau 1934; Russell 1948; Ruud 1948; Solemdal 1997; and Went 1972, pp. 169–72.

30 Solemdal 1997 evokes the work of this group and Hjort's working style. See also Rollefsen 1966.

31 Hjort 1914. Based on the ability to age herring using scale rings, this work served as an important stimulus to the study of fish populations – and was considered suspect by those who did not accept that scale rings could be used to age fish.

32 Lea 1919, pp. 81–2.

33 Hjort 1914, pp. 217–22.

34 Hjort 1914, p. 179.

35 Helland-Hansen and Nansen 1909b, pp. iv, 216–18; 1909a; and treated at greater length in joint papers in 1917 and 1920.

36 Hjort 1914, pp. 179–92.

37 Hjort 1914, p. 202.

38 Hjort 1914, p. 203.

39 Hjort 1914, p. 204.

40 Hjort 1914, p. 205. This, later christened the 'match-mismatch hypothesis' by the English biological oceanographer David Cushing (Cushing 1972, pp. 216–17; 1973, pp. 400–8; 1975, pp. 180–2), became the subject of intense research efforts from the 1960s until the present.

41 Hjort 1914, p. 206.

42 Hjort 1914, p. 209.

43 It is likely that the close similarity in volume-transport and other oceanographic features between the St Lawrence River – Gulf of St Lawrence system and the Baltic Sea outflow (both areas with major herring fisheries) made the opportunity to work in Canada especially appealing to Hjort. I am indebted to Dr Artur Svansson of Göteborg University for this suggestion.

44 Large amounts of material are available documenting the organization of the Canadian Fisheries Expedition, especially in Library and Archives Canada (LAC) (esp. RG 23, vol. 1204), the University of Toronto Archives (Huntsman Collection), the Harvard University Archives (H.B. Bigelow General Correspondence, HUG 4212.5), and the University of Oslo Library Manuscript Collection (Johan Hjort: Ms. 4to 2911. Etterlatte papirer). The definitive study of the expedition based upon archival records is Hubbard 1993, ch. 3, and 2006, ch. 3.

45 Hjort 1919, p. xi.

46 Certainly the most significant figure in Canadian east coast marine biology from 1914 through the 1950s, Huntsman's career was deeply influenced by his early work with Hjort. On his career, see Dickie 1985; Johnstone 1977, pp. 305–6; Needler 1975; and many mentions in Hubbard 2006.

47 Hjort 1915.

48 Johan Hjort to A.B. Macallum, 25 November 1914, LAC, RG 23, vol. 1204, File 726-2-4 [1]. See also Hjort 1915, pp. 5–7, and p. 24, where he said, 'The gulf [sic] of St. Lawrence seems to provide the most excellent conditions for investigations of this kind.' The Gulf also had similar features to the herring-producing regions of southern Sweden and Norway (see note 43).

49 Dealt with in detail by Hubbard 1993, pp. 140–5, and 2006, pp. 80–2.

50 Hjort 1915, p. 6.

51 Brought from Scotland in 1904 in aid of J.J. Cowie's attempts to modernize the East Coast herring fishery.

52 Johan Hjort to G.J. Desbarats, 16 March 1915, 'MEMORANDUM regarding hydrographical and biological investigations in the Gulf of St. Lawrence and adjacent waters. May to August 1915,' LAC, RG 23, vol. 1204, file 726-2-4 [1].

53 Hjort 1919, pp. xvii-xviii; Bjerkan 1919, pp. 350–2. Nansen had been experimenting with designs for water bottles for several years, and was promoting the use of German-made Richter reversing thermometers allowing water temperature to be read to at least 0.1 °C. See Carpine 1993, 1997, on these technological changes.

54 Rollefsen 1970. See Bjerkan 1919 on the chemical techniques.

55 On Willey, see Hubbard 1993, p. 133, and 2006, p. 77; Hutchinson 1944; and Johnstone 1977, pp. 70–1. Mavor is less well known. The son of the distinguished University of Toronto political economist James Mavor (1854–1925), after graduate work at Harvard and a short period at the University of Wisconsin, he spent most of his career at Union College, Schenectady, New York. He was curator of the Biological Station at St Andrews, temporarily replacing Huntsman, in 1914 and 1917 (Hubbard 1993, p. 133, and 2006, p. 77; for a contemporary photograph, see Johnstone 1977, p. 87).

56 H.B. Bigelow of Harvard, with whom Hjort was in close contact, had begun limited oceanographic studies in the Gulf of Maine for the U.S. Fish Commission in 1912. These developed into definitive studies of the area long after the Canadian Fisheries Expedition (Bigelow 1926, 1927; Brosco 1989).

57 As late as 1950, Hjort's obituarist, the English biological oceanographer A.C. Hardy, called the report 'among the rarest and most prized of publications that an oceanographer can possess.' (Hardy 1950, p. 171).

58 Lea 1919; Dannevig 1919; Huntsman 1919; Willey 1919; Gran 1919. At this time, Gran was still struggling to understand the control of phytoplankton production in nature, attributing the onset of the spring bloom to some factor from land carried into the sea during the spring melt. It was a further fifteen years before he and Trygve Braarud recognized that a combination of high levels of dissolved nutrients and stratification of the water column

resulted in the beginning of phytoplankton growth in spring (Mills 1989, ch. 5).

59 One might have expected Bjerknes's principal Norwegian disciple, Helland-Hansen, to have tackled this part of the report. But by 1914, Hjort and Helland-Hansen, collaborators in the early years, had fallen out in a turf war between the biological station of Bergens Museum, directed by Helland-Hansen from 1906 to 1917, and Hjort's fisheries institute in Bergen. Under the circumstances, there was no one but Sandström to do the job.

60 Hjort 1919, p. xxiv. The mention of sections taken transverse to the direction of current movement as part of the cruise plan indicates that Hjort had a dynamic analysis in mind from the beginning. And the mention of ice-melting indicates that he was also giving thought to Otto Pettersson's theory of temperature-induced thermohaline circulation (Pettersson 1899, 1900a, 1907) as well as to the European debate over the control of herring abundance by alternations of Atlantic and Baltic water (Svansson 1965, 1999).

61 Hjort 1923, p. 45, and the lengthy discussion following on pp. 45–8.

62 Hjort 1923, p. 30.

63 Sandström 1919, p. 221.

64 Sandström 1919, pp. 221–2.

65 Sandström 1919, p. 288.

66 Sandström 1919, pp. 261, 288–90. According to Artur Svansson (*in litt.*), Sandström had been greatly influenced by his mentor Otto Pettersson's interest in the effect of the meeting of Atlantic and Baltic water masses on the abundance of herring in southern Scandinavian waters.

67 Sandström 1919, p. 267.

68 Sandström 1919, p. 267. Throughout his career, Sandström carried out laboratory experiments in hydrodynamics and maintained an interest in instruments. Although an able theoretician, he was unusual in taking hydrodynamic theory into the laboratory.

69 Sandström 1919, Tables 1–5, Plates XIV, XV.

70 Sandström 1919, p. 225.

71 Surfaces of equal specific volume.

72 Sandström 1919, p. 228, figs. 4–6.

73 Sandström 1919, p. 231.

74 Sandström 1919, p. 238.

75 Density surfaces slope upward toward the centre of cyclonic gyres and downward toward the centre of anticyclonic gyres (see Sandström 1919, pp. 239–41, figs. 22–4).

76 Sandström 1919, p. 251. Lying below the surface water of the Gulf of St Lawrence is a very cold (0°C or less) layer of saline water 20 to more than 90 me-

tres thick atop the warmer and even more saline bottom layer derived from
the Atlantic (Hachey 1961, pp. 82–4). The intermediate layer is now believed
to be formed mainly by winter cooling and sinking *in situ* in the Gulf.

77 Sandström 1919, pp. 253–5, 275–8.

78 Sandström 1919, p. 255.

79 Sandström 1919, p. 257. Here he mentions, in particular, the Gulf Stream,
   which he regarded as partially wind-driven, in that it was part of the Sargasso Sea circulation, but resulting mainly from the buoyancy forces of
   warm water occurring at great depths on its western margin. Sandström had
   a long-standing interest in the Gulf Stream (see a preliminary treatment
   in Sandström and Helland-Hansen 1905, pp. 162–3, pp. 257–3 of the 1919
   monograph, and especially Liljequist 1993, ch. 41). On modern interpretations of the dynamics of the Gulf Stream, see Stommel 1965 and Open
   University Course Team 2001, pp. 107–13.

80 Sandström 1919, p. 263. It seems surprising that at this stage of his career
   Sandstöm did not see any linkage between wind stress on the ocean and the
   distribution of properties in the water column, even though Mohn before
   him had recognized such a relationship.

81 Sandström 1919, p. 267.

82 Sandström 1919, pp. 268–72, and Table 5, pp. 317–41.

83 Sandström 1919, Plate XIV.

84 Helland-Hansen 1905, p. 6.

85 Proudman 1953, p. 72. Sandström was quite aware that the calculated current velocities were relative, not absolute, because the water at the deepest
   part of the section might be in motion too, rather than being immobile.
   This problem could be minimized by selecting the reference level at the
   greatest depth, where the current velocity was likely to be small (1919,
   pp. 273–4). Friction could be ignored, and other problems eliminated, by
   choosing a section across the current (1919, p. 274) – indicating that some
   knowledge of the current had to be available from the start if a correct
   result was to be achieved.

86 Sandström rarely referred in his publications to the work of others, or even
   to his own work, as my colleague Dr Artur Svansson has pointed out (*in
   litt.*).

87 Dr R.W. Trites of the Bedford Institute of Oceanography told me that even
   during the 1950s Sandström's techniques – and the monograph – were still
   in use in H.B. Hachey's Atlantic Oceanographic Group based in the Biological Station at St Andrews, New Brunswick. By 1961, Defant, in his textbook
   of dynamic oceanography, mentions only the experimental results in the
   1919 monograph (Defant 1961, p. 486).

88 Sandström 1919, p. 221.

89 See the epigraph to this chapter.

90 On 'inventory science,' see Zeller 1987. Further background is given by De Vecchi 1984 a,b and 1985.

91 See, for example, any of Prince's cited publications, 1913–24.

92 In the case of physics, see Gingras 1991.

93 Johnstone 1977, pp. 136–9; Hubbard 1993, pp. 397–403, and 2006, pp. 192–213.

94 Hubbard 1993, pp. 160–9. The Belle Isle Strait Expedition of 1923 was the last to be patterned on the Canadian Fisheries Expedition. Thereafter Board research took less ambitious form. On the results of some of these expeditions, see Craigie and Chase 1918; Huntsman 1924; Mavor 1920, 1922, 1923; and Vachon 1918.

95 For example, for many years the Biological Station at St Andrews, New Brunswick, had only one Nansen bottle.

96 James Mavor's (1922, 1923) papers on the hydrography of the lower Bay of Fundy were to be completed by a dynamic analysis of the currents based on Sandström's monograph (see Mavor 1922, p. 16), but it was never published.

5. 'Physische Meereskunde': From Geography to Physical Oceanography in Berlin, 1900–1935

1 Paffen and Kortum 1984.

2 The change of spelling occurs in the late 1880s, as a result of 'Germanization' of spelling.

3 See chapter 3 on applications of Scandinavian dynamical oceanography by Krümmel's students.

4 Defant 1928b, p. 459.

5 Anonymous 1906, Probst 1995.

6 Beckinsale 1975; Defant 1933; Kortum 1983, 1987; Lüdecke 1995, pp. 23–5; Stahlberg 1929.

7 Richthofen 1904.

8 Probst 1995, p. 11; Röhr 1981.

9 Richie 1998, p. 228. Based solely upon the exhibits in the sections dealing with trade and naval warfare, Richie's generalization is correct – but the Museum had significant scientific content as well, as a few of the illustrations in Röhr (1981) indicate.

10 Beckinsale 1974.

11 Penck 1910, p. 3.

12 Penck 1912, p. 419.

13 Lehmann 1965, Penck 1912.

14 One of the large lakes just west of Berlin – used by Grund and his successor, Alfred Merz, as surrogate oceans.

15 See Merz 1910. The *Habilitationsschrift* represented research beyond the doctorate, qualifying the candidate for university teaching.

16 Penck 1926, Priesner 1993. Penck's lengthy obituary of Merz is especially poignant. When Merz died in 1925 in Buenos Aires during the *Meteor* expedition, Penck had lost two sons, a biological one, Walther Penck, also a distinguished geologist, in 1923, and only two years later a scientific one, Merz.

17 Merz 1912, 1913; Wendicke 1912.

18 Merz 1915.

19 Merz 1911.

20 On the laboratory at the time, see Kofoid 1910, pp. 261–6.

21 Merz 1915b, p. 111.

22 Ekman 1902, 1905, 1906.

23 Nansen 1912.

24 Helland-Hansen and Sandström 1903, 1905.

25 Merz 1912, p. 179.

26 Penck 1926, p. 87. Not only had Merz been one of Penck's favourite students, he was his son-in-law. See Beckinsale 1974, pp. 502–3, on Penck's political philosophy; also MacLeod 1995, p. 431.

27 Richie 1998, p. 251. There are several other analyses of the German professoriate before the 1930s; this one is brief and to the point, making specific reference to Berlin.

28 Outlined by Laqueur 1974, especially chs 1, 6; Forman 1971, 1973. On the resulting antipathy to German scientists, see Kevles 1971.

29 Among these was certainly *not* Albert Einstein. See especially Stern 1999, chs 2 and 3, on the reaction of this individualist, and the ambivalent reactions of Fritz Haber and Max Planck, eminent members of the University of Berlin, to the intellectual attitudes there.

30 For his views after the Great War, see Merz 1922a.

31 On pan-Germanism, see especially Chickering 1984; and on other ideological and intellectual movements of early twentieth-century Germany, Meyer 1955 and Richie 1998, chs 6 and 7.

32 Forman 1973, p. 152.

33 The situation in Germany in 1918 and after is outlined by Bessel 1993; Craig 1981, chs 11, 12; Kevles 1971; and Richie 1998, ch. 8.

34 Roll 1987.

35 Archiv der Humboldt Universität, Berlin, Aktenverzeichnis Institut für
    Meereskunde, Nr 256, 1–2, 57–58 (examined in Museum für Verkehr und
    Technik, Berlin).

36 Wüst 1920.

37 Statsarkiv I Bergen, Norway, Dep. 827, Prof. Bjørn Helland-Hansen arkiv,
    Boxes 6, 9, Correspondence from Georg Wüst. Wüst arrived in Bergen
    on 25 August 1913, too late to take the annual summer course given by
    Helland-Hansen, Hjort, and others. The outbreak of war a year later ended
    this important series of classes permanently.

38 Brennecke 1921. Wilhelm Brennecke (1875–1924) was an employee of the
    Deutsche Seewarte in Hamburg and oceanographer on the voyages of *Planet*
    in 1906–7 and *Deutschland* in 1911–12 (the German Antarctic Expedition,
    directed by Wilhelm Filchner). On Brennecke, see G. Schott 1924 and W.
    Schott 1987. On the expeditions, see Reichs-Marine Amt 1909, Watermann
    1989, W. Schott 1987, and Filchner 1994.

39 Mills 2005, pp. 253–60.

40 See G. Schott 1902. Gerhard Schott (1866–1961) was a long-serving em-
    ployee of the Deutsche Seewarte (Dunbar 1961, Lenz 1986, W. Schott 1987,
    and Schulz 1936). The *Valdivia* expedition under Carl Chun (1852–1914)
    was mainly biological; Schott was its oceanographer.

41 G. Schott 1902, p. 165.

42 E. Lenz 1830, 1831, 1832, 1845, 1847, 1848. See also chapter 2.

43 Drygalski 1902, p. 47. On the expedition, its work, and its leader, see Dryg-
    alski 1902, 1903, 1904, 1989, and Fels 1971.

44 Brennecke 1909, pp. 71–5. The expedition on *Planet* is described by
    Stutzbach-Michelsen 1989.

45 Brennecke 1909, p. 95.

46 Brennecke 1909, p. 98.

47 Brennecke 1911a,b, 1913, 1914.

48 Brennecke 1915.

49 For some of the evidence and early ideas on the origin of North Atlantic
    Deep Water, see Nansen 1912 (who suggested its origin mainly Southeast of
    Greenland), Brennecke 1915, p. 58, and the discussion in Brennecke 1921.
    Recent views are outlined by Bacon 2002.

50 Schott 1924, p. 50.

51 An abbreviated version is given by Wüst 1968.

52 Merz and Wüst 1922, p. 25.

53 John Young Buchanan (1844–1925) was the chemist on the *Challenger* voy-
    age. He had a long and distinguished career in oceanography after *Challeng-
    er* (Kutzbach 1970), beginning with summaries of the physical and (some of

the) chemical results. Alexander Buchan (1829–1907), a meteorologist of the British climatological school (Kutzbach 1979, pp. 227–8), was assigned the task of summarizing the physical oceanography.

54 Brennecke and Schott 1922, Merz 1922b.

55 Both accuracy and precision of deep-sea thermometers improved strikingly – better than an order of magnitude – around the turn of the century.

56 It is easy to agree with Brennecke and Schott. Although the charts and data presented by Buchanan and Buchan show some hydrographic evidence of AABW, AAIW, and NADW, neither discusses their implications.

57 Brennecke and Schott 1922, p. 285.

58 Merz 1922b.

59 Pettersson 1923.

60 Pettersson 1904, Sandström 1904. This was an early attempt by Sandström to apply the new mathematical technique on a large scale, much as he and Helland-Hansen had attempted to do at the end of their 1903 publication reformulating Bjerknes's theorem for use in the oceans.

61 Pettersson 1923, pp. 69–70.

62 Petterson 1923, p. 71.

63 Merz 1922a; Schott 1923.

64 Merz accused Brennecke of allowing a misunderstanding to arise with Pettersson over the interpretation of vertical circulation in the Sargasso Sea. Correspondence with Brennecke (as editor of the *Annalen der Hydrographie*), Archiv der Humboldt-Universität, Berlin, Aktenverzeichnis Institut für Meereskunde: Meereskundliche Untersuchungen und Expeditionen, Nr 15 (examined in Museum für Verkehr und Technik, Berlin).

65 Merz and Wüst 1923a,b.

66 Merz and Wüst's replies do not throw much light on their understanding of the dynamic method in 1923. But it appears that neither understood Pettersson's point, despite Wüst's study with Helland-Hansen in Bergen a decade earlier.

67 As chapter 3 shows, Mohn's calculation of the circulation of the Norwegian Sea involved considerably more than merely calculating the pressure field – although this misconception seems to have been common.

68 Merz 1925, p. 563.

69 Alfred Merz to Otto Pettersson and Johan Sandström, May-June 1923, Archiv der Humboldt-Universität, Berlin, Aktenverzeichnis Institut für Meereskunde: Meereskundliches Untersuchungen und Expeditionen, Nr 153 (examined in Museum für Verkehr und Technik, Berlin). The gist of the reply from Sandström (which is not included) is that he regarded the 1904 calculation as a methodological exercise, not a definitive work on ocean circulation.

70 Emery 1980; Penck 1925; W. Schott 1987, pp. 25–7; Spiess 1926, 1985. A definitive, historically rigorous account of the events surrounding the origin of the *Meteor* expedition – and of the expedition itself – remains to be written. Contemporary accounts, like those of Penck and Spiess, disguise or omit some of the political rationale and underpinnings. Emery's short outline in English is useful as an introduction to some of the factual material but accepts contemporary accounts at face value.

71 Economic circumstances – the incredible inflation of the Mark in 1922/23 from less than 200 to the U.S. dollar to 4.2 European billion (10^{12}) to the dollar by late in 1923 (Craig 1981, pp. 448–56, Richie 1998, pp. 320–4) – nearly scuttled the *Meteor* expedition. Even when the Mark stabilized and inflation ended, the expedition had to be restricted in time and to one ocean, the Atlantic.

72 Spiess 1926, pp. 4–5, 1985, p. 379; Wüst 1928a, pp. 67, 82–3; Defant 1927, pp. 365–6.

73 The portion of the Gulf Stream system passing through the Strait of Florida and up the east coast of Florida.

74 Wüst 1924, p. 41.

75 Wüst 1924, p. 41. The primary comparison was between Pillsbury's current measurements and dynamic calculations. The 'third method' was Pillsbury's estimate of the ship's drift due to currents. For a modern viewpoint, see Warren 2006, pp. 7–8.

76 Pillsbury's success in direct current measurements was certainly due to his (and his crew's) skill in using current meters and in deep-sea mooring, but it was aided by the high velocity of the Florida Current where he worked; in modern terms, he found a high signal-to-noise ratio, a situation rare in the open ocean and one that *Meteor* was unlikely to encounter.

77 Wüst 1928a, p. 83.

78 Fritz Spiess (1881?–1959), a highly competent naval officer trained in hydrography and the related sciences, later rose to the rank of rear-admiral and directed the Deutsche Seewarte (the German Marine Observatory) from 1934 to 1945 (Errulat 1960; W. Schott 1987, p. 7; Schumacher 1959).

79 Indeed, in all of Germany. Both Penck and Merz had held the chair of geography. On Defant, see Böhnecke 1976; W. Schott 1987, pp. 30–1; Thorade 1944; and Wüst 1964a,b. *En route* to the Berlin chair, Defant joined *Meteor* for its last two transects – the importance of which he described in 1936.

80 Defant 1928b, p. 459.

81 Defant 1928a.

82 Defant 1928b, p. 460.

83 Defant 1928b, pp. 470–3.

84 Defant 1929b.

85 Defant 1929c, 1934.
86 The effect of *Dynamische Ozeanographie* was felt rapidly. Margaret Deacon
tells me that her father, George Deacon, a young hydrologist (physical oceanographer) with the British *Discovery* Investigations in the early 1930s, took
the time and trouble to translate it into English for his own use.
87 Notably Gertrud Kobe's (1934) on the currents of the Skagerrack and
Günther Dietrich's (1935) on the Agulhas Current.
88 Wüst 1928b.
89 Dietrich 1972. In later analyses of the *Meteor* results, he made important
uses of the dynamic method.
90 Defant 1928b, p. 492.

6. 'Découverte de l'océan': Monaco and the Failure of French Oceanography

1 This is an extensively explored subject: see especially Crosland 1992; Fox
1992; Fox and Weisz 1980; Nye 1986; Paul 1985; and Weisz 1983, among
many others.
2 Kofoid 1910 describes these and more, indicating their links with the
universities. See also Paul 1985, pp. 103–17, notably on the influence of the
famous Stazione Zoologica at Naples, founded in 1872. A recent important
work is Fischer 2002.
3 Kofoid 1910, p. 35.
4 Shinn 1979.
5 Biographical material is abundant, but with a few exceptions (Albert I[er]
1998; Carpine-Lancre 1991; 1993, 1998; Carpine-Lancre and Saldanha 1992;
Mills and Carpine-Lancre 1992) repetitive, often hagiographic, and not
analytical (Anonymous ca. 1920; Anonymous 1923a; Damien 1964; Fontaine
1972; Herdman 1922; Mill 1922; and Petit 1970). Albert's own account of
his early life at sea is in Albert I[er] 1966 (first published in 1902).
6 Anonymous ca. 1920, p. 20; Richard 1934a, pp. 454–5; Carpine-Lancre
1993, p. 122; Carpine-Lancre 1998, pp. 24–5.
7 Albert I[er] 1892a,b,c; 1898.
8 Carpine 2002, pp. 57–83, on the Prince himself; other chapters deal with
the range of instruments on the ships and onshore in Monaco. A complete
list of stations is given by Richard 1934b.
9 Electric lighting probably first appeared on the U.S. research vessel *Albatross*
about 1882 (Tanner 1897, pp. 287–8; Mills 1983, p. 38).
10 Lacroix et al. 1946; Portier 1945; Rouch 1948.
11 H. Pettersson 1953, p. 120; Mills 1983, p. 50.

12 Albert Ier 1896, p. 440.

13 See chapter 7; also Brosco 1989.

14 Albert Ier 1910, Ier 1932c, pp. 47–8; Richard 1910, p. 72; Mills 1983, p. 50; Carpine-Lancre and Saldanha 1992, p. 146.

15 Albert Ier 1998, pp. 45–87; Carpine-Lancre 1998, pp. 4–12.

16 Albert Ier 1904, p. 8; 1932a, p. 212.

17 Anonymous 1910, pp. 7–8.

18 Carpine-Lancre 1989.

19 Richard 1932; Rouch 1958.

20 Albert Ier 1932b (1912), p. 335.

21 Rabot 1987, p. 37.

22 Carpine-Lancre 2003, pp. 32–3.

23 Anonymous 1911; Idrac 1934; Joubin 1934.

24 Vallaux 1936, p. 1. Jacqueline Carpine-Lancre (*in litt.*) has emphasized that Vallaux's account is full of historical errors.

25 Thoulet 1905b.

26 Biographical details and much else are in Carpine 2002, pp. 159–78; Carpine-Lancre 2003; Jamieson 2005; Portier 1936; and Vallaux 1936.

27 Equivalent to a lectureship.

28 Nye 1986, ch. 2, deals *in extenso* with science and politics at Nancy. See also Carpine-Lancre 2003, p. 20.

29 Thoulet 1887, 2005. Until the Anglo-French Entente (*Entente cordiale*) of 1904, France maintained an active fishery along the north and west coasts of the British colony of Newfoundland.

30 Thoulet 1889a,b.

31 Thoulet 1905b. Vallaux gives 1894 as the date of the Sorbonne lectures, but this is incorrect, according to Jacqueline Carpine-Lancre (*in litt.*).

32 Ritchie 2003, 1967. See also McConnell 1990.

33 Carpine-Lancre 2003, p. 19; see pp. 15–19 on background.

34 Carpine-Lancre 2003, p. 23.

35 Carpine-Lancre 2003, pp. 27–8. Now widely referred to as GEBCO, the General Bathymetric Chart of the Oceans, the production of this major contribution was transferred to the newly founded International Hydrographic Bureau (IHB), based in Monaco, in 1921 (Carpine-Lancre 2003, pp. 45–6). Further information on GEBCO and the IHB is in Scott 2003. For a contemporary account of the IHB, see Poilleux 1924.

36 Carpine-Lancre 2003, pp. 30–2.

37 Thoulet 1893, pp. 265–6; 1905c.

38 Thoulet 1901, p. 198.

39　Thoulet 1890b, p. vii.

40　Thoulet 1899, p. 408, only a portion of which is quoted.

41　Thoulet 1890b, p. 1.

42　Thoulet 1895b, pp. 268–9.

43　Thoulet 1904b, p. vi.

44　Thoulet 1893; 1895b, p. 259; 1904a-c are among the first.

45　Especially in Tizard et al. 1885. The section cited was written by Buchanan.

46　Zöppritz 1878a,b; 1879a,b.

47　Thoulet 1893, p. 261. His first references to the calculations of Zöppritz are Thoulet 1891b, p. 327; 1893 pp. 260–1;1895b, p. 268; and 1896, p. 104.

48　For all the relevant ideas, see Thoulet 1890a,c,d and 1896, pp. 116–23.

49　Thoulet 1890a, p. 500.

50　Thoulet went to some trouble to correct reported deep-water densities for the effect of pressure on recorded temperatures (to show, for example, that some apparent inversions in the *Challenger* data were removed once the effects of pressure on temperature measurements were taken into account). He appears to have been the first to recognize the significance of this correction, which is discussed at length in Thoulet 1895b, pp. 266–7, and much later in Thoulet 1930a.

51　Thoulet 1890a, pp. 502–3.

52　Thoulet 1891a, p. 1146.

53　Thoulet 1891a,b, 1892, 1893, 1895b, 1896, 1902b, 1903.

54　Knudsen 1901. On the development of this important concept in oceanography, see Wallace 1974; and on ICES's Central Laboratory and Knudsen's role in it, see Smed 2005.

55　Chevallier 1905.

56　Thoulet 1902a; 1902b, p. 552; 1902c, p. 1460; 1903, p. 98; 1905a, pp. 98–100; Wallace 1974, ch. 8; Smed 1992.

57　Pettersson 1907, p. 290. The controversy is discussed in detail by Smed 1992, pp. 80–5. See also the comments by Menaché 1952, especially on pp. 11, 13, and 19. Knudsen had never claimed the complete accuracy of his tables to several decimal places, and had discussed the meaning of the awkward constant in the equation relating salinity and chlorine content (Wallace 1974, pp. vii–viii, 146–9).

58　Thoulet 1895b, p. 267; 1903.

59　Helland-Hansen 1916. Mamayev 1987, p. 666, gives a translation into English of the relevant pages.

60　Thoulet 1926a,b; 1927a,b,c; 1930a,b; 1933. The quotation is from 1927c, p. 866.

61　Thoulet 1930b, p. 37. He would have been pleased by the discovery of deep-sea vents in the 1970s and their implication in the chemical composition of seawater.

62 Thoulet to Jules Richard, 20 October 1921. Letter in Archives, Musée océanographique de Monaco.

63 Thoulet 1890b, p. v. On the 'ingénieurs hydrographes de la marine' Dortet de Tessan and Bouquet de la Grye, see Gougenheim 1968; also Rollet de l'Isle 1951.

64 Thoulet 1921a, p. 4.

65 Gougenheim 1968 pp. 90–1; quoted at length in Thoulet 1922, pp. 259–60. The origin is Dortet de Tessan 1842–4.

66 Bouquet de la Grye 1882; see also Gougenheim 1968, pp. 92–5.

67 Bouquet de la Grye 1882, pp. 468–9.

68 Bouquet de la Grye 1882, p. 470.

69 Bouquet de la Grye 1882, pp. 471–5; Gougenheim 1968, pp. 94–5. Modern calculations, involving dynamic height calculations rather than simple levelling, show that the Sargasso Sea is approximately 30 centimetres higher than the average height of the North Atlantic as a result of the piling up of water caused by wind stress around the North Atlantic meteorological high pressure system and the effect of the Earth's rotation.

70 Thoulet 1890b, p. 351.

71 Thoulet 1890b, pp. 369–70; 1896, pp. 108–9; 1921a, pp. 11, 13, 18; 1921b, p. 862, 1921c, p. 2; 1925b, pp. 420–1. But the descent was not into the abyss, only to the level at which the sinking water met water of its own density, normally within the top few hundred metres.

72 Thoulet 1896, p. 113. He included a discussion, with calculations, of the potential Coriolis effect on the slope of the ocean surface, even though he believed it was of little relevance to the currents themselves. The inherent paradox in this seems not to have occurred to Thoulet.

73 Thoulet 1904c.

74 Chevallier 1906; Thoulet 1904a,c; Thoulet 1904b, pp. 377–9; Thoulet 1908; Thoulet and Chevallier 1906.

75 Because of Thoulet's belief that the composition of seawater was variable and could not be characterized using the abundance of a single ion, like chloride (which, based on the theory of constancy of composition, was the basis of Knudsen's Tables).

76 Thoulet 1905a, p. 100.

77 Thoulet 1905a, p. 104. The symbols represent the density of seawater at the *in situ* temperature corrected for pressure and the same property at a given depth, n.

78 Thoulet 1905a, p. 107.

79 Thoulet 1905a, p. 111.
80 Information thanks to Jacqueline Carpine-Lancre.
81 Otto Pettersson to Martin Knudsen, 1 May 1905, quoted by Smed 1992, p. 81.
82 Notably the Darmstadt physicist Carl Forch (1909), who showed that to get correct results using Thoulet's method it was necessary to calculate the weight of the whole water column. When one did so, Thoulet's results (in 1904a and 1905a, also Chevallier 1906) proved to be wrong. Forch, too, despite his close contact with Otto Krümmel, ignored the Earth's rotation, making his corrections of limited long-term interest.
83 Thoulet 1924d, p. 337.
84 Thoulet 1921a–c; 1922, p. 241ff; 1924a–e; 1925a–d; 1926a; 1927c; 1930a,b.
85 Buchanan 1884 and 1919, pp. 105–12; Tizard et al. 1885.
86 Summarized in Murray and Hjort 1912.
87 Thoulet 1921c, p. 5.
88 Julien Thoulet to Jules Richard, 31 March 1921. Letter in Archives, Musée océanographique de Monaco. Ekman was, of course, a Swede.
89 Thoulet 1925a, p. 650; 1926a, p. 139; 1927b, p. 265. The work referred to was Sandström 1923.
90 Le Danois 1924 (in which there is no mention of Thoulet); also later publications, including Le Danois 1937a,b.
91 Vallaux 1927. After Thoulet's death, Vallaux (1937a,b, 1938) wrote the best-informed reviews of and commentaries on physical oceanography for the French scientific community, including in his ambit American and British work.
92 Vallaux 1936, p. 1.
93 Jacqueline Carpine-Lancre tells me (*in litt.*) that the Spanish oceanographer Raphael de Buen spent some time in Thoulet's laboratory in Nancy during 1911–12 and referred to Thoulet in later works, mostly, it appears, in connection with sedimentology.
94 Rozwadowski 2002, pp. 71–3.
95 Richard 1907, pp. 176–7.
96 Richard 1907, p. 179.
97 Rouch 1922, p. 2.
98 Rouch 1922, pp. 200, 203; Schott 1902.
99 Rouch 1943–8.
100 Berget 1931, pp. 182–5; later in the same book (pp. 199–215), he outlined Ekman's work on the wind-driven circulation in considerable detail without placing it in meaningful context.
101 Berget 1931, pp. 268, 270.

102 Berget 1931, pp. 273–4. In this, Berget closely follows Richard 1907, which appears to have been the basis of this section.

103 Lacombe 1965. Henri Lacombe held the chair of physical oceanography created in 1955 in the Muséum d'histoire naturelle in Paris.

7. Slipping away from Norway: Dynamic Oceanography Comes to the United States

1 Woods Hole Oceanographic Institution (WHOI) Archives, Office of the Director, 1935–1940, Folder: Columbus O'D. Iselin 1936–1940. Iselin, a staff member of Woods Hole Oceanographic Institution, was writing to his director from the congress of the International Association of Physical Oceanography being held in Edinburgh. Carl-Gustaf Rossby (1898–1957), a former member of the Bergen School of Meteorology, was teaching at the Massachusetts Institute of Technology (see especially Bergeron 1959, Byers 1960, and Phillips 1998).

2 William Emerson Ritter papers, SIO Archives, La Jolla, California. James Gillespie Blaine (1830–1893), representative from Maine and Republican senator, was secretary of state under Presidents Garfield (1881) and Harrison (1889–92). He was an unsuccessful presidential candidate in 1884.

3 On McEwen, see especially Mills 1990, 1991 and on SIO history, Raitt and Moulton 1967 and Mills 1993. The first sections of this chapter have been adapted from Mills 1990.

4 SIO Archives, George F. McEwen papers, Folders 4, 6, 44.

5 For a recent summary of Ritter's ideas and their development, see Pauly 2000, pp. 201–13.

6 Raitt and Moulton 1967, pp. 1–66; Shor 1981.

7 Ritter 1912b, p. 147.

8 The success of astronomy, and especially observatories, was highly influential throughout late nineteenth–early twentieth-century science (see Mills 2001).

9 Ritter 1905b, pp. 49–53; 1905a, pp. viii–ix.

10 Pauly 2000, pp. 207–10.

11 Ritter 1908, pp. 330–1.

12 Ritter 1912b, p. 219.

13 Ritter 1912b, p. 140; 1912a.

14 Ritter 1912b, p. 227.

15 Ritter 1916a, pp. 261–4.

16 Ritter and Bailey 1918.

17 Ritter 1912b, pp. 223–4.

18 Ritter 1916b, p. 23.
19 Ritter 1915, p. 231. On Ritter and Loeb, see Pauly 1987, pp. 110, 133, and 140.
20 Ritter 1916b, p. 24.
21 Thone and Bailey 1927, pp. 259–60.
22 Ritter 1908, pp. 332–3.
23 McEwen 1910, pp. 98–201; Thorade 1909; V.W. Ekman 1905, 1906. On Thorade, see W. Schott 1987, p. 40.
24 McEwen 1929b, p. 199.
25 McEwen 1912, pp. 265, 268–79.
26 McEwen 1915, p. 139; 1912, p. 282.
27 Ritter 1913, p. 118.
28 McEwen 1929b, p. 199.
29 McEwen 1915, 1922, 1924, 1927, 1929b,c,d, 1937a,b, 1938.
30 McEwen 1937b, 1938.
31 Zöppritz 1878a,b, 1879a,b.
32 Ekman 1905, pp. 3–40.
33 Jacobsen 1913. See discussion in McEwen 1922, pp. 79–80.
34 McEwen 1929a, pp. 259–61.
35 McEwen 1929b, p. 259.
36 The change of name took effect in 1925, but the transition, which apparently originated with Ritter, began at least two years earlier (Ritter 1923, pp. 44–5).
37 On the long-range forecasting, not described in detail here, see Mills 1990, pp. 287–92.
38 Among many biographical notices, see especially Devik 1966; Fjeldstad 1959; Friedman 1994, 2001, 2002; and Revelle and Munk 1948. Important light on his career at SIO is provided by Oreskes and Rainger 2000.
39 Sverdrup 1926, 1927.
40 Sverdrup 1931.
41 Interviews with Walter Munk and Robert S. Arthur at Scripps Institution of Oceanography, 11 and 15 June 1987.
42 McEwen, 'Seminar outline in dynamical oceanography, November to December 1929,' (redated 1928–1935), typescript, George F. McEwen papers, SIO Archives.
43 University of California, Berkeley, *General Catalog 1937–38*, Scripps Institution of Oceanography, 1.
44 Sverdrup to T.W. Vaughan, 11 April 1936 (Shor 1978, p. 11.)
45 SIO Archives, George F. McEwen Biographical File, folder 350.
46 McEwen 1921b, p. 493.

47 Vaughan et al. 1937, p. xiii.
48 Sverdrup, Johnson, and Fleming 1942, p. 1.
49 McEwen 1937c, p. 11.
50 McEwen 1921a, p. 600.
51 McEwen 1922, p. 80; 1929b, p. 261.
52 Vaughan 1926, p. 70.
53 McEwen 1929a,c.
54 McEwen 1930a; McEwen, Thompson, and Van Cleve 1930.
55 McEwen 1930b, 1932.
56 McEwen 1921a, pp. 605–6.
57 Sverdrup 1938b,c, 1939a,b.
58 Recollections by his daughter in George F. McEwen Personnel File, University of California at San Diego.
59 Vaughan correspondence in George F. McEwen Personnel File, University of California at San Diego.
60 Sumner 1945, p. 206.
61 Sargent 1979, Prologue. Roger Revelle (1909–1991) completed a Ph.D. under Vaughan in 1936, as Richard Fleming (1909–1990) had done under Erik Moberg in 1935. Revelle, who was to become one of the most significant figures in the entire history of SIO, followed a well-blazed scientific trail to work in Helland-Hansen's Geophysical Institute in Bergen during 1936–7.
62 Revelle 1978, p. 69.
63 Schlee 1980, p. 55.
64 Brosco 1985, p. 55. On Bigelow, see especially Anonymous 1955; Brosco 1985, 1989; Graham 1968; Redfield 1976; and the journal *Oceanus* 14 (2) (1968, ('H.B. Bigelow' issue). His autobiography (Bigelow 1964) provides much interesting detail.
65 Hedgpeth 1945, 1946, 1974.
66 Redfield 1976, p. 54; Brosco 1985, pp. 10–11; 1989, pp. 239, 245–57.
67 Especially Bigelow 1926, 1927a,b, 1928; and Bigelow and Welsh 1925. In addition there was a regular series of publications on the work in the *Bulletin of the Museum of Comparative Zoology*. Despite the lack of ship time, Bigelow remained an advisor to the Bureau for many years.
68 Bigelow 1927b, p. 855. See also Bigelow 1927a for his reservations about the use of the dynamic method.
69 Bigelow 1927b, p. 513. Smith also figures in Bigelow's summary of the Gulf of Maine work (1927a).
70 Bigelow and Leslie 1930, pp. 478–83.
71 Edward H. Smith to H.B. Bigelow, 16 January 1928, Harvard University Ar-

chives, HUG 4212.5, H.B. Bigelow General Correspondence, Box 5, Folder S. Critical biographical information on Smith is sparse. See Anonymous 1962, Dinsmore and Strobridge 1998, and Driggers 1980.

72 Bigelow was appointed curator of oceanography at the MCZ in 1927, the year he became associate professor of zoology, and remained in that position until his retirement in 1950.

73 Matthews 1914; Smith et al. 1937, p. 6. On Matthews's career (1873–1956), see Carruthers 1956. His report on the *Scotia* results included dynamic calculations (see Matthews 1914, pp. 11–12, 37–46).

74 Its formal name was the International Service of Ice Observation and Ice Patrol.

75 Brosco 1985, pp. 36–9.

76 Smith 1926a, p. 107.

77 Sandström 1919. See chapter 4.

78 E.H. Smith to H.B. Bigelow, 2 March 1922, Harvard University Archives, HUG 4212.7, H.B. Bigelow General Correspondence, Box 5, Ice Patrol.

79 Smith 1926a, 1927.

80 Smith 1926a, p. 109.

81 E.H. Smith to H.B. Bigelow, 30 October 1924, Harvard University Archives, HUG 4212.7, H. B. Bigelow General Correspondence, Box 5, Ice Patrol.

82 E.H. Smith to H.B. Bigelow, 5 January 1925, Harvard University Archives, HUG 4212.7, H.B. Bigelow General Correspondence, Box 5, Ice Patrol.

83 Smith 1926a.

84 E.H. Smith to Commandant, Ice Patrol, 5 August, 19 October, and 13 November 1925, Harvard University Archives, HUG 4212.5, H.B. Bigelow General Correspondence, Box 4, Folder S. Smith spent several weeks in the British Meteorological Office before returning home in the autumn of 1925, examining air pressure records in relation to iceberg drift, a subject that I do not deal with here.

85 Smith's Ph.D. thesis, dated 1930, was titled 'Arctic Ice: With Reference to Its Distribution to the North Atlantic Ocean.'

86 E.H. Smith to H.B. Bigelow, 8 May 1926, Harvard University Archives, HUG 4212.5, H.B. Bigelow General Correspondence, Box 4, Folder S. This was the area in which *Titanic* had come to grief in 1912.

87 E.H. Smith to Commandant, Ice Patrol, 28 September 1924, Harvard University Archives, HUG 4212.5, H.B. Bigelow General Correspondence, Box 4, Folder S.

88 Ricketts 1932.

89 Riis-Karstensen 1931. There was some overlap in the Labrador Sea.

90 A press release made much of the modern nature of the short-wave radio

equipment aboard, including plans to maintain contact with the United States via radio amateurs (1928 'MARION EXPEDITION Under the direction of the United States Coast Guard,' Archive, Geophysical Institute, University of Bergen, Folder 27, USA res. 1930–1940).

91 Riis-Karstensen 1936, Kiilerich 1939, on the Danish results; Smith et al. 1937, on *Marion*. Ice in the area was covered in a separate part of the *Marion* results (Smith 1931).

92 Biographical information from Mrs Mette Mosby Haugan, 10 September 2002. See also Anonymous 1942, 1965. Mosby was appointed upon the recommendation of Nansen and Smith in 1929, following a meeting of the Board of the Ice Patrol in which Smith and Bigelow were appointed to recommend the appointment of a civilian physical oceanographer (E.H. Smith to H.B. Bigelow, 23 February 1929, Harvard University Archives, HUG 4212.5, H.B. Bigelow General Correspondence, Box 6, Folder S).

93 Anonymous 1969; Barnes 1969; also information in Edward H. Smith Biographical File, WHOI Archives, Woods Hole, Massachusetts.

94 Helland-Hansen and Nansen 1909, 1926; Nansen 1913.

95 McEwen 1926b, p. vi.

96 Woods Hole Oceanographic Institution was founded in 1930, with Bigelow, who retained his positions at Harvard, as director. See especially Burstyn 1980, Revelle 1980, and Schlee 1980.

97 *Time*, 84, No. 1, 6 July 1959. See the associated article, pp. 44–54.

98 For biographical information, see Revelle 1978; Stommel 1994; and *Oceanus* 16(2) (1971) ('C. O'D. Iselin' issue), which includes a bibliography. His importance in promoting oceanography is dealt with by Weir 2001.

99 See the quotation from Iselin in 1936 at the opening of this chapter.

100 Iselin 1927, 1930; Bigelow and Iselin 1927.

101 Iselin 1929.

102 On the new ship, see Schlee 1978, especially pp. 16–60, and Schlee 1980.

103 Woods Hole Oceanographic Institution Archives, Iselin Collection A. 1926–1943.

104 Bigelow to Helland Hansen, 17 July 1929, Harvard University Archives, HUG 4212.5, H.B. Bigelow General Correspondence, Box 7, Folder H.

105 For example, Iselin 1929, figures 7, 8.

106 Mainly the contribution of Iselin's colleague Fritz Fuglister (1909–1987), according to one of my reviewers.

107 Iselin 1933b, p. 261. He described the complexity of the Gulf Stream and its definitions in 1933a.

108 For early expressions of these ideas, see Defant 1928a, pp. 628ff, and especially Defant 1929b. By analogy with the atmosphere, Defant, who

had started his career in geophysics as a meteorologist, divided the oceans vertically into an upper well-mixed layer delimited by the thermocline (the 'troposphere') and a deep, quieter region extending to the bottom (the 'stratosphere'). These distinctions were embodied later in two major monographs (Wüst 1935, Defant 1936b).

109 Iselin published little or nothing of this critique. But his notes for WHOI staff meetings from 1933 through 1936 are full of it (WHOI Archives, Iselin Collection A. 1926–1943, various folders).

110 Iselin 1936, pp. 94–7. He seems to have gotten the idea from C.-G. Rossby and meteorology; see Iselin 1939a. It was an important topic of research at WHOI, especially in the hands of R.B. Montgomery.

111 Iselin 1937, 1938b,c, 1939b, 1940.

112 A later, much-cited paper (Iselin and Fuglister 1948), using results obtained using the newly developed bathythermograph and Loran positioning to show meanders and eddies of the Gulf Stream, makes similar use of dynamic calculations, again without comment.

113 Stommel 1994, pp. 169–70.

114 Henry Stommel (1920–1992) and Iselin were colleagues for many years. Their social and scientific backgrounds were very different, as were their mathematical abilities. Stommel became one of the foremost, and probably the most insightful and creative, theoretical physical oceanographers of his generation. On his career and influence, see Luyten and Schmitz 1992, Ryan 1984, Veronis 1992, and Warren and Wunsch 1981.

115 Woods Hole Oceanographic Institution Archives, Iselin Collection A. 1926–1943, Folder 'Iselin, C. Manuscripts 1939' and Folder 6. All quotations come from these documents.

116 Especially true of the Sargasso Sea–Gulf Stream areas because of their high vertical stability.

117 The whole history of the Institution, to the time of Iselin's talk in 1939, had taken place during the Great Depression. Only the Second World War, on the apparently distant horizon in the United States in 1940, brought an end to the small-scale, day-to-day financing of oceanography.

118 At this time, Iselin was in close contact with Johan Hjort in debating the ability of physical oceanography to contribute to fisheries problems (Rozwadowski 2002, p. 117).

119 Iselin 1942, pp. 14–15.

120 Revelle 1978, p. 68.

121 At the same time, he was an expert micromanager. One of my reviewers, who knew him, aptly comments that 'Iselin's great contribution was his strong support for WHOI scientists, warmly encouraging them, interest-

ing himself in their work, finding money for them, and not pestering them.'

8. Facing the Atlantic and the Pacific: Dynamic Oceanography Re-emerges in Canada, 1930–1950

1 On Huntsman and his career, see note 46 in chapter 4.
2 Hubbard 1993, pp. 160–9, and 2006, pp. 85–7, give an account of the expeditions. In 1924 Huntsman's energies were redirected into the directorship of the Biological Board's new technological station in Halifax, while he retained the directorship at St Andrews.
3 An evocative phrase borrowed from Jennifer Hubbard's early (1993) study of the development of the marine sciences on the East Coast of Canada.
4 A.G. Huntsman, 'The *Cheticamp Expedition* of 1917 of the biological boat 'Prince,' ca. 1925, A.G. Huntsman Papers, University of Toronto Archives, UTA B78 0010 0151/14 (incomplete manuscript, 8 pages).
5 Edward E. Prince, note dated 8 August 1917, Library and Archives Canada (LAC), RG 23, Vol. 1204, File 726-2-4 [3].
6 Craigie 1916a, p. 151. E.H. Craigie (1897–1989), who took a B.A. at the University of Toronto in 1916, became the first person awarded a Ph.D. in zoology at the university and had a long career there in anatomy and neurology rather than in the marine sciences (see Gourley 2002). Mavor, described earlier in chapter 4, joined the faculty of Union College in 1913, where he had a long career teaching biology to undergraduates.
7 Vachon 1918; Craigie 1916b; Craigie and Chase 1918. Alexandre Vachon (1885–1953), a chemist, first came to St Andrews in 1916 to study the relation of the plankton (and thus the distribution of herring) to temperature, salinity, and density. He later was involved in the founding of marine stations on the St Lawrence estuary and the Gulf of St Lawrence. His relation with St Andrews ended when he became a senior administrator at Université Laval and later Archbishop of Ottawa (see Anonymous 1953 and Saint-Pierre 1994).
8 Mavor 1920a,b, 1922, 1923.
9 Mavor 1922, 1923.
10 Huntsman 1924a, 1925. The purpose of the expedition was specifically to determine the pattern of circulation north of Newfoundland through the Strait of Belle Isle and the sources of water reaching the Gulf of St Lawrence and the Scotian Shelf in relation to the cod fishery.
11 See Huntsman 1924b and 1927.
12 Hubbard 1993, pp. 329–43; 2006, pp. 173–82.

13 Huntsman 1927b.

14 Bigelow 1927a,b. It was the work leading to this monograph that led to Edward H. Smith's work with Bigelow, as described in chapter 7. The NACFI was established (as the International Commission on Marine Fisheries Investigations) in 1920, bringing together representatives from Canada, the United States, Newfoundland, and France to promote the exchange of information on fisheries and to encourage international cooperation in research and the sharing of data (NACFI 1932, pp. 3–4). For a detailed discussion, see Hubbard 1993, pp. 314–29, and 2006, pp. 163–72. Huntsman and Bigelow were long-time members and soon became friends.

15 Canada 1928, p. 79. Hubbard (1993, pp. 246–56, and 2006, pp. 132–8) discusses the background and results of the Maclean Commission.

16 On Fraser's life and career, see Arai 2004; Clemens 1947; Foerster 1948; Johnstone 1977, pp. 92–3; and Schmitt 1948.

17 Johnstone 1977, pp. 167–8; Reed 1948. Cameron became the chairman of the Biological Board of Canada in 1932, overseeing its transition into the Fisheries Research Board of Canada in 1937.

18 Fraser and Cameron 1915, pp. 133, 136; Fraser 1921, p. 38.

19 Cameron and Mounce 1922; Mounce 1922.

20 Mounce's later career was as a plant pathologist with Agriculture Canada (Estey 1994, pp. 141–3; Ginns 1988). Lucas became a noted medical researcher in the Banting Institute, Toronto, and McPhail, a pharmacologist and physiologist, had a distinguished career in Canadian universities, the pharmaceutical industry, and defence research.

21 Mills 1989, chs 8, 9.

22 Lucas and Hutchinson 1927; Hutchinson 1928, 1929; Hutchinson, Lucas, and McPhail 1929, 1930; Lucas 1929; Hutchinson and Lucas 1931.

23 Hutchinson and Lucas 1931, p. 231.

24 On Clemens's career, see Clemens 1958, 1968. Needler 1958 gives an account of PBS as Clemens found it.

25 Clemens 1958, p. 794.

26 Clemens 1958, pp. 794–5. On Carter, see Johnstone 1977, pp. 132–3, and Ricker 1979.

27 Johnstone 1977, p. 133.

28 Campbell 1985, 1989; Johnstone 1977, pp. 121–2. The most interesting account of some aspects of his career is autobiographical; see Hachey 1983. Hachey never completed a Ph.D., but he received honorary doctorates from St Thomas University in 1950 and from St Francis Xavier University in 1969.

29 While he was a graduate student at McGill, probably in 1925, Hachey heard

Huntsman speak on the Gulf of St Lawrence, although he claims to have been unimpressed (Hachey 1983).

30 Anonymous 1940; Johnstone 1977, pp. 78–84. Cox became a member of the Biological Board of Canada in 1923 (Johnstone 1977, p. 103). Harry Hachey married Katherine Cox in 1930.

31 In 1928, Bigelow was secretary of the first National Academy of Sciences Committee on Oceanography, then preparing to recommend funding for oceanography in the United States, including the founding of Woods Hole Oceanographic Institution (of which he became first director in 1930), and preparing its report, including a very influential abridgement (Bigelow 1931).

32 NACFI 1932, pp. 28–30.

33 Hachey 1929a; 1961, p. 19; NACFI 1932, p. 28.

34 Hachey 1929b.

35 Zaslow 1971, pp. 218–23, 255–6; 1988, pp. 38–9.

36 Huntsman 1931b; Hachey 1931c. Documentary material on the Hudson's Bay Fishery Expedition and an earlier investigation in the Bay is in Library and Archives Canada, RG 23, Volume 1207, 726–4–1 through 726–4–7 and 762–42–1 through 762–42–2.

37 Hachey 1931c, p. 468.

38 Huntsman 1931c, p. 462; Hachey 1931b,c, 1935c.

39 Hachey 1931d, Sandström 1919, Smith 1926.

40 Hachey 1931d, p. 115. See Mills 1989, pp. 167–8, for a discussion of the use of kite diagrams by Braarud to link stability with the beginning of the spring bloom. Gran and Braarud were in St Andrews individually or together from about 1930 through 1932. Did they learn the kite diagram technique from Hachey, or he from them? Or (it seems unlikely) were their uses not related?

41 Huntsman 1930, p. 3; 1931a,c.

42 Hachey 1932a,b, 1933b, 1934a, 1935a.

43 Hachey 1934b, p. 331. This paper appears to have sunk without trace; I have never seen a reference to it in the oceanographic literature.

44 Hachey 1933a, 1934c, 1935b, d,e, 1936a,b, 1937b,d; Leim and Hachey 1935.

45 Hachey 1937c, 1938a. The first WHOI / Biological Board collaborative cruises were during the winter of 1936. Thereafter they were apparently limited by lack of money on the Canadian side (Hachey 1937c, p. 10).

46 Hachey 1937a, 1938a,b. Helland-Hansen introduced T-S analysis in 1916, but it was not widely used until the 1930s, when Georg Wüst used the technique with data from the *Meteor* Expedition.

47 McEwen 1912; Hachey 1937b, pp. 271–4.

48 Montgomery 1938; Hachey 1947, pp. 7–11 (this paper, delayed by the war, was begun about 1939). On the very influential work of R.B. Montgomery (1910–1988), see Boicourt 1975, and Boicourt and Cannon 1988.
49 Hachey 1938b; Helland-Hansen 1934; Parr 1937.
50 Hachey 1939, pp. 344–6; Montgomery 1937; Rossby 1938. Hachey met Carl-Gustaf Rossby (1898–1959) in Woods Hole in 1937 (Hachey 1937c) and was impressed by his theoretical work On Rossby, see especially Bergeron 1959; Byers 1959, 1960; Lewis 1996; and Phillips 1998.
51 McKenzie and Hachey 1939; Hachey 1941.
52 Hachey 1942, 1947. Both papers were submitted to the *Journal of the Fisheries Research Board of Canada* in 1940. Huntsman, the editor, asked Columbus Iselin to assess them. Iselin's response was mixed: the 1942 paper was acceptable, but according to Iselin, the observations in the paper on dynamics 'hardly warrant the sort of analysis Hachey has attempted' (WHOI Archives, Columbus Iselin correspondence, letter to A.G. Huntsman, 4 Sept. 1940). Hachey seems to have revised and expanded the paper in 1946 after his war service.
53 Grier 1969 (1941); Vaughan et al. 1937. Most of this section is based directly on Mills 2002.
54 Carter 1931a,b, 1932a,b, 1933, 1934.
55 For biographical information, see Johnstone 1977, pp. 152–3; Levings 2000; Tabata 1987a,b; and J.H. Tully 1988.
56 Tully 1936d.
57 Carter and Tully 1932.
58 R.H. Herlinveaux, personal communication, 9 November 1991.
59 Tully 1933, 1937e.
60 Tully 1935.
61 Tully 1936a,b, 1937a–d,f,g, 1938a,b, 1942.
62 Tully 1937a, p. 1.
63 Marmer 1926.
64 Tully's description of the method (1936c) is a hybrid, owing its origin to these authors but especially to Sandström 1919 and Smith 1926.
65 Tully 1936b; 1937b, c.
66 Tully 1937d, p. 1.
67 The Biological Board of Canada was renamed the Fisheries Research Board of Canada in 1937.
68 Tully 1937d, p. 2.
69 Tully 1937d, p. 7.
70 Tully 1937d, p. 8.
71 Tully 1937d, p. 17. He reiterated these aims two years later (Tully 1939b).

72 Bjørn Helland-Hansen correspondence, Geophysical Institute, Bergen, Norway (now in the Staatsarchiv in Bergen). I thank Professor Odd H. Saelen for his help in allowing me to work with this correspondence.

73 Financial problems probably prevented Tully from going to Bergen. Perhaps, too, his marriage in 1938 affected his choice of a graduate school. T.G. Thompson had a close relationship with Clemens and PBS, so Seattle must have seemed an attractive alternative.

74 He completed the work for his Ph.D. in Seattle during the academic year 1946–7. On the Alberni Inlet work, see Tully 1939a, 1949.

75 Much of this section is taken from Mills 1995. Eggleston 1950, pp. 116–50, Goodspeed 1958, pp. 207–9, 211, and Zimmerman 1989, p. 96, give information on the beginning of Canadian defence research. A more general account is that of Avery 1998, although he ignores the war at sea.

76 Later called the Defence Research Establishment Atlantic (DREA) and currently DRDC (Defence Research and Development Canada) Atlantic; for details, see Goodspeed 1958, pp. 211–14, and Longard 1993. A Pacific counterpart was opened in Esquimalt, British Columbia, in 1948 (Goodspeed 1958, pp. 214ff; Chapman 1998).

77 A not unmixed blessing, according to Zimmerman 1988. See also Eggleston 1950, pp. 135–6; Goodspeed 1958, pp. 210–11; Middleton 1979, pp. 87–8, 92–3; Zimmerman 1989, pp. 94–5. The NRC's formal link with the RCN lasted until 1943, when the RCN's Research Division was established.

78 Underwater sound is greatly affected by variations in temperature and salinity, especially in the region of sharp temperature change below the surface called the thermocline. The BT was developed by Athelstan Spilhaus before the war to enable rapid measurement of vertical temperature change (Spilhaus 1938; Spilhaus and Miller 1948). The U.S. Navy rapidly began use of BTs to increase the effectiveness of their work on undersea warfare.

79 Milner 1988, p. 150; Zimmerman 1989, pp. 132–3.

80 Hachey 1983, section XVII.

81 Zimmerman 1989, p. 132. On similar work in 1943, see Hachey 1983, section XVII.

82 Middleton 1978, p. 91; Zimmerman 1989, pp. 132–3.

83 Leary 1999, pp. 9–11. The name of the laboratory was changed to U.S. Navy Electronics Laboratory (USNEL) in November 1945.

84 Tully 1951, p. 10.

85 W.M. Cameron 1948a,b, 1950; Trites 1956.

86 Hachey 1948; W.M. Cameron 1949.

87 Campbell 1958, 1971; Collin and Dunbar 1964. Little formal documentation of this work exists in open literature, although brief accounts are found

in POG and PBS reports and in contemporary newspapers during the 1950s. Collin and Dunbar used some of the data from the cruises in their survey of Arctic oceanography.

88 Doe 1950a,b, 1951, 1952, 1955; Waldie, Doe et al. 1950.

89 Dodimead 1956; Dodimead and Tully 1958; Oceanic Observations ... 1955; Tully 1956, 1958; Tully and Dodimead 1956.

90 Anonymous 1957; Campbell 1976, pp. 2159–60; Lauzier 1953; Lauzier and Trites 1959; McLellan 1957; McLellan, Lauzier, and Bailey 1953.

91 Hachey 1950, p. 30; 1961, facing p. 21.

92 Campbell 1976, p. 2160; Ford, Longard, and Banks 1952; Fuglister and Worthington 1951; Hachey 1950, p. 31; Lauzier 1950.

93 The POG retained responsibility for the Western Arctic – even though U.S. interest in Arctic oceanography was waning, cutting off the supply of ship-time to the POG (Mills 2001, p.4).

94 Dunbar 1951, 1982; Hachey 1961, p. 22; Campbell 1958, 1976, p. 2161; Mills 2001. An RCN mission to Hudson Bay in 1948 provided the first new oceanographic information from there since Hachey's visit in 1930 (Dunbar 1951, p. 19). Under M.J. Dunbar of McGill University, the Fisheries Research Board of Canada had begun biological oceanographic work in the Eastern Arctic in 1949, concentrating mainly on oceanic production and marine mammals (Dunbar 1949, 1951).

95 Mills 1995, p. 10. In 1959 the JCO was reorganized and renamed the Canadian Committee on Oceanography. Its influence in promoting and coordinating Canadian oceanography was considerable through the 1960s. Thereafter its members were increasingly minor administrators rather than deputy-ministers or senior scientists (there were some notable exceptions), and with the increase of Canadian oceanographic laboratories and personnel by the 1970s, it became largely functionless. It was disestablished as a result of a government program review in the 1980s. On early organization of the JCO and CCO, see Hachey 1965, pp. 295–305.

96 J.P. Tully, 'Oceanography in Relation to University Training,' undated typescript, Dalhousie University Archives, Presidents' Office, Institute of Oceanography 1948–1959, MS1–3, A678. There is no indication who was intended to read Tully's memo; it was probably prepared for the JCO.

97 The Scripps Institution of Oceanography in La Jolla, California, expanded dramatically after the Second World War, capitalizing at first on the scientists who had found their way into oceanographic research during the war, but who lacked formal qualifications in the field; see Shor 1978, ch. 2.

98 N.A.M. MacKenzie, memo 'Institute of Oceanography,' accompanying letter to registrar of UBC, 29 July 1949, Archives, University of British

Columbia, Senate documents, OCEANOGRAPHY – Institute of. 31–4. SENATE. Letters. 4. From the Board of Governors. (a) Approval of recommendations of Senate. (1) Re. Institute of Oceanography.

99 *Proceedings and Transactions of the Royal Society of Canada*, Third Series, 42 (1948): p. 57.

100 University of British Columbia Archives, President's Office, General Correspondence, microfilm roll 122, Minutes of Committee on Oceanography, 6 October 1949, Appendix A, Note by W.A. Clemens on background of Institute of Oceanography, UBC.

101 Scientific Archives, Pacific Biological Station, Nanaimo, BC, File No. BZ/1/7/. 1-4-1, Correspondence to and from Dr F.H. Sanders, George S. Field to J.P. Tully, 20 September 1948.

102 Dalhousie's strongest suit was in biology, not in physical oceanography, which was of greatest interest to Field. On the competition between UBC and Dalhousie, see Mills 1995.

103 Dalhousie University in Halifax had made a similar bid, but was not successful until 1959 (see Mills 1995).

104 W.M. Cameron's role, then and later, was very important. As a federal government employee, first with the Defence Research Board and later with the Department of Mines and Technical Surveys, he played a major role in the expansion of government oceanography, notably the founding of the Bedford Institute of Oceanography in the early 1960s.

9. Studying *The Oceans* and the Oceans

1 Lectures on oceanography had been inaugurated by Prince Albert in Paris in 1903.

2 To mention only strictly oceanographic laboratories and departments. A host of marine laboratories, not mentioned here, and mostly devoted to marine biology, were founded or expanded in this period. An incomplete but useful account is given by Limburg (1979).

3 The striking exceptions include, for example, the books by Leary (1999) and Weir (2001), along with a smattering of articles in periodicals.

4 See Hamblin 2005

5 On Sverdrup and his career, see especially Friedman 1994, 2001. Other useful accounts of his life and scientific approach include Devik 1966, Munk and Day 2003, Nierenberg 1996, and Revelle and Munk 1948.

6 Munk and Day 2003, pp. 180–84; Rainger 2003, pp. 465–75. Its early links had been with Berkeley, but by the time of Sverdrup's appointment, SIO was formally a branch of UCLA.

7 In a letter to Helland-Hansen, 17 September 1936 (quoted in Friedman 1994, pp. 29–30).

8 Rainger 2003, p. 475; McGowan 2004, pp. 106–7.

9 Munk 1992, pp. 156–7; Munk and Day 2003, pp. 184–5.

10 Munk 1992, p. 156.

11 Munk and Day 2003, pp. 184, 198n31.

12 Baker 1992, p. 154.

13 Sverdrup 1928, pp. 1–3. The deliberations of the first NASCO led to the foundation of Woods Hole Oceanographic Institution in 1930 and the funding of oceanography elsewhere in the United States, including SIO (see Burstyn 1980).

14 Rainger 2003, pp. 476–9.

15 Sverdrup 1939a, pp. 8–22. SIO was between ships shortly after Sverdrup arrived in La Jolla, the largely unsuitable *Scripps* having been destroyed in a fire late in 1936; it was replaced, thanks to a gift from Robert P. Scripps, by the much more seaworthy *E.W. Scripps* in December 1937.

16 Sverdrup 1940, p. 4.

17 Sverdrup 1948, p. 14.

18 Sverdrup 1928, p. 6.

19 Sverdrup 1928, pp. 6–8. Defant, by 1928 the director of the Institut für Meereskunde of the University of Berlin, was an early exponent of the transformation of oceanography into a geophysical science. Helland-Hansen, directing the work of the Geophysical Institute in Bergen and of its ship *Armauer Hansen* at sea, was an evangelist of simultaneous systematic surveys by small-to-modest-sized vessels. Neither, however, showed the same interest as Sverdrup in bringing together the physics, biology, chemistry, and geology of the seas.

20 Sverdrup 1940, p. 3.

21 Munk and Day 2003, p. 198n31.

22 Sverdrup et al. 1942, p. v.

23 A rationale for the synthesis of oceanographic knowledge was set out in the first chapter, the only part of the book that took an explicitly historical approach, making the case that a unified science of oceanography could only arise through the 'correlation' of individual pre-existing sciences (Sverdrup *et al.* 1942, pp. 1–7). The other disciplines have not been reviewed recently, but McGowan (2004) emphasizes the importance of biology to Sverdrup.

24 Sverdrup et al. 1942, p. 69; see also pp. 61, 89–93. The use of eddy coefficients in oceanography was introduced by the Dane J.P. Jacobsen in 1913 and expanded by V.W. Ekman (1927) and G.I. Taylor (1931). Sverdrup cited only the last.

25 Sverdrup et al. 1942, p. 90.

26 Sverdrup et al. 1942, p. 138. It is doubtful that Sverdrup knew of the Carpenter-Croll controversy. His speculations were based on newer information on the distribution of temperature and salinity, and partly on Nansen's (1906) mistaken belief that deep-water formation took place south of Greenland. Direct evidence of deep-water formation in the Norwegian and Greenland Seas and to a lesser degree in the Labrador Sea began to accumulate only in the 1950s and '60s (for a review, see Bacon 2002).

27 At this point in *The Oceans,* Sverdrup did not cite Deacon's work (Deacon 1937), although he used it later in the book in discussions of ocean circulation. For the significance of Deacon's work on Sverdrup and later oceanography, see Mills 2005, p. 260.

28 Sverdrup et al. 1942, pp. 141–6. Although, as Pantiulin (2002) points out, the term 'water mass' was introduced by Helland-Hansen and Nansen in 1909 in their great work 'The Norwegian Sea,' it had not been used regularly and synoptically until Sverdrup applied it in 1942.

29 Sverdrup et al. 1942, p. 146. The method originated with Wüst's (1935) analysis of the results of the *Meteor* expedition,1925–1927.

30 Sverdrup et al. 1942, p. 163.

31 Sverdrup et al. 1942, p. 394.

32 Sverdrup et al. 1942, p. 399.

33 See chapter 7.

34 Sverdrup et al. 1942, pp. 395–8. This was largely a discussion of the wind-induced Ekman spiral, as developed by V.W. Ekman (1902, 1905, and later papers). Later in his American career, Sverdrup developed this approach in a highly influential way, as discussed subsequently in this chapter.

35 Sverdrup et al. 1942, p. 393.

36 Sverdrup et al. 1942, pp. 391–2.

37 Sverdrup et al. 1942, p. 395.

38 See chapter 7.

39 Warren 1992, p. 158.

40 Sverdrup et al. 1942, p. 431.

41 Sverdrup et al. 1942, p. 431.

42 Ekman's original formulation had assumed constant density with depth.

43 Sverdrup et al., 1942, pp. 489–503. In a few sentences (p. 501), Sverdrup mentioned the likelihood that upwelling and coastal currents would be accompanied by the production of eddies – a subject that dominated oceanography beginning in the late 1950s.

44 Notably Sandström 1908, an experimental investigation of the effects of

heating and cooling in the sea. For a recent evaluation of Sandström's experiments, see Coman et al. 2006.

45 Sverdrup et al. 1942, p. 508.

46 Sverdrup et al. 1942, p. 509.

47 Sverdrup et. al. 1942, p. 510.

48 Sverdrup et al. 1942, p. 139.

49 Nor indeed in much later work. Carl Wunsch has reopened the debate in a modern framework, showing the necessity of wind energy for the formation of cooled deep waters (Wunsch 2002, Wunsch and Ferrari 2004).

50 By Columbus Iselin, quoted by Warren 1992, p. 157.

51 Warren 1992, p. 158.

52 Sverdrup called it the Antarctic Circumpolar Ocean; in 2000 the International Hydrographic Organization formally named the circumpolar ocean south of 60°S the Southern Ocean.

53 On Deacon's work in the Antarctic and the relationship between his work and *The Oceans*, see Mills 2005.

54 Sverdrup et al. 1942, p. 625.

55 Sverdrup et al. 1942, p. 742.

56 Sverdrup et al. 1942, p. 747.

57 Carruthers 1944. J.N. Carruthers (1895–1973) knew European oceanography well and was responsible in a major way for the rebuilding of that science in Europe after the Second World War (Böhnecke 1973, G.E.R. Deacon 1973, Ramster 1975).

58 G.E.R. Deacon 1945, pp. 652, 653.

59 Warren 1992, p. 158.

60 Sverdrup 1938a, pp. 5–7, expanded upon in Sverdrup 1939a, pp. 5–15. In the latter, he also, seemingly presciently, pointed out the ubiquity of eddies, 'the nature of which we have yet to explore and the dynamic consequences of which we can at present only vaguely perceive' (p. 15).

61 Sverdrup 1947, p. 319.

62 Based on the equations of motion – see Open University Course Team 2001, pp. 98–102, for an explanation of how the standard equations of motion are applied to the oceans.

63 Sverdrup 1947, p. 325.

64 For the whole subject of the effect of the Earth's rotation, wrapped up in the concept of vorticity, including the effect of the wind in Sverdrup's model, see especially Open University Course Team 2001, pp. 85–92.

65 For a modern explanation of Sverdrup's approach and it significance, see Open University Course Team 2001, pp. 90–2.

66 Walter Munk's pervading influence in geophysical oceanography is shown

in detail in Garrett and Wunsch 1984, Graham 1987, Munk 1980, Revelle 1990, and Sargent 1983.

67 Henry Stommel (1948) had already explored the startling implications of latitudinal variation in the Coriolis effect, as discussed shortly. Munk was indebted to Stommel for this insight.

68 Munk 1950, p. 83.

69 Munk 1950, p. 88; see also Munk 1955.

70 Munk 1950, p. 92.

71 His career, with many commentaries, as befits such a colourful and influential character, is outlined in Luyten and Hogg 1992, Luyten and Schmitz 1992, Ryan 1984, Sullivan 1992, Veronis 1992, Warren and Wunsch 1981, and Wunsch 1997.

72 Wunsch 1997, p. 331.

73 Warren 1981, p. 26. 'Merzian,' of course, refers to Alfred Merz, whose work with Georg Wüst in Berlin in the 1920s first established the reality of global meridional circulation (see chapter 5).

74 Montgomery 1981, pp. xxiv–xxvi; Veronis 1981, p. xix; Wunsch 1997, pp. 337–8. See also Richardson 2005 and Stommel 1954 on the background of work on the Gulf Stream based at Woods Hole.

75 Stommel 1948, p. 202.

76 Stommel 1948, p. 205.

77 Stommel 1950b, 1951; see also the very clear exposition in these terms in Open University Course Team 2001, pp. 92–7.

78 Stommel 1950a, p. 252; 1950b. See also his popular outline of the problem in 1955 and reviews for practitioners in 1957.

79 Stommel 1958c, pp. 154, 163.

80 Swallow 1955; Swallow and Worthington 1957, 1961. Stommel's development of the idea of a deep western countercurrent to the Gulf Stream spread widely earlier, beginning late in 1954, according to John Swallow (see Luyten and Hogg 1992, p. 22), but was not mentioned formally until the 1957 meeting. It is developed in a theoretical framework in his paper 'A Survey of Ocean Current Theory' (Stommel 1957b) and presented explicitly in Stommel 1958a.

81 One of the investigators, Val Worthington, commented of this result, 'I was dumfounded' (McCartney et al. 1982, p. xv).

82 Stommel 1958a,b.

83 Woods Hole Oceanographic Institution 2001. Arnold Arons entered the story in part because of one important driving force at WHOI in the 1950s, the U.S. government's interest in using the deep sea to dispose of nuclear wastes – and their willingness to fund research on the deep sea,

including Stommel's. This is not made clear in the sparse biographical material available on Arons.

84 Stommel, Arons, and Faller 1958.

85 Robinson and Stommel 1959, p. 307.

86 Stommel and Arons 1960a. Their approach was in part to explore the implications of having 'sources,' that is, the locations of deep water formation, in a variety of places, not just at high latitudes, and the effects of various kinds of marginal barriers.

87 Using information on Antarctic Deep Water and North Atlantic Deep Water provided by Wüst 1938 and 1951.

88 Stommel and Arons 1960b, pp. 224–5. For an account of more recent thermocline theory, see Pedlosky 2006.

89 Stommel and Arons 1960b, pp. 222–5.

90 Bolin and Stommel 1961, Arons and Stommel 1967, Stommel and Arons 1972.

91 Robinson and Stommel 1959, p. 295, footnote 3. Stommel returned to Carpenter and Croll in 1980 and 1987, further emphasizing the futility of the debate.

92 My choice would be the Open University Course Team's *Ocean Circulation*, second edition, published in 2001, because of its clarity, comprehensiveness, and an uncharacteristic, although modest, attention to the history of the field.

93 Stommel 1989 (also in Luyten and Hogg 1992, pp. 51–63).

94 He may have been referring here primarily to his friend Carl-Gustav Rossby (1898–1957) of MIT rather than to Vilhelm Bjerknes and his school.

95 Stommel 1989, p. 49.

96 Richardson 2005, pp. 4–7; Open University Course Team 2001, pp. 71–4.

97 Stommel 1989, p. 49. During the winter of 1969, Stommel was involved in the study of deep-water convection forming Mediterranean deep water during an expedition called MEDOC 69.

98 Eddies entering the Atlantic from the Mediterranean and propagating at mid-depths as rotating lenses of water of 40–150 km in diameter into the Central Atlantic and even into the Western Atlantic (Open University Course Team 2001, pp. 223–4).

99 Stommel 1989, p. 49.

100 Stommel 1989, p. 51.

REFERENCES

Adams, J. 1995. Historical studies in the Faroe-Shetland Channel: The Scottish perspective. *Ocean Challenge* 6(1): 14–17.

Adhémar, J.A. 1842. *Révolutions de la mer.* Paris: Privately published.

Aitken, H.J. 1985. *Syntony and spark: The origins of radio technology.* New York: John Wiley. xvi + 347pp.

Albert Ier, Prince de Monaco. 1892a. Sur une nouvelle carte des courants de l'Atlantique Nord. *Comptes rendus hebdomadaires des séances de l'Académie des sciences* 114(6): 264–8.

– 1892b. A new chart of the currents of the North Atlantic. *Proceedings of the Royal Geographical Society* (New Series), 14(9): 619–22.

– 1892c. A new chart of the currents of the North Atlantic. *Scottish Geographical Magazine* 8(10): 528–31.

– 1896. Voyages scientifiques du yacht *Princesse Alice. Report of the Sixth International Geographical Congress held in London, 1895*, pp. 437–41.

– 1898. Oceanography of the North Atlantic. *Geographical Journal* 12(5): 445–69.

– 1904. Les progrès de l'océanographie. *Monaco. Bulletin du Musée océanographique*, 1(6), 13pp.

– 1910. Sur la douzième campagne scientifique de la *Princesse-Alice. Comptes rendus hebdomadaires des séances de l'Académie des sciences* 150(22): 1396–7.

– 1932a (1904). Les progrès de l'océanographie: Recueil de travaux publiés sur ses campagnes scientifiques. *Résultats des campagnes scientifiques, accomplies sur son yacht par Albert Ier, Prince souverain de Monaco*, 84: 207–15.

– 1932b (1912). Les progrès de l'océanographie: Recueil de travaux publiés sur ses campagnes scientifiques. *Résultats des campagnes scientifiques, accomplies sur son yacht par Albert Ier, Prince souverain de Monaco*, 84: 323–36.

– 1932c. Recueil des travaux publiés sur ses campagnes scientifiques. *Résultats* *des campagnes scientifiques, accomplies sur son yacht par Albert Ier, Prince souverain de Monaco*, 84: 369pp.

- 1966. *La carrière d'un navigateur.* 6th edn. Monaco: Éditions des Archives du Palais princier. xxii + 239pp.
- 1998. *Des œuvres de science, de lumière et de paix.* Ed. J. Carpine-Lancre. Monaco: Palais de S.A.S. le Prince. 205pp.

Aleem, A.A 1967. Concepts of currents, tides and winds among medieval Arab geographers in the Indian Ocean. *Deep-Sea Research* 14: 459–63.

- 1968. Ahmad Ibn Majid. Arab navigator of the XVth century and his contributions to marine sciences. *Monaco. Bulletin de l'Institut océanographique, No. spécial 2 (Premier Congrès international d'histoire de l'océanographie) 2*: 565–80.
- 1980. On the history of Arab navigation. In M. Sears and D. Merriman, eds, *Oceanography: The past*, pp. 582–95. New York: Springer-Verlag.

Allen, P. 1992. *The atlas of atlases: The map maker's vision of the world: Atlases from the Cadbury Collection, Birmingham Central Library.* New York: Harry N. Abrams. 160pp

Anderson, F. 1984. The demise of the Fisheries Research Board of Canada: A case study of Canadian research policy. *Scientia Canadensis* 8: 151–6.

Andersson, K.A. 1949. Johan Hjort, 1869–1948. *Journal du Conseil International pour l'Exploration de la Mer* 16: 3–8.

Andrewes, W.J.H., ed. 1996. *The quest for longitude.* Collection of Historical Instruments. Cambridge, MA: Harvard University. 437pp.

Anonymous. 1871. The general oceanic circulation. *Nature* 4: 97–8.

- 1885. Karl Zöppritz: Nachruf. *Meteorologische Zeitschrift* 2: 277–8.
- 1895. Obituary, Admiral Jansen. *Geographical Journal* 2(5): 465–8.
- 1906. Das Institut und Museum für Meereskunde an der Universität Berlin. *Zeitschrift der Gesellschaft für Erdkunde zu Berlin* 1906(5): 1–11.
- 1910. *Discours prononcés à l'occasion des fêtes d'inauguration du Musée océanographique de Monaco, 29 mars 1910 – 1er avril 1910* . Monaco. 87pp.
- 1911. Berget (Alphonse). Paris. *Dictionnaire biographique* 1: 228–9.
- 1912. Prince, Edward Ernest. In H.J. Morgan, ed., *The Canadian men and women of the times: A handbook of Canadian biography of living characters*, pp. 918–19. Toronto: William Briggs.
- 1914. Who's who in the fishing world. *The Canadian Fisherman* 1 (June 1914): 165.
- ca. 1920. *Notes biographiques sur S.A.S. le Prince Albert Ier de Monaco.* Monaco. 29pp.
- 1923a. Le Prince Albert de Monaco, 1848–1922. *Conseil international de recherches. Union géodésique et géophysique internationale, Section d'océanographie physique, Bulletin*, No. 3, pp. 5–13.

- 1923b. *Festskrift Tillägnad Professor Otto Pettersson den 12 Februari 1923. Sven Otto Pettersson en Havets Arbetare.* Helsingfors: Holger Schildts Tryckeri. 114pp.
- 1940. Professor Philip Cox. *Journal of the Fisheries Research Board of Canada* 5: i.
- 1941. Otto Pettersson. 12 February 1848 – 17 January 1941. *International Hydrographic Bulletin (Bulletin hydrographique international)* 9–10: 51.
- 1942a. *Festskrift Tillägnad Vilhelm Bjerknes den 14 Mars 1942.* Stockholm: Centraltryckeriet. xvi + 146pp.
- 1942b. Mosby, Olav. In *Studentene fra 1915. Biografiske Opplysninger Samlet til 25–årsjubileet 1940.* Oslo: J. Chr. Gundersen.
- 1943. J.J. Cowie. *Fisheries Research Board of Canada, Progress Reports (Pacific),* No. 54: 3.
- 1951. Vilhelm F.K. Bjerknes, For. Mem. R.S. *The Meteorological Magazine* 80 (949): 181–4.
- 1953. Comité de perfectionnement de l'Institut océanographique. Décès de Monseigneur Alexandre Vachon, member du Comité. *Les amis du Musée océanographique de Monaco,* No. 27 (July), p. 16.
- 1955. Foreword. In *Papers in Marine Biology and Oceanography. Dedicated to Henry Bryant Bigelow by his former students and associates on the occasion of the twenty-fifth anniversary of the founding of the Woods Hole Oceanographic Institution 1955,* pp. xi–xvii. Deep-Sea Research, Supplement to Volume 3.
- 1957. Oceanography. In Fisheries Research Board of Canada, *Annual Report of the Fisheries Research Board of Canada 1955,* pp. 115–26.
- 1962. Rear-Admiral Edward H. Smith, USCG. *The Polar Record* 11: 345.
- 1965. *Studentene fra 1915. Biografiske Opplysninger Samlet til 50–årsjubileet 1965.* Oslo.
- 1969. Capt. Floyd M. Soule, USCGR (Ret.) (1901–1968). *U.S. Coast Guard, Oceanographic Report* 19 (CG373–19): ii–iii.
Arago, D.J.F. 1838. Instructions concernant la météorologie et la physique du globe, par M. Arago. Courants sous-marins. *Comptes rendus hebdomadaires de l'Académie des sciences de Paris,* 1837, 7: 212.
Arai, M.N. 2004. Charles McLean Fraser (1872–1946) – his contributions to hydroid research and to the development of fisheries biology and academia in British Columbia. *Hydrobiologia* 530/531(1): 3–11.
Aristotle. 1937a. *Problems I: Books I-XXI, with an English translation by W.S. Hett, M.A.* London: William Heinemann. x + 461pp.
- 1937b. *Problems II: Books XXII–XXXVIII, with an English translation by W.S. Hett, M.A.,* with *Rhetorica ad Alexandrium, with an English translation by H. Rackham, M.A.* London: William Heinemann. vi + 456pp.
- 1952. *Meteorologica, with an English translation by H.D.P. Lee.* London: Heinemann. xxx + 433pp.

Arons, A., and H. Stommel. 1967. On the abyssal circulation of the world ocean – III. An advective-lateral mixing model of the distribution of a tracer property in an ocean basin. *Deep-Sea Research* 14: 441–57.

Attlmayr, F., ed.. 1883. *Handbuch der Oceanographie und maritimen Meteorologie. Im Auftrage des K.K. Reichs-Kriegs-Ministeriums (Marine-Section)*. Wien: Kaiserlich-Königlichen Hof-und Staatsdruckerei. 2 Bande.

Austin, J., and A. McConnell. 1980. *The construction of a thermometer. By James Six. Prefaced by an account of his life and works and the use of his thermometer over two hundred years by Jill Austin and Anita McConnell*. London: Nimbus. 28pp. + xiv + 64 pp + xxiv (Appendix).

Avery, D.H. 1998. *The science of war: Canadian scientists and Allied military technology during the Second World War*. Toronto: University of Toronto Press. xv + 406pp.

Bacon, S. 2002. The dense overflows from the Nordic Seas into the deep North Atlantic. *ICES Marine Science Symposia* 215: 148–55.

Baird, D., R.I.G. Hughes, and A. Nordmann, eds.. 1998. *Heinrich Hertz: Classical physicist, modern philosopher*. Boston Studies in the Philosophy of Science 198. Boston/Dordrecht: Kluwer. xii + 318pp.

Baker, D.J. 1992. *The Oceans* 50th anniversary. *Oceanography* 5(3): 154–5.

Ball, R. 1891. *The cause of an ice age*. New York: D. Appleton and Co. xi + 180pp.

Barlaup, A., ed. 1966. *Det Norske Meteorologiske Institutt, 1866–1966*. Oslo: Fabritius.

Barnes, C.A. 1969. Obituary.– Floyd Melville Soule. *Deep-Sea Research* 16(5): 399–403.

Beckinsale, R.P. 1974. Penck, Albrecht. *Dictionary of Scientific Biography* 10: 501–6.

– 1975. Richthofen, Ferdinand von. *Dictionary of Scientific Biography* 11: 438–41.

Beechey, F.W. 1974 (1851). Hydrography. In: D.M. Knight, ed., *Admiralty manual of scientific inquiry*, pp. 52–107. London: Dawson.

Bergeron, T. 1959. The young Carl-Gustaf Rossby. In B. Bolin, ed., *The atmosphere and the sea in motion: Scientific contributions to the Rossby memorial volume*, pp. 51–5. New York: Rockefeller Institute Press.

Bergeron, T., O. Devik, and C.L. Godske. 1962. Vilhelm Bjerknes, March 14, 1862 – April 9, 1951. *Geofysiske Publikasjoner* 24: 7–25.

Berget, A.. 1930. Leçons d'océanographie physique professées à l'Institut océanographique de Paris. 1ère partie. Généralités. Océanographie physique. *Paris. Annales de l'Institut océanographique, nouvelle série*, 9, 352pp.

– 1931. Leçons d'océanographie physique professées à l'Institut océanographiqe de Paris. 2e partie. L'océan et l'atmosphère. *Paris. Annales de l'Institut océanographique, nouvelle série*, 11, 396pp.

Berghaus, H. 1845. *Physikalischer Atlas.* Erster Band. Gotha: Verlag J. Perthes. 234pp.

Biermann, K.-R. 1972. Humboldt, Friedrich Wilhelm Heinrich Alexander von. *Dictionary of Scientific Biography* 6: 549–55.

Bigelow, H.B. 1926. Plankton of the offshore waters of the Gulf of Maine. *Bulletin of the United States Bureau of Fisheries* 40, Part II (Document No. 968): 1–509.

– 1927a. Dynamic oceanography of the Gulf of Maine. *Transactions of the American Geophysical Union* 8: 206–11.

– 1927b. Physical oceanography of the Gulf of Maine. *Bulletin of the United States Bureau of Fisheries* 40, Part II (Document No. 969): 511–1027.

– 1928. Exploration of the waters of the Gulf of Maine. *Geographical Review* 18(2): 232–60.

– 1931. *Oceanography: Its scope, problems, and economic importance.* Boston: Houghton Mifflin. v + 262pp.

– 1964. *Memories of a long and active life.* Cambridge, MA: Cosmos Club.

Bigelow, H.B., and C. O'D. Iselin. 1927. Oceanographic reconnaissance of the northern sector of the Labrador Current. *Science* 65: 551–2.

Bigelow, H.B., and M. Leslie. 1930. Reconnaissance of the waters and plankton of Monterey Bay, July 1928. *Bulletin of the Museum of Comparative Zoology* 70(5): 430–581.

Bigelow, H.B., and W.W. Welsh. 1925. Fishes of the Gulf of Maine. *Bulletin of the United States Bureau of Fisheries* 40(1): 1–567.

Biswas, A.K. 1970a. Edmond Halley, F.R.S., hydrologist extraordinary. *Notes and Records of the Royal Society of London* 25: 47–57.

– 1972b. *History of hydrology.* Amsterdam: North Holland. xii + 336pp.

Bjerkan, P. 1919. Results of the hydrographical observations made by Dr. Johan Hjort in the Canadian Atlantic waters during the year 1915. In Canada, Department of the Naval Service, *Canadian Fisheries Expedition, 1914–1915: Investigations in the Gulf of St. Lawrence and Atlantic waters of Canada under the direction of Dr. Johan Hjort, leader of the expedition, director of fisheries for Norway,* pp. 349–403. Ottawa: King's Printer.

Bjerknes, V.F.K. 1898a. Über die Bildung von Cirkulationsbewegungen und Wirbeln in reibungslosen Flüssigkeiten. *Videnskabsselskabets Skrifter I. Mathematisk – naturvidenskabelig Klasse,* No. 5, 29pp.

– 1898b. Über einen hydrodynamischen Fundamentalsatz und seine Anwendung besonders auf die Mechanik der Atmosphäre und des Weltmeeres. *Kongliga Svenska Vetenskaps-Akademiens Handlingar* 31(4), 35pp.

– 1900a. Das dynamische Princip der Cirkulationsbewegungen in der Atmosphäre. *Meteorologische Zeitschrift* 17: 97–106.

– 1900b. *Vorlesungen über hydrodynamische Fernkräfte nach C.A. Bjerknes's Theorie.* Leipzig: J.A. Barth. xvi + 388pp.

– 1901. Cirkulation relativ zu der Erde. *Öfversigt af Kongliga Vetenskaps-Akademiens Förhandlingar* 58(10): 739–57.

– 1902. *Vorlesungen über hydrodynamische Fernkräfte nach C.A. Bjerknes's Theorie.* Leipzig: J.A. Barth. xvi + 316pp.

– 1906. *Die isobaren Flächen der Atmosphäre und des Meeres dargestellt nach den Veröffentlichungen der internationalen Commission für wissenschaftliche Luftschiffahrt und den Bulletin du Conseil Permanent International pour l'Exploration de la Mer.* Stockholm: P. Palmquist's Aktiebolag. 9pp.

– 1921. Fluides baroclines. In P. Apell, *Traité de méchanique rationelle,* vol. III: *Équilibre et mouvement des milieux continus,* Ch. 37, pp. 562–96. Paris: Gauthier-Villars.

– 1925. *C.A. Bjerknes: Hans liv og arbeide: Traek av norsk kulturhistorie i det nittende aarhundre.* Oslo. 240pp.

– 1933. *C.A. Bjerknes. Sein Leben und seine Arbeit.* Aus dem Norwegischen ins Deutsche übertragen von Else Wegener-Köppen. Berlin. 218pp.

Bjerknes, V.F.K, J. Bjerknes, H. Solberg, and T. Bergeron. 1933. *Physikalische Hydrodynamik mit Anwendung auf die dynamische Meteorologie.* Berlin: Julius Springer. xvii + 797pp.

Bjerknes, V.F.K., and J.W. Sandström. 1901. Über die Darstellung des hydrographischen Beobachtungsmateriales durch Schnitte, die als Grundlage die theoretischen Diskussion der Meerescirkulationen und ihrer Ursachen dienen können. *Göteborgs Kongliga Vetenskaps – och Vitterhetssammhälles Handlingar* 4:e Földjen, Häfte 3, 19pp.

– 1903. Ueber Drachen für meteorologische Untersuchungen bei hydrographischen Expeditionen. *Svenska Hydrografisk-biologiska Kommissionens Skrifter,* 5pp.

– 1910. *Dynamic meteorology and hydrography. Part 1. Statics.* Publication no. 88. Washington, DC: Carnegie Institution of Washington. 234pp.

Blindheim, J. 1995. Historical studies in the Faeroe-Shetland Channel: The Norwegian perspective. *Ocean Challenge* 6(1): 4–8.

Böhnecke, G. 1973. In memoriam: Dr. J.N. Carruthers. *Deutsche Hydrographische Zeitschrift* 26(3): 119.

– 1976. In memoriam Albert Defant, 1884–1974. *Meteor Forschungsergebnisse* A(18): 1–8.

Boguslawski, G. von. 1884. *Handbuch der Oceanographie. Band I. Räumliche, physikalische und chemische Beschaffenheit der Ozeane.* Stuttgart: Verlag von J. Engelhorn. xviii + 400 pp.

Boguslawski, G. von, and O. Krümmel. 1887. *Handbuch der Ozeanographie. Band*

II. Die Bewegungsformen des Meeres, von Dr Otto Krümmel. Mit einem Beitrage von Prof. Dr K. Zöppritz. Stuttgart: J. Engelhorn. xvi + 592pp.

Boicourt, W.C. 1975. Raymond Braislin Montgomery. *Journal of Marine Research* 33, Supplement, III-V.

Boicourt, W.C., and G.A. Cannon. 1988. Raymond B. Montgomery (1910–1988). *Oceanography* 1(2): 59.

Bolin, B., and H. Stommel. 1961. On the abyssal circulation of the world ocean – IV. Origin and rate of circulation of deep ocean water as determined with the aid of tracers. *Deep-Sea Research* 8: 95–110.

Bonney, T.G. 1901. Prestwich, Sir Joseph. *Dictionary of National Biography, Supplement*, III: 284–7.

Bouquet de la Grye, A. 1882. Recherches sur la chloruration de l'eau de mer. *Annales de chemie et de physique, 5ᵉ série*, 25: 433–77.

Bourne, W. 1578. *A booke called the treasure for traveillers*. London: Thomas Woodcocke. Variably paged.

Bowen, M.J. 1970. Mind and nature: The physical geography of Alexander von Humboldt. *Scottish Geographical Magazine* 86: 222–33.

Bravo, M. 1993. James Rennell: Antiquarian of ocean currents. *Ocean Challenge* 4(1–2): 41–50.

Brennecke, W. 1909. Ozeanographie. *Forschungsreise SMS "Planet" 1906/07*, III, x + 154 pp.

– 1911a. Ozeanographischen Arbeiten der Deutschen Antarktischen Expedition. II. Bericht. *Annalen der Hydrographie und maritimen Meteorologie* 39: 464–71.

– 1911b. Ozeanographischen Arbeiten der Deutschen Antarktischen Expedition. III. Bericht. *Annalen der Hydrographie und maritimen Meteorologie* 39: 642–7.

– 1913. Ozeanographischen Arbeiten der Deutschen Antarktischen Expedition. V. Bericht. *Annalen der Hydrographie und maritimen Meteorologie* 41: 134–44.

– 1914. Deutsche Antarktische Expedition. Die ozeanographische Arbeiten im Weddell-Meer. *Zeitschrift der Gesellschaft für Erdkunde zu Berlin* 1914: 118–29.

– 1915. Aufgaben und Probleme der Ozeanographie. *Annalen der Hydrographie und maritimen Meteorologie* 33: 49–62.

– 1921. Die ozeanographischen Arbeiten der Deutschen Antarktischen Expedition 1911–1912. *Archiv der deutschen Seewarte* 39(1), vi + 216 S.

Brennecke, W., and G. Schott. 1922. 'Die atlantische Vertikalzirkulation:' Eine Entgegnung auf die Abhandlung von A. Merz und G. Wüst. *Zeitschrift der Gesellschaft für Erdkunde zu Berlin* 7–8: 277–88,

Broch, H. 1954. Sars, Georg Ossian, 1837–1927. *Norsk Biografisk Leksikon* 12: 260–4.

Brosco, J.P. 1985. Henry Bryant Bigelow and transformation of American oceanography. B.A. thesis, Department of History and Sociology of Science, University of Pennsylvania. 80pp.

– 1989. Henry Bryant Bigelow, the U.S. Bureau of Fisheries, and intensive area study. *Social Studies of Science* 19: 239–64.

Brosse, J. 1983. *Great voyages of discovery: Circumnavigators and scientists, 1764–1843*. New York: Facts on File. 232pp.

Brown, G.I. 1999. *Count Rumford: The extraordinary life of a scientific genius*. London: Sutton. 182pp.

Brown, R.C., and R. Cook. 1974. *Canada, 1896–1921: A nation transformed*. Toronto: McClelland and Stewart. xiv + 412pp.

Brown, S.C. 1967. *Benjamin Thompson – Count Rumford: Count Rumford on the nature of heat*. New York: Pergamon Press. 205pp.

– 1979. *Benjamin Thompson, Count Rumford*. Cambridge, MA: MIT Press. 361pp.

Buchan, A. 1895. Report on oceanic circulation, based on the observations made on board H.M.S. *Challenger*, and other observations. *Report on the scientific results of the voyage of H.M.S. Challenger during the years 1873–76. Summary, Part 2, Appendix (physics and chemistry,* Part viii), vii +38pp.

Buchanan, J.Y. 1874. Note on the vertical distribution of temperature in the ocean. *Proceedings of the Royal Society of London* 23: 123–7.

– 1884. Report on the specific gravity of samples of ocean water, observed on board H.M.S. *Challenger* during the years 1873–76. *Report on the scientific results of the voyage of H.M.S. Challenger during the years 1873–76. Physics and chemistry* 1(2), 46pp.

– 1885. *Report on the scientific results of the voyage of H.M.S. Challenger during the years 1873–76. Narrative,* 1(2), Ch. 22, pp. 948–1003.

– 1896. Specific gravities and oceanic circulation. *Transactions of the Royal Society of Edinburgh* 38(2): 317–42.

– 1919. *Accounts rendered of work done and things seen*. Cambridge: Cambridge University Press. lvii + 435pp.

Buchwald, J.Z. 1985. *From Maxwell to microphysics: Aspects of electromagnetic theory in the last quarter of the nineteenth century*. Chicago: University of Chicago Press. xv + 339pp.

– 1990. The background to Heinrich Hertz's experiments in electrodynamics. In T.H. Levere and W.R. Shea, eds, *Nature, experiment, and the sciences*, pp. 275–306. Boston/Dordrecht: Kluwer.

– 1992. The training of German research physicist Heinrich Hertz. In M.J. Nye et al., eds, *The invention of physical science*, pp. 119–45. Boston/Dordrecht: Kluwer.

Buff, H. 1850. *Zur Physik der Erde*. Braunschweig: Friedrich Vierweg und Sohn. 251pp.

– 1851. *Familiar letters on the physics of the earth, treating of the chief movements of the land, the waters, and the air, and the forces that give rise to them.* Edited by A.W. Hoffman. London: Taylor, Walton, and Moberly. xiii + 273pp.

Bullen, K.E. 1973. Lamb, Horace. *Dictionary of Scientific Biography* 7: 594–5.

Bullough, V.L. 1976. Varenius, Bernardus (Bernhard Varen). *Dictionary of Scientific Biography* 13: 583–4.

Burstyn, H.L. 1965. The deflecting force of the earth's rotation from Galileo to Newton. *Annals of Science* 21: 47–80.

– 1966a. The deflecting force and Coriolis. *Bulletin of the American Meteorological Society* 47: 890–1.

– 1966b. Early explanations of the role of the earth's rotation in the circulation of the atmosphere and ocean. *Isis* 57: 167–87.

– 1971a. Croll, James. *Dictionary of Scientific Biography* 3: 470–1.

– 1971b. Ferrel, William. *Dictionary of Scientific Biography* 4: 590–3.

– 1971c. Theories of winds and ocean currents from the discoveries to the end of the seventeenth century. *Terrae Incognitae* 3: 7–31.

– 1980. Reviving American oceanography: Frank Lillie, Wickliffe Rose, and the founding of Woods Hole Oceanographic Institution. In M. Sears and D. Merriman, eds, *Oceanography: The past*, pp. 57–66. New York: Springer-Verlag.

– 1984. William Ferrel and American science in the centennial years. In E. Mendelsohn, ed., *Transformation and tradition in the sciences: Essays in honor of I. Bernard Cohen*, pp. 337–51. Cambridge: Cambridge University Press.

Byers, H.R. 1959. Carl-Gustaf Rossby, the organizer. In B. Bolin, ed., *The atmosphere and the sea in motion: Scientific contributions to the Rossby Memorial Volume*, pp. 56–9. New York: Rockefeller University.

– 1960. Carl-Gustaf Arvid Rossby. December 28, 1898 – August 19, 1957. *National Academy of Sciences of the U.S.A., Biographical Memoirs* 34: 249–70.

Byrne, D.J. 1916. Practical problems in the fish business. In Canada, Commission of Conservation, *Conservation of fish, birds and game*, pp. 10–27. Toronto: Methodist Book and Publishing House.

Calman, W.T. 1927. Georg Ossian Sars. Obituary. *Proceedings of the Linnean Society of London*, 139th Session, pp. 98–100.

Camerini, J.R. 1993a. Evolution, biogeography, and maps: An early history of Wallace's Line. *Isis* 84: 700–27.

– 1993b. The *Physical Atlas* of Heinrich Berghaus: Distribution maps as scientific knowledge. In R.G. Mazzolini, ed., *Non-verbal communication in science prior to 1900*, pp. 479–512. Firenze: Olschi.

Cameron, A.T., and I. Mounce. 1922. Some physical and chemical factors influencing the distribution of marine flora and fauna in the Strait of Georgia and adjacent waters. *Contributions to Canadian Biology and Fisheries (New Series)* 1(4): 39–72.

Cameron, W.M. 1948a. Fresh water in Chatham Sound. *Fisheries Research Board of Canada. Progress Reports of Pacific Coast Stations* 76: 72–5.

– 1948b. Oceanography of Chatham Sound. *Fisheries Research Board of Canada. Annual Report of the Pacific Biological Station for 1948,* Appendix No. 99, p. 103.

– 1949. Submarine target studies. Appendix No. 5. Project Nodales (1). *Fisheries Research Board of Canada. Annual Report, Pacific Oceanographic Group,* p. 13.

– 1950. Oceanography of Chatham Sound. *Fisheries Research Board of Canada. Annual Report for 1950 of the Pacific Biological Station,* Appendix No. 117, p. 147.

Camões, Luis Vaz de. 1997. *The Lusíads.* Translated with an Introduction and Notes by Landeg White. Oxford and New York: Oxford University Press. xxvi + 258pp.

Campbell, N.J. 1958. Recent oceanographic activities of the Atlantic Oceanographic Group in the Eastern Arctic. Atlantic Oceanographic Group, Biological Station, St Andrews, N.B. *Fisheries Research Board of Canada. Progress Reports of the Atlantic Coast Stations,* No. 69, pp. 18–21.

– 1976. An historical sketch of physical oceanography in Canada. *Journal of the Fisheries Research Board of Canada* 33: 2155–67.

– 1985. Henry Benedict Hachey, 1901–1985. *Canadian Journal of Fisheries and Aquatic Sciences* 42: 1850–1.

– 1989. Henry Benedict Hachey, 1901–1985. *Transactions of the Royal Society of Canada,* Series V, 4: 359–63.

Canada. Royal Commission Investigating the Fisheries of the Maritime Provinces and the Magdalen Islands. 1928. *Report of the Royal Commission Investigating the Fisheries of the Maritime Provinces and the Magdalen Islands.* Ottawa: King's Printer. 125pp.

Cannon, S.F. 1978. *Science in culture: The early Victorian period.* New York: Science History Publications. xii + 296pp.

Cantor, G.N., and M.J.S. Hodge, eds. 1981. *Conceptions of ether: Studies in the history of ether theories, 1740–1900.* Cambridge: Cambridge University Press. x + 351pp.

Carpenter, J.E. 1888. William Benjamin Carpenter. A memorial sketch. In W.B. Carpenter, *Nature and man: Essays scientific and philosophical,* pp. 3–152. London: Kegan Paul, Trench and Co.

Carpenter, W.B. 1850. On the mutual relations of the vital and physical forces. *Philosophical Transactions of the Royal Society of London* 140: 727–57.

– 1868. Preliminary report of dredging operations in the seas to the north of the British islands, carried on in H.M.S. *Lightning,* by Dr. Carpenter and Dr. Wyville Thomson, Professor of Natural History in Queen's College, Belfast. *Proceedings of the Royal Society of London* 17: 168–200.

- 1869. On the temperature and animal life of the deep sea. *Proceedings of the Royal Institution* 5: 503–10.
- 1870a. On the temperature and animal life of the deep sea. *Nature* 1: 488–90, 540–2, 563–6.
- 1870b. On the temperature and animal life of the deep sea. *Proceedings of the Royal Institution* 6: 63–82.
- 1871a. The Gibraltar Current. *Nature* 4: 468.
- 1871b. Oceanic circulation. *Nature* 5: 59–60.
- 1871c. On the Gibraltar Current, the Gulf Stream, and the general oceanic circulation. *Proceedings of the Royal Geographical Society* 15: 54–88.
- [Anonymous]. 1872a. Oceanic circulation. *Edinburgh Review* 135: 430–71.
- 1872b. Oceanic circulation. *Nature* 6: 473.
- 1872c. On the thermo-dynamics of the general oceanic circulation. *Report of the British Association for the Advancement of Science, Edinburgh 1871*, Transactions, p. 51.
- 1872d. Report of scientific researches carried on during the months of August, September, and October, 1871, in H.M. Surveying-ship *Shearwater*. *Proceedings of the Royal Society of London* 20: 535–644.
- 1873. Address of William B. Carpenter, M.D., LL.D., F.R.S., President [Man the interpreter of nature]. *Report of the British Association for the Advancement of Science*, Brighton 1872, lxix–lxxxiv. (Reprinted in J.E. Carpenter 1888.)
- 1874a. Further inquiries on oceanic circulation. *Proceedings of the Royal Geographical Society* 18: 301–407.
- 1874b. Lenz's doctrine of ocean circulation. *Nature* 10: 170–1.
- 1874c. Ocean circulation. *Nature* 10: 62.
- 1874d. Ocean currents. *Nature* 9: 423–4.
- 1874e. On the physical cause of ocean currents. *London, Edinburgh and Dublin Philosophical Magazine and Journal of Science*, Series 4, 47: 359–62.
- 1875a. Ocean circulation. *Nature* 12: 454–5.
- 1875b. Ocean circulation. *Nature* 12: 533.
- 1875c. Ocean-circulation: Researches in the 'Challenger' and 'Tuscarora.' *Contemporary Review* 26: 565–89.
- 1875d. Remarks on Mr. Croll's 'crucial-test' argument. *London, Edinburgh and Dublin Philosophical Magazine and Journal of Science*, Series 4, 50: 402–4.
- 1875e. Summary of recent observations on ocean temperature made in H.M.S. *Challenger* and U.S.S. *Tuscarora*; with their bearing on the doctrine of a general oceanic circulation sustained by difference of temperature. *Proceedings of the Royal Geographical Society* 19: 493–514.
- 1876a. Oceanic circulation. *The Athenaeum*, No. 2533, pp. 666–7.
- 1876b. Report on the physical investigations carried on by P. Herbert Carpen-

ter, B.A., in H.M.S. 'Valorous' during her return voyage from Disco Island in August 1875. *Proceedings of the Royal Society of London* 25: 230–7.

– 1880. The deep sea and its contents. *The Nineteenth Century* 7: 593–618.

Carpenter, W.B., and J.G. Jeffreys. 1870. Report on deep-sea researches carried on during the months of July, August, and September 1870 in H.M. Surveying-ship 'Porcupine.' *Proceedings of the Royal Society of London* 19: 146–221.

– 1871. Report on deep-sea researches carried on during the months of July, August, and September, 1870, in H.M. Surveying-ship 'Porcupine.' *Nature* 3: 334–9, 415–17, 454–7.

Carpenter, W.B., J.G. Jeffreys, and C.W. Thomson. 1870. Preliminary report of the scientific exploration of the deep sea in H.M. Surveying-vessel 'Porcupine,' during the summer of 1869. *Proceedings of the Royal Society of London* 18: 397–492.

Carpine, C. 1993. Catalogue des appareils d'océanographie en collection au Musée océanographique de Monaco. 4. Bouteilles de prélèvement d'eau. *Monaco. Bulletin de l'Institut océanographique* 75 (1440), 175pp.

– 1997. Catalogue des appareils d'océanographie en collection au Musée océanographique de Monaco. 6. Thermomètres. *Monaco. Bulletin de l'Institut océanographique* 76 (1442), 211pp.

– 2002. *La pratique de l'océanographie au temps du Prince Albert Ier*. Monaco: Musée océanographique. iii + 331pp.

Carpine-Lancre, J. 1989. Le Prince Albert de Monaco et l'Exposition universelle de 1889. *Annales monégasques, revue d'histoire de Monaco* 13: 7–42.

– 1991. Le Prince Albert Ier de Monaco et le Muséum national d'histoire naturelle. *Les amis du Muséum national d'histoire naturelle* 167: 33–6.

– 1993. Le Prince Albert Ier de Monaco, marin et océanographe: chronologie sommaire. *Océanis* 19(4): 121–35.

– 1998a. *Albert Ier: Prince of Monaco (1848–1922)*. Monaco: Éditions EGC. 32pp.

– 1998b. *Albert Ier, Prince de Monaco: Des oeuvres de science, de lumière et de la paix*. Monaco: Palais de S.A.S. le Prince. 205pp.

– 2003. The origin and early history of 'la Carte générale bathymétrique des océans.' In D. Scott, ed., *The history of GEBCO, 1903–2003: The 100-year story of the General Bathymetric Chart of the Oceans*, pp. 15–51. Lemmer, Netherlands: GITC bv.

Carpine-Lancre, J., and L. Saldanha. 1992. *Dom Carlos I, Roi de Portugal, Albert Ier, Prince de Monaco: Souverains océanographes*. Lisbonne: Fondation Calouste Gulbenkian. 178pp.

Carruthers, J.N. 1944. The oceans: Their physics, chemistry and general biology. *Quarterly Journal of the Royal Meteorological Society* 70: 159–60.

– 1956. Donald John Matthews, 1873–1956. *Journal du Conseil international pour l'Exploration de la Mer* 22: 3–8.

Carter, N.M. 1931a. An oceanographical investigation of certain types of fiords. *Biological Board of Canada. PBS Summary Reports. Investigations in Fisheries and Oceanography*, pp. 1–3.

– 1931b. Oceanographical investigations in the Strait of Georgia. *Biological Board of Canada. PBS Summary Reports. Investigations in Fisheries and Oceanography*, pp. 1–3.

– 1932a. Fjords and fjord formation. *Biological Board of Canada. PBS Summary Reports. Investigations in Fisheries and Oceanography*, No. 3, p. 1.

– 1932b. The oceanography of the fiords of southern British Columbia. *Biological Board of Canada. Progress Reports of Pacific Coast Stations*, No. 12, pp. 7–11.

– 1933. Fiords and fiord formation. *Biological Board of Canada. PBS Summary Reports. Investigations in Fisheries and Oceanography*, No. 40, p. 1.

– 1934. The physiography and oceanography of some British Columbia fiords. *Proceedings of the 5th Pacific Science Congress*, 1: 721–33.

Carter, N.M., and J.P. Tully. 1932. Oceanographical investigations in the Strait of Georgia. *Biological Board of Canada. PBS Summary Reports. Investigations in Fisheries and Oceanography*, No. 8, p. 1.

Castens, G. 1905a. *Untersuchungen über die Strömungen des Atlantischen Ozeans: Die Dichte- und Windverhältnisse*. Inaugural-Dissertation zur Erlangung der Doktorwürde der hohen philosophischen Fakultät der Königlichen Christian-Albrechts-Universität zu Kiel. 38pp.

– 1905b. Untersuchungen über die Strömungen des Atlantischen Ozeans: Wissenschaftliche Meeresuntersuchungen, Abteilung Kiel, N.F. 8: 261–76.

Chapman, R.P., ed. 1998. *Alpha and omega. An informal history of the Defence Research Establishment Pacific, 1948–1995*. Dartmouth, NS: Defence Research Establishment Atlantic. xx + 283pp.

Charnock, H., and M. Deacon. 2001. Introduction. Ocean circulation. In M. Deacon, T. Rice, and C. Summerhayes, eds, *Understanding the oceans: A century of ocean exploration*, pp. 155–7. London and New York: UCL Press.

Chevallier, A. 1905. Relation entre la densité et la salinité des eaux de mer. *Monaco. Bulletin du Musée océanographique* 2(31), 12pp.

– 1906. Courants marins profonds dans l'Atlantique Nord. *Monaco. Bulletin du Musée océanographique* 3(63), 16pp.

Chickering, R. 1984. *We men who feel most German: A cultural study of the Pan-German League, 1886–1914*. London: George Allen and Unwin. xii + 365pp.

Christiansen, M.E. 1993. Georg Ossian Sars (1837–1927), the great carcinologist of Norway. *Crustacean Issues 8* [*History of Carcinology*, ed. Frank Truesdale], pp. 143–64.

Clemens, W.A. 1947. Charles McLean Fraser (1872–1946). *Proceedings and Transactions of the Royal Society of Canada*, Series 3, 41: 127–9.

– 1958. Reminiscences of a director. *Journal of the Fisheries Research Board of Canada* 15: 779–96.

– 1968. Education and fish: An autobiography of Wilbert Amie Clemens. *Fisheries Research Board of Canada. Manuscript Report Series*, No. 974, ii + 102pp.

Colding, L.A. 1870. Om strømningsforholdene i almindelige ledninger og i havet. *Danske Videnskabernes Selskabs Forhandlinger, Skrifter*, Ser. 5, 9 (III) (1869): 81–214.

– 1871. Nogle bemaerkingen om luftens stromningsforhold. *Danske Videnskabernes Selskabs Forhandlinger, Oversigt* (1871): 89–108.

– 1877. Some remarks concerning the nature of currents of air. In Short Memoirs on Meteorological Subjects (trans. E.C. Abbe). *Annual Report of the Smithsonian Institution for 1877*, pp. 447–62.

Collin, A.E., and M.J. Dunbar. 1964. Physical oceanography in Arctic Canada. *Oceanography and Marine Biology Annual Reviews* 2: 45–75.

Coman, M.A., R.W. Griffiths, and G.O. Hughes. 2006. Sandström's experiments revisited. *Journal of Marine Research* 64: 783–96.

Cook, A. 1998. *Edmond Halley: Charting the heavens and the seas*. Oxford: Clarendon Press. xvi + 540pp.

Cook, R. 1987. The triumphs and trials of materialism. In: R.C. Brown, ed. *The illustrated history of Canada*, pp. 375–466. Toronto: Lester and Orpen Dennys.

Cowie, J.J. 1912. Sea-fisheries of Eastern Canada. In: Canada, Commission of Conservation, *Sea-fisheries of Eastern Canada*, pp. 94–119. Ottawa: Mortimer.

C.P. 1893. Maclaurin, Colin. *Dictionary of National Biography* 35: 196–8.

Craig, G.A. 1981. *Germany, 1866–1945*. Oxford: Oxford University Press. xv + 825pp.

Craigie, E.H. 1916a. Hydrographic investigations in the St Croix River and Passamaquoddy Bay in 1914. *Contributions to Canadian Biology and Fisheries* 1914–1915 (15): 151–61.

– 1916b. A hydrographic section of the Bay of Fundy in 1914. *Contributions to Canadian Biology and Fisheries* 1914–1915 (16): 163–7.

Craigie, E.H., and W.H. Chase. 1918. Further hydrographic investigations in the Bay of Fundy. *Contributions to Canadian Biology and Fisheries* 1917–1918 (7): 127–48.

Croll, J. 1864. On the physical cause of the change of climate during geological epochs. *Philosophical Magazine*, 4th Series, 28: 121–37.

– 1867. On the eccentricity of the earth's orbit, and its physical relations to the glacial epoch. *Philosophical Magazine*, 4th Series, 33: 119–31.

– 1870a. On ocean-currents. I. Ocean-currents in relation to the distribution of heat over the globe. *Philosophical Magazine*, 4th Series, 39: 81–106.

– 1870b. On ocean-currents. Part II. Ocean-currents in relation to the physical theory of secular changes of climate. *Philosophical Magazine,* 4th Series, 39: 180–94.

– 1870c. On ocean currents. Part III. On the physical cause of ocean-currents. *London, Edinburgh and Dublin Philosophical Magazine and Journal of Science,* 4th Series, 40: 233–59.

– 1871a. Ocean currents. *Nature* 3: 247.

– 1871b. Ocean currents. *Nature* 4: 304.

– 1871c. On ocean currents. Part III. On the physical cause of ocean-currents. *London, Edinburgh and Dublin Philosophical Magazine and Journal of Science,* 4th Series, 42: 241–80.

– 1872a. Ocean currents. *Nature* 5: 201–2.

– 1872b. Ocean currents. *Nature* 5: 399.

– 1872c. Ocean currents. *Nature* 5: 502–3.

– 1872d. Ocean currents. *Nature* 6: 240–1.

– 1872e. Kinetic energy. *Nature* 6: 324.

– 1872f. Oceanic circulation. *Nature* 6: 453–4.

– 1874a. Ocean currents. *Nature* 10: 52–3.

– 1874b. On ocean currents – part III. On the physical causes of ocean-currents. *London, Edinburgh and Dublin Philosophical Magazine and Journal of Science,* 4th Series, 47: 94–122, 168–90.

– 1874c. On the physical cause of ocean currents. *London, Edinburgh and Dublin Philosophical Magazine and Journal of Science,* 4th Series, 47: 434–7.

– 1875a. The 'Challenger''s crucial test of the wind and gravitation theories of oceanic circulation. *London, Edinburgh and Dublin Philosophical Magazine and Journal of Science,* 4th Series, 50: 242–50.

– 1875b. *Climate and time in their geological relations: A theory of secular changes of the earth's climate.* New York: Appleton. xvi + 577pp.

– 1875c. Further remarks on the 'crucial test' argument. *London, Edinburgh and Dublin Philosophical Magazine and Journal of Science,* 4th Series, 50: 489–91.

– 1875d. Oceanic circulation. *Nature* 12: 494.

– 1875e. Oceanic circulation. *Nature* 13: 66–7.

– 1875f. The wind theory of oceanic circulation. – Objections examined. *London, Edinburgh and Dublin Philosophical Magazine and Journal of Science,* 4th Series, 50: 286–90.

– 1876. On the *Challenger's* crucial test of the wind and gravitation theories of oceanic circulation. *Report of the British Association for the Advancement of Science,* Bristol 1875, 45(2): 191–3.

– 1879. Zöppritz on ocean currents. *Nature* 19: 202–4.

Crosland, M. 1992. *Science under control. The French Academy of Sciences, 1795–1914.* Cambridge: Cambridge University Press. xix + 454pp.

Cushing, D.H. 1972. The production cycle and numbers of marine fish. *Symposia of the Zoological Society of London* 29: 213–32.

– 1973. The natural regulation of fish populations. In: F.R. Harden-Jones, ed., *Sea fisheries research*, pp. 399–411. London and New York: John Wiley and Sons.

– D.H. 1975. *Marine ecology and fisheries*. Cambridge: Cambridge University Press. xiv + 278pp.

Cuvier, G.R. 1810. *Rapport historique sur les progrès des sciences naturelles depuis 1789, et sur leur état actuel*. Paris: Imprimerie impériale. 395pp. (Reprint 1968, *Culture et Civilisation* 115.)

Damien, R. 1964. *Albert 1er, Prince souverain de Monaco, précédé de l'historique des origines de Monaco et de la dynastie des Grimaldi*. Villemomble: Institut de Valois. 518pp.

Dampier, W. 1697. *A new voyage around the world. Vol. I*. London: James Knapton. 550pp.

– 1699. *Discourse of winds, breezes, storms, seasons of the year, tides and currents of the torrid zone throughout the world*. Supplement to *A new voyage around the world* (1697). London: James Knapton. 112pp.

Dannevig, A. 1919. Biology of Atlantic waters of Canada: Canadian fish eggs and larvae. In Canada, Department of the Naval Service, *Canadian Fisheries Expedition, 1914–1915: Investigations in the Gulf of St. Lawrence and Atlantic waters of Canada under the direction of Dr. Johan Hjort, leader of the expedition, director of fisheries for Norway*, pp. 1–49. Ottawa: King's Printer.

Darrigol, O. 1998. From organ pipes to atmospheric motions: Helmholtz on fluid mechanics. *Historical Studies in the Physical and Biological Sciences* 29(1): 1–51.

– 2005. *Worlds of flow: A history of hydrodynamics from the Bernoullis to Prandtl*. Oxford: Oxford University Press. xiv + 300pp.

Deacon, G.E.R. 1937. The hydrology of the Southern Ocean. *Discovery Reports* 15: 1–124.

– 1945. Oceanographic investigations. *Nature* 155: 652–4.

– 1973. James Norman Carruthers [24 Nov. 1895–18 Mar. 1973]. *Polar Record* 16 (105): 873–5.

Deacon, G.E.R., and M. Deacon. 1969. Captain Cook as a navigator. *Notes and Records of the Royal Society of London* 24: 33–42.

Deacon, M. 1968. Some early investigations of the currents in the Strait of Gibraltar. *Monaco. Bulletin de l'Institut océanographique, No. spécial 2 (Premier Congrès internationale d'histoire de l'océanographie)* 1: 63–75.

– 1977. Staff-Commander Tizard's journal and the voyages of H.M. ships *Knight Errant* and *Triton* to the Wyville Thomson Ridge in 1880 and 1882. In M. Angel, ed., *A voyage of discovery: George Deacon 70th anniversary volume*, pp. 1–14, Oxford: Pergamon Press.

– ed. 1978a. *Oceanography: Concepts and history*. Benchmark Papers in Geology, vol. 35. Stroudsburg, PA: Dowden, Hutchinson and Ross. xvii + 394pp.

– 1978b. Vice-Admiral T.A.B. Spratt and the development of oceanography in the Mediterranean 1841–1873. National Maritime Museum, London. *Maritime Monographs and Reports*, No. 37, v + 74pp.

– 1985a. An early theory of ocean circulation: J.S. von Waitz and his explanation of the currents in the Strait of Gibraltar. *Progress in Oceanography* 14: 89–101.

– 1985b. George Strong Nares (1831–1915). *Arctic* 38: 148–9.

– 1986a. The contribution of Edmond Halley to meteorology and oceanography. *Weather* 41: 246–50.

– 1986b. Early knowledge of the Bosporus undercurrent. *Mariner's Mirror* 72: 207–10.

– ed. 1993a. *A treatise concerning the motions of the seas and winds* (1677), together with *Isaac Vossius, De motu marium et ventorum* (1663). Delmar, NY: Scholars Facsimiles and Reprints.

– 1993b. Wind pressure versus density differences: a 19th century controversy about ocean circulation. *Ocean Challenge* 4(1–2): 53–8.

– 1996. How the science of oceanography developed. In C.P. Summerhayes and S.A. Thorpe, eds, *Oceanography. An illustrated guide*, New York: John Wiley and Sons.

– 1997. *Scientists and the sea, 1650 -1900: A study of marine science*. 2nd edn. Aldershot, England: Ashgate. xi + 459pp.

– 1998. From salt manufacture to the circulation of the ocean: J.S. von Waitz and his 1755 paper on the water budget of the Mediterranean Sea. *Historisch-Meereskundliches Jahrbuch* 5: 75–91.

Deacon, M., and A. Savours. 1976. Sir George Strong Nares (1831–1915). *Polar Record* 18(113): 127–41.

Deacon, M., and C. Summerhayes. 2001. Introduction. In M. Deacon, T. Rice, and C. Summerhayes, eds, *Understanding the oceans: A century of ocean exploration*, pp. 1–23. London and New York: UCL Press.

Defant, A. 1927. Über die wissenschaftlichen Aufgaben und Ergebnisse der Expedition. Die Deutsche Atlantische Expedition auf dem Vermessungs- und Forschungsschiff 'Meteor.' Festsitzung zur Begrüssung der Expedition am 24. Juni 1927. *Zeitschrift der Gesellschaft für Erdkunde zu Berlin* 7/8: 359–69.

– 1928a. Physik des Meeres. In Wien und Harms, eds, *Handbuch der Experimentalphysik* 25(2), pp. 569–686. Leipzig.

- 1928b. Die systematische Erforschung des Weltmeeres. Jubiläums-Sonderband 1928. *Zeitschrift der Gesellschaft für Erdkunde zu Berlin* 1928: 459–505.
- 1929a. *Dynamische Ozeanographie* (III Band der *Einführung in der Geophysik*). Berlin: Julius Springer. x + 222S.
- 1929b. Stabile Lagerung ozeanischer Wasserkörper und dazu gehörige Stromsysteme. *Veröffentlichungen des Instituts für Meereskunde an der Universität Berlin. Neue Folge. A. Geographisch-naturwissenschaftliche Reihe* 19: 1–33.
- 1929c. Die Theorie der Meeresströmungen und die ozeanische Zirkulation. *Geografiska Annaler* 3(4): 268–93.
- 1933. Ferdinand von Richthofen als Begründer des Instituts und Museums für Meereskunde. *Berliner geographische Arbeiten* 5: 10–14.
- 1934. Ist die Tiefsee in Ruhe? In *Tiefseebuch: Ein Querschnitt durch die neuere Tiefseeforschung*, pp. 46–63. Das Meer in volkstümlichen Darstellung. Dritter Band. Berlin: E.G. Mittler.
- 1936a. Antrittsrede des Hrn Defant. *Sitzungsberichten der Preussischen Akademie der Wissenschaften, Öffentliche Sitzung vom 2. Juli 1936*, pp. 1–3.
- 1936b. Schichtung und Zirkulation des Atlantischen Ozeans. Die Troposphäre. *Deutsche Atlantische Expedition, 1925–1927. Wissenschaftliche Ergebnisse* 6(1).
- 1939. Deutsche meereskundliche Forschungen 1928 bis 1938. *Zeitschrift der Gesellschaft für Erdkunde zu Berlin* 1939, 3/4: 81–102.
- 1960. Die meereskundliche Erkenntnisse Alexander von Humboldts im Lichte der modernen Ozeanographie. *Deutsche Geographentag Berlin 1959. Tagungs – Berichte und wissenschaftliche Abhandlungen*. Wiesbaden. pp. 84–94.
- 1961. *Physical oceanography. Volume 1*. London and New York: Pergamon. xvi + 729pp.
Department of the Naval Service. 1919. *Canadian Fisheries Expedition, 1914–1915: Investigations in the Gulf of St. Lawrence and Atlantic waters of Canada under the direction of Dr. Johan Hjort, head of the expedition, director of fisheries for Norway*. Ottawa: King's Printer. xxviii + 495pp.
Despretz, 1837. Recherches sur le maximum de densité de l'eau pure et des dissolutions aqueuses. *Annales de chimie* 70: 5.
De Vecchi, V. 1984a. The dawning of a national scientific community in Canada, 1878–1896. *Scientia Canadensis* 18(26): 32–58.
- 1984b. Science and scientists in government, 1878–1896 – Part I. *Scientia Canadensis* 8(27): 112–42.
- 1985. Science and scientists in government, 1878–1896. – Part II. *Scientia Canadensis* 9(29): 97–113.
Devik, O. 1956. Professor Bjørn Helland-Hansen. In H. Dahl, ed., *Festskrift til Professor Bjørn Helland-Hansen*, pp. 9–23. Bergen: A.S. John Griegs Boktrykkeri.

– 1958. Bjørn Helland-Hansen. *Forhandlinger ved Det Kongelige Norske Videnskabers* Selskab 30 (1957): 128–36.

– 1966. Sverdrup, Harald Ulrik. *Norsk Biografisk Leksikon* 15: 379–87.

Dickie, L.M. 1985. Huntsman, Archibald Gowanloch. *The Canadian Encyclopedia II*: 850.

Dickson, H.N. 1901. The circulation of the surface waters of the North Atlantic Ocean. *Philosophical Transactions of the Royal Society of London* A, 196: 61–203.

Dietrich, G. 1935. Aufbau und Dynamik des südlichen Agulhas-Stromgebietes. *Veröffentlichungen des Instituts für Meereskunde an der Universität Berlin, Neue Folge. A. Geographisch-naturwissenschaftliche Reihe* 27: 1–79.

– 1958. Bjørn Helland-Hansen. *Deutsche Hydrographische Zeitschrift* 11(1): 36–27.

– 1970. Alexander von Humboldts 'Physische Weltbeschreibung' und die moderne Meeresforschung. *Deutsche Geographentag Kiel 1969. Tagungs-Berichte und wissenschaftliche Abhandlungen.* Wiesbaden. pp. 105–22.

– 1972. Georg Wüst's scientific work, dedication to his eightieth birthday. In A.L. Gordon, ed., *Studies in physical oceanography*, pp. xi–xx. New York: Gordon and Breach.

Dinsmore, R.B., and T.B. Strobridge. 1998. Rear Admiral Edward 'Iceberg' Smith. United States Coastguard Academy home page: http://www.cga.edu/acd/science/ehs.htm

Dodimead, A.J. 1956. Project Norpac. *Fisheries Research Board of Canada. Progress Reports of Pacific Coast Stations* 105: 16–18.

Dodimead, A.J., and J.P. Tully. 1958. Canadian oceanographic research in the Northeast Pacific Ocean. *Proceedings of the Ninth Pacific Science Congress* 16: 180–95.

Doe, L.A.E. 1950a. The offshore project. *Fisheries Research Board of Canada. Pacific Oceanographic Group, Annual Report for 1950*, pp. 16–19.

– 1950b. The offshore project. *Fisheries Research Board of Canada. Pacific Oceanographic Group, Annual Report for 1950*, Appendix No. 120, p. 149.

– 1951. Project Offshore. *Fisheries Research Board of Canada. Annual Report for 1951 of the Pacific Biological Station*, p. 187.

– 1952. Offshore project. *Fisheries Research Board of Canada. Pacific Oceanographic Group, Annual Report for 1952*, pp. 23–4.

– 1955. Offshore waters of the Canadian Pacific Coast. *Journal of the Fisheries Research Board of Canada* 12(1): 1–34.

Dortet de Tessan, U. 1844. Physique. In A du Petit-Thouars, *Voyage autour du monde sur la frégate la Vénus, pendant les années 1836–1839* (1840–55), vol. 9. Paris: Gide.

– 1842–4. *Physique.* In A. du Petit Thouars, *Voyage autour du monde sur la frégate la Vénus pendant les années 1836–39, (1840–64).* Paris: Gide.

Driggers, V.W. 1980. Edward H. Smith and the 1928 *Marion* Expedition revisited: a compilation. In M. Sears and D. Merriman, eds, *Oceanography: The past*, pp. 114–126. New York: Springer-Verlag.

Drygalski, E. von. 1903. Die Deutsche Südpolar Expedition auf dem Schiff 'Gauss.' *Bericht über die wissenschaftlichen Arbeiten. Veröffentlichungen des Instituts für Meereskunde und die Geographisches Instituts an der Universität Berlin* 5, iv + 181S

– 1904. *Zum Kontinent des eisigen Südens: Deutsche Südpolarexpedition, Fahrter und Forschung des 'Gauss' 1901–1903*. Berlin: Georg Reimer. xiv + 668S.

– 1989. *The southern ice-continent: The German South Polar Expedition aboard the Gauss, 1901–1903*. Translated by M.M. Raraty. Huntingdon: Bluntisham Books and Erskine Press. xxi + 373pp.

Drygalski, E. von, et al. 1902. Die Deutsche Südpolar Expedition auf dem Schiff 'Gauss.' *Bericht über die wissenschaftlichen Arbeiten. Veröffentlichungen des Instituts für Meereskunde und die Geographisches Instituts an der Universität Berlin* 2 , vi + 73S.

Du Buat, P.-L.-G. 1816. *Principes d'hydraulique et de pyrodynamique*. 3 vols. Paris.

Dumont d'Urville, J. 1830–3. *Voyage de la corvette l'Astrolabe executé par ordre du Roi, pendant les années 1826–1829*. 5 vols. Paris.

– 1833. *Voyage de découvertes de l'Astrolabe pendant les années 1826–1829: Observations nautiques, météorologiques, hydrographiques et de physique*. Paris.

– 1834–5. *Voyage pittoresque autour du monde*. 2 vols. Paris: Chez L. Terré. vii + 576pp, 584pp.

Dunbar, M.J. 1949. *Calanus:* New Arctic research vessel. *Arctic* 2: 56–7.

– 1951. Eastern Arctic waters. *Fisheries Research Board of Canada Bulletin* 88, 131pp.

– 1961. Obituary. Gerhard Schott. *Geographical Review* 51(4): 590.

– 1982. The history of oceanographic research in the waters of the Canadian Arctic islands. In M. Zaslow, ed., *A century of Canada's Arctic islands, 1880–1980*, pp. 141–52. Ottawa: Royal Society of Canada.

Eckert, M. 1913. Otto Krümmel. *Geographische Zeitschrift* 19(10): 545–54.

Eddington, A.S. 1942. Obituary of Joseph Larmor. *Obituaries of Fellows of the Royal Society* 4: 192–207.

Eggleston, W. 1950. *Scientists at war*. Toronto: Oxford University Press. x + 291pp.

Ekman, F.L. 1870. Om salthalten i hafsvattnet utmed Bohuslänska kusten. *Kongliga Svenska Vetenskaps-Akademiens Handlingar* 9(4), 44pp.

– 1875a. Notice sur les mouvements de l'eau de la mer, dans la voisinage de l'embouchure des fleuves, pour servir à la connaissance de la nature des courants marins. *Archives des Sciences physiques et naturelles* 54: 62–71.

– 1875b. Om de strömningar som uppstå i närheten af flodmynningar: Ett bidrag till kännedomen af hafsströmmarnes natur. *Öfversigt af Kongliga Vetenskaps-Akademiens* Förhandlingar, No. 7.

– 1876. On the general causes of the ocean currents. Nova Acta Regiae Societas, *Scientiarum Upsaliensis,* Ser. III, 1879, 10(6): 1–52.

– 1881. *Appareils hydrographiques exposés par le Professeur F.-L. Ekman au Congrès géographique de Venise, 1881.* Stockholm: P.A. Nordstedt and Söner. 19pp.

Ekman, F.L., and O. Pettersson. 1893. Den Svenska hydrografiska expeditionen år 1877 under ledning af F.L. Ekman. I. Första afdelningen af F.L. Ekman. II.Andra afdelningen efter F.L. Ekmans död utarbetad af O. Pettersson. *Kongliga Svenska Vetenskaps-Akademiens Handlingar* 25(1): 1–163.

Ekman, V.W. 1902. Om jordrotationens inverkan på vindströmmar i hafvet. Kristiania, *Nyt Magazin for Naturvidenskab* 40(1): 1–27.

– 1904. On dead-water: Being a description of the so-called phenomena often hindering the headway and navigation of ships in Norwegian fjords and elsewhere, and an experimental investigation of its causes. *Norwegian North Polar Expedition 1893–1896, Scientific Results* 5(15): 1–152.

– 1905. On the influence of the earth's rotation on ocean currents. *Arkiv för Matematik, Astronomi och Fysik* 2(11): 1–52.

– 1906. Beiträge zur Theorie der Meeresströmungen. *Annalen der Hydrographie und maritimen Meteorologie* 34: 423–30, 472–84, 527–40, 566–83.

– 1927. Meeresströmungen. In F. Auerbach and W. Hort, eds, *Handbuch der Physikalischen und Technischen Mechanik* 5 (1), pp. 177–206.

– 1930a. Dr Gustaf Ekman 29/8 1852 – 26/2 1930. *Svensk geografisk årsbok* 6: 148–51.

– 1930b. Dr Gustaf Ekman 1852–1930. *Meddelanden från Lunds universitets geografiska institution, Ser. C,* 60: 148–51.

– 1939. Neuere Ergebnisse und Probleme zur Theorie der Konvektionsströme im Meere. *Gerlands Beiträge zur Geophysik, Supplement, Band* 4, *Ergebnisse der kosmischen Physik IV, Physik der Hydro- und Lithosphäre.* Leipzig. 74pp.

– 1941. Otto Pettersson 12/2 1848 – 17/1 1941. *Svensk geografisk årsbok* 17: 85–92.

Ekman, V.W., and H. Pettersson. 1949. Fredrik Laurents Ekman. *Svenskt biografiskt lexikon* 13: 111–16.

Eliassen, A. 1982. Vilhelm Bjerknes and his students. *Annual Review of Fluid Mechanics* 14: 1–11.

Emery, W.J. 1980. The *Meteor* Expedition: An ocean survey. In M. Sears and D. Merriman, eds, *Oceanography: The past,* pp. 690–702. New York: Springer-Verlag.

Engelhardt, R. 1899a. Untersuchungen über die Strömungen der Ostsee: Die Dichtigkeitsfläche. *Aus dem Archiv dem Deutschen Seewarte* 22(6): 1–27.

– 1899b. Untersuchungen über die Strömungen der Ostsee: Die Dichtigkeitsfläche. *Inaugural-Dissertation zur Erlangung der Doktorwürde der hohen philosophischen Fakultät der Christian-Albrechts-Universität in Kiel.* Altona: Hammerich and Lesser. 27pp.

- 1964. Der *Physikalische* Atlas des Heinrich Berghaus und Alexander Keith Johnston's *Physical Atlas. Petermanns geographische Mitteilungen* 108: 133–49.
- 1969. A. v. Humboldt's Abhandlung über die Meeresströmungen. *Petermanns geographische Mitteilungen* 113: 100–10.
- 1976. *Heinrich Berghaus: Der Kartograph von Potsdam.* Acta Historica Leopoldina, No.10, 411pp.

Erman, G.A. 1828. Nouvelles recherches sur le maximum de densité de l'eau salée. *Annales de chimie* 38: 287.

Errulat, F. 1960. In Memoriam. Dr h.c. F. Spiess zum Gedächtnis. *Deutsche Hydrographische Zeitschrift* 13(2): 85–9.

Estey, R.H. 1994. *Essays on the early history of plant pathology and Mycology in Canada.* Montreal and Kingston: McGill-Queen's University Press. xi + 384pp.

Euler, H. von. 1941. Sven Otto Pettersson. In memoriam. *Svensk Kemisk Tidskrift* 53: 28–32.

Evans, J. 1897. Sir Joseph Prestwich. *Proceedings of the Royal Society of London* 60: xii–xvi.

Eyles, J. 1975. Rennell, James. *Dictionary of Scientific Biography* 11: 376.

Fels, E. 1971. Drygalski, Erich von. *Dictionary of Scientific Biography* 4: 193–4.

Ferrel, W. 1856. An essay on the winds and currents of the ocean. *Nasheville Journal of Medicine and Surgery* 6: 287–301, 375–89.

- 1859. The motions of fluids and solids relative to the earth's surface. *Mathematical Monthly* 1: 140–8, 210–16, 300–7, 366–73, 379–406.
- 1859–60. The motions of fluids and solids relative to the earth's surface. *Mathematical Monthly* 2: 89–97, 339–46, 374–90.
- 1861. The motions of fluids and solids relative to the earth's surface. *American Journal of Science,* 2nd Series, 31: 27–51.
- 1872a. Ocean currents. *Nature* 5: 384–5.
- 1872b. Ocean currents. *Nature* 6: 120.
- 1874. Relation between the barometric gradient and the velocity of the wind. *American Journal of Science,* 3rd Series, 8: 343–62.

Filchner, W. 1994. *To the sixth continent: The second German South Polar Expedition.* With the collaboration of Alfred King and Eric Przybylloch. Trans. and ed. William Barr. Huntingdon: Bluntisham Books; Banham, Norfolk, Erskine Press. 42 + viii + 253pp.

Findlay, A.G. 1853. Oceanic currents, and their connection with the proposed Central-American canals. *Journal of the Royal Geographical Society* 23: 217–42.

Findlen, P. 2004. Introduction. The last man who new everything ... or did he? Athanasius Kircher, S.J. (1602–1680) and his world. In P. Findlen, ed., *Athanasius Kircher: The last man who knew everything*, pp. 1–48. New York and London: Routledge.

Fischer, J.L. 2002. Créations et fonctions des stations maritimes françaises. *Revue pour l'histoire du CNRS*, No. 7, pp. 26–31.

Fjeldstad, J.E. 1959a. Harald Ulrik Sverdrup, 15 November 1888 – 21 August 1957. *Norsk Geografisk Tidsskrift* 16(1–8): 5–23.

– 1959b. Harald Ulrik Sverdrup, 15 November 1888 – 21 August 1957. *Norsk Polarinstitutt Meddelelser* 82, 23pp.

Foerster, R.E. 1948. Charles McLean Fraser. *Anatomical Record* 100: 398–9.

Fontaine, M. 1972. *Albert Ier, chef et propagateur de l'océanographie, précurseur de l'océanologie.* Monaco: Commission nationale pour l'UNESCO. 19pp.

Forbes, E.G. 1974. The birth of scientific navigation: The solving in the 18th century of the problem of finding longitude at sea. National Maritime Museum, Greenwich, England. *Maritime Monographs and Reports*, No.10, vii + 25pp.

Forch, C. 1906. Zur Theorie der Meeresstrømungen. *Annalen der Hydrographie und maritimen Meteorologie* 15(10): 114–22.

– 1909. Der Druckgradient im Meerwasser in seiner Abhängigkeit von Temperatur- und Salzverteilung. *Annalen der Hydrographie und maritimen Meteorologie* 37(11): 492–500.

– ed. 1911. *Kayser's Physik des Meeres.* Zweite Auflage. Paderborn: Ferdinand Schöningh. vi + 384pp.

Ford, W.L., J.R. Longard, and R.E. Banks. 1952. On the nature, occurrence and origin of cold low salinity water along the edge of the Gulf Stream. *Journal of Marine Research* 11: 281–93.

Forman, P. 1971. Weimar culture, causality, and quantum theory, 1918–1927: Adaptation by German physicists and mathematicians to a hostile intellectual environment. *Historical Studies in the Physical Sciences* 3: 1–115.

– Scientific internationalism and the Weimar physicists: The ideology and its manipulation in Germany after World War I. *Isis* 64: 151–80.

Fox, R. 1992. *The culture of science in France, 1700–1900.* Brookfield, VT: Variorum. xiii + 335pp.

Fox, R., and G. Weisz, eds. 1980. *The organization of science and technology in France, 1808–1914.* Cambridge: Cambridge University Press. x + 355pp.

Frängsmyr, T., J.L. Heilbron, and R.E. Rider, eds. 1990. *The quantifying spirit in the 18th century.* Berkeley: University of California Press. viii + 411pp.

Fraser, C.M. 1921. Temperature and specific gravity variations in the surface waters of Departure Bay, B.C. *Contributions to Canadian Biology and Fisheries* 1918–1920 (III): 35–47.

Fraser, C.M., and A.T. Cameron. 1915. Variations in density and temperature in the coastal waters of British Columbia – preliminary notes. *Contributions to Canadian Biology and Fisheries* 1914–1915 (XIII): 133–43.

Frey, D.G. 1982. G.O. Sars and the Norwegian Cladocera: A continuing frustration. *Hydrobiologia* 96: 267–93.

Friedman, R.M. 1989. *Appropriating the weather: Vilhelm Bjerknes and the construction of a modern meteorology.* Ithaca, NY: Cornell University Press. xx + 251pp.

– 1994. *The expeditions of Harald Ulrik Sverdrup: Contexts for shaping an ocean science.* La Jolla, CA: Scripps Institution of Oceanography. 49pp.

– 2001. Polar dreams and California sardines: Harald Ulrik Sverdrup and study of ocean circulation prior to World War II. In M. Deacon, T. Rice, and C. Summerhayes, eds, *Understanding the oceans: A century of ocean exploration,* pp. 158–72. London and New York: UCL Press.

– 2002. Contexts for constructing an ocean science: The career of Harald Ulrik Sverdrup (1888–1957). In K.R. Benson and P.F. Rehbock, eds, *Oceanographic history: The Pacific and beyond,* pp. 17–27. Seattle: University of Washington Press.

F.R.S. 1874. Ocean circulation – Dr. Carpenter and Mr. Croll. *Nature* 10: 83–4.

Fuglister, F.C., and L.V. Worthington. 1951. Hydrography of the Western Atlantic: Some results of a multiple ship survey of the Gulf Stream. *Woods Hole Oceanographic Institution Technical Report,* No. 18, Reference No. 51–9.

Gareis, A., und A. Becker. 1867. *Zur Physiographie des Meeres: Ein Versuch.* Triest: H.F. Schimpf's Buchhandlung. viii + 135SS.

Garrett, C., and C. Wunsch, eds. 1984. *'It's the water that makes you drunk.' A celebration in geophysics and oceanography – 1982. In honor of Walter Munk on his 65th birthday, October 19, 1982, at Scripps Institution of Oceanography, University of California, San Diego.* La Jolla, CA: Scripps Institution of Oceanography. SIO Reference Series 84–5, 118pp.

Gill, A. 1997. *The devil's mariner: A life of William Dampier, pirate and explorer, 1651–1715.* London: Michael Joseph. xx + 396pp.

Gill, A.E. 1982. *Atmosphere – ocean dynamics.* New York: Academic Press. xv + 662pp.

Gingras, Y. 1991. *Physics and the rise of scientific research in Canada.* Montreal and Kingston: McGill-Queen's University Press. xii + 203pp.

– 2001. What did mathematics do to physics? *History of Science* 39: 383–416.

Ginns, J. 1988. Irene Mounce, 1894–1987. *Mycologia* 80: 607–8.

Gold, E. 1947. J.W. Sandström. *Nature* 159: 395–6.

Goodspeed, D.J. 1958. *A history of the Defence Research Board of Canada.* Ottawa: Queen's Printer. xi + 259pp.

Gougenheim, A. 1968. Deux ingénieurs hydrographes du XIXe siècle, précurseurs en matière de dynamique des mers. *Monaco. Bulletin de l'Institut océanographique, No. spécial 2 (Premier Congrès international d'histoire de l'océanographie)* 1: 87–97.

Gough, J. 1988. Fisheries history. *The Canadian Encyclopedia II*: 781–4.

– 1991. Fisheries management in Canada, 1880–1910. Fisheries and Oceans Canada. *Canadian Manuscript Report of Fisheries and Aquatic Sciences* 2105, iv + 196pp.

– 1993. Fisheries management in Canada: A historical sketch. In L.S. Parsons and W.H. Lear, eds, Perspectives in Canadian marine fisheries management, *Canadian Bulletin of Fisheries and Aquatic Sciences* 226, pp. 5–53.

Gould, W.J. 1993. James Rennell's view of the Atlantic circulation: A comparison with our present knowledge. *Ocean Challenge* 4(1–2): 26–32.

– 2001. Direct measurement of subsurface ocean currents: A success story. In M. Deacon, T. Rice, and C. Summerhayes, eds, *Understanding the oceans: A century of ocean exploration,* pp. 173–92. London and New York: UCL Press.

Gourley, D. 2002. Recollections of Professor E. Horne Craigie. *ZooNews* (Department of Zoology, University of Toronto), July 2002. http://www.zoo.utoronto. ca/zfa/Newsletter-htm/jul-02/craigie.htm

Graham, D.M. 1987. Dr. Walter H. Munk: Oean science 'generalist.' *Sea Technology* 28(7): 40.

Graham, M. 1968. Obituary of Henry Bryant Bigelow. *Deep-Sea Research* 15: 125–8.

Gran, H.H. 1919. Quantitative investigations as to phytoplankton and pelagic protozoa in the Gulf of St. Lawrence and outside the same. In: Canada, Department of the Naval Service, *Canadian Fisheries Expedition, 1914–1915: Investigations in the Gulf of St. Lawrence and Atlantic waters of Canada under the direction of Dr. Johan Hjort, leader of the expedition, director of fisheries for Norway,* pp. 487–95. Ottawa: King's Printer.

Granatstein, J.L., I.M. Abella, D.J. Bercuson, R.C. Brown, and H.B. Neatby. 1983. *Twentieth century Canada.* Toronto: McGraw-Hill Ryerson. ii + 440pp.

Grant, E. 1996. *The foundations of modern science in the Middle Ages: Their religious, institutional, and intellectual contexts.* Cambridge: Cambridge University Press. xiv +247pp.

Grier, M.C. 1969 (1941). *Oceanography of the North Pacific Ocean, Bering Sea and Bering Strait: A contribution toward a bibliography.* New York: Greenwood Press. xxii + 290pp.

Griffiths, G. 1993. James Rennell and William Scoresby: Their separate quests for accurate current data. *Ocean Challenge* 4(1–2): 34–8.

Günther, S. 1900. Zöppritz, Karl Jacob. *Allgemeine deutsche Biographie* 45: 434–7.

Guldberg, C.M., and H. Mohn. 1876. *Études sur les mouvements de l'atmosphère.* Part 1. Christiania: A.W. Brøgger. 39pp.

– 1877. Über die gleichförmige Bewegung der horizontalen Luftströme. *Zeitschrift der Österreichischen Gesellschaft für Meteorologie* 12(4): 49–60.

– 1880. *Études sur les mouvements de l'atmosphère.* Part 2. Christiania: A.W. Brøgger. 53pp.

Gunther, E.R. 1936. A report on oceanographical investigations in the Peru coastal current. *Discovery Reports* 13: 107–276.

Gurney, A. 1997. *Below the convergence: Voyages toward Antarctica, 1699–1839.* New York: W.W. Norton. ix + 315pp.

Hachey, H.B. 1929a. Effective mixing in the Passamaquoddy region. *Biological Board of Canada. Report of the Atlantic Biological Station for the Quarter Ending Sept. 30, 1929.* In *Atlantic Biological Station, Annual Report 1929–1930,* by A.G. Huntsman. Toronto: mimeographed, 4pp.

– 1929b. Seeing the world in a catsup bottle: The reminiscences of post cards that have covered many miles of ocean waters in the interest of science. *The Glass Container* 9(1): 3pp.

– 1931a. Appendix No. 1. The general hydrography of the waters of the Bay of Fundy region. In *Biological Board of Canada. Report of the Atlantic Biological Station for 1931,* by A.G. Huntsman. Toronto: mimeographed, p. 1.

– 1931b. Appendix No. 3. The general hydrography and hydrodynamics of the waters of the Hudson Bay region. In *Report of the Atlantic Biological Station for 1931.* Typescript, p. 1.

– 1931c. Biological and oceanographic conditions in Hudson Bay. 2. Report on the Hudson Bay Fisheries Expedition of 1930. A. Open water investigations with the S.S. *Loubyrne. Contributions to Canadian Biology and Fisheries, New Series,* 6 (23): 465–71.

– 1931d. The general hydrography and hydrodynamics of the waters of the Hudson Bay region. *Contributions to Canadian Biology and Fisheries, Series D,* 7: 91–118.

– 1932a. Appendix No. 1. Hydrography of the waters of the south coast of Nova Scotia. In *Biological Board of Canada. Report of the Atlantic Biological Station for 1932,* by A.G. Huntsman. Toronto, mimeographed, p. 1.

– 1932b. Contributions to the hydrography of the waters of the Scotian Shelf. Hydrodynamics of the waters – 1932. *Biological Board of Canada. Manuscript Reports of the Biological Stations,* No. 105, 7pp.

– 1933a. Appendix No. 36. Hydrography of the waters of the Scotian Shelf. In *Biological Board of Canada. Report of the Atlantic Biological Station for 1933,* by A.G. Huntsman. Toronto: mimeographed, unpaged.

– 1933b. Appendix No. 37. Hydrodynamics of waters of Scotian Shelf. In *Biological Board of Canada. Report of the Atlantic Biological Station for 1933,* by A.G. Huntsman. Toronto: mimeographed, unpaged.

– 1934a. Appendix No. 26. Hydrodynamics of the waters of the Scotian Shelf – 1933. In *Biological Board of Canada. Report of the Atlantic Biological Station for 1933,* by A.H. Leim. St Andrews, NB: mimeographed, unpaged.
– 1934b. A derivation of Bjerknes's fundamental hydrodynamical concept. *Internationale Revue der gesamten Hydrobiologie* 31: 331–6.
– 1934c. North Atlantic cyclones and coastal waters. *Biological Board of Canada. Fisheries Experimental Station (Atlantic), Note No. 37. Progress Reports of Atlantic Biological Stations,* No. 11, pp. 6–9.
– 1935a. Appendix No. 21. Hydrodynamics of the waters of the Scotian Shelf – 1934. In *Biological Board of Canada. Report of the Atlantic Biological Station for 1935,* by A.H. Leim. St Andrews, NB: mimeographed, unpaged.
– 1935b. Appendix No. 22. General hydrography of the waters of the Scotian Shelf - 1935. In *Biological Board of Canada. Report of the Atlantic Biological Station for 1935,* by A.H. Leim. St Andrews, NB: mimeographed, unpaged.
– 1935c. The circulation of Hudson Bay water as indicated by drift bottles. *Science* 82 (2125): 275–6.
– 1935d. The effect of a storm on an inshore area with markedly stratified waters. *Journal of the Biological Board of Canada* 1(4): 227–37.
– 1935e. The waters of the Scotian Shelf. *Biological Board of Canada. Progress Reports of Atlantic Coast Stations,* No. 16, pp. 3–6.
– 1936a. Appendix No. 21. Ekman's theory applied to water replacement on the Scotian shelf. In *Biological Board of Canada. Report of the Atlantic Biological Station for 1936,* by A.H. Leim. St Andrews, NB: mimeographed, p. 1.
– H.B. 1936b. Transgressions in the Atlantic Ocean. *Science* 83(2154): 349–50.
– 1937a. Appendix No. 20. Temperature – salinity diagrams for the Scotian Shelf. In *Fisheries Research Board of Canada. Report of the Atlantic Biological Station for 1937,* by A.H. Leim. St Andrews, NB; mimeographed, p. 1.
– 1937b. Ekman's theory applied to water replacements on the Scotian Shelf. *Proceedings of the Nova Scotian Institute of Science* 19: 264–76.
– 1937c. The status of physical oceanography in the Western North Atlantic. *Fisheries Research Board of Canada. Manuscript Reports of the Biological Stations,* No. 165, 10pp.
– 1937d. The submarine physiography and oceanographical problems of the Scotian Shelf. *Transactions of the American Fisheries Society* 66: 237–41.
– 1938a. The cold water layer of the Scotian Shelf. *Science* 88: 307–8.
– 1938b. The origin of the cold water layer of the Scotian Shelf. *Transactions of the Royal Society of Canada,* Section III, 29(210): 31–42.
– 1939. Temporary migrations of Gulf Stream waters on the Atlantic seaboard. *Journal of the Fisheries Research Board of Canada* 4(5): 339–48.
– 1941. Winter flooding of the Scotian Shelf. Atlantic Biological Station Note

No. 81. *Fisheries Research Board of Canada. Progress Reports of Atlantic Biological Stations,* No. 29, pp. 3–5.

– 1942. The waters of the Scotian Shelf. *Journal of the Fisheries Research Board of Canada* 5(4): 377–97.

– 1947. Water transports and current patterns for the Scotian Shelf. *Journal of the Fisheries Research Board of Canada* 7(1): 1–16.

– 1948. Report for 1948 of the Canadian Joint Committee on Oceanography. *Fisheries Research Board of Canada. Annual Report for 1948,* Appendix 11, pp. 94–5.

– 1950. *Canadian Joint Committee on Oceanography. Annual Report – 1950.* St Andrews, NB, 34pp + appendices.

– 1961. Oceanography and Canadian Atlantic waters. *Fisheries Research Board of Canada Bulletin,* No. 134, vi + 120pp.

– 1965. History of the Fisheries Research Board of Canada. *Fisheries Research Board of Canada. Manuscript Report Series (Biological),* No. 843, 499pp.

– 1983. 'Grandpa's golden years (for my grandchildren).' Charlotte County Archives, St Andrews, NB, unpaged manuscript.

Hadley, G. 1735. Concerning the cause of the general trade winds. *Philosophical Transactions of the Royal Society* 39: 58–62.

Hall, M.V.D. 1979. The contribution of the physiologist, William Benjamin Carpenter (1813–1885), to the development of the principles of the correlation of forces and the conservation of energy. *Medical History* 23: 129–55.

Hall, R. 1996. *Empires of the monsoon. A history of the Indian Ocean and its invaders.* London: HarperCollins. xxxiii+575pp.

Halley, E. 1686. An historical account of the trade winds and monsoons, observable in the seas between and near the tropicks, with an attempt to assign the physical cause of the said winds. *Philosophical Transactions of the Royal Society* 16: 153–168.

– 1687. An estimate of the quantity of vapour raised out of the sea by the warmth of the sun: Derived from an experiment shown before the Royal Society, at one of their late meetings. *Philosophical Transactions of the Royal Society of London* 16: 366–70.

– 1691. An account of the circulation of the watry vapours of the sea, and of the cause of springs. *Philosophical Transactions of the Royal Society* 16: 468–73.

Halvorsen, J.B., ed. 1896. *Norsk Forfatter-Lexikon 1814–1880.* Fjerde Bind M - R. Kristiania: Den Norske Forlagsforening.

Hamblin, J.D. 2005. *Oceanographers and the Cold War: Disciples of marine science.* Seattle: University of Washington Press. xxix + 346pp.

Hamlin, C. 1982. James Geikie, James Croll and the eventful ice age. *Annals of Science* 39: 565–83.

Hansen, W. 1954. Vagn Walfrid Ekman. *Deutsche hydrographische Zeitschrift*
7(1/2): 65–6.

Hardy, A.C. 1950. Johan Hjort, 1869–1948. *Obituary Notices of Fellows of the Royal
Society* 7: 167–81.

Harman, P.M. 1982. *Energy, force, and matter: The conceptual development of Nineteenth-century physics.* Cambridge: Cambridge University Press. ix + 182pp.

– 1987. Mathematics and reality in Maxwell's dynamical physics. In: R. Kargon
and P. Achinstein, eds, *Kelvin's Baltimore Lectures and modern theoretical physics,*
pp. 267–97. Cambridge, MA: MIT Press.

– 1998. *The natural philosophy of James Clerk Maxwell.* Cambridge: Cambridge
University Press. xiv + 232pp.

Hattersley-Smith, G. 1976. The British Arctic Expedition, 1875–76. *Polar Record*
18(113): 117–26.

Hedgpeth, J.W. 1945. The United States Fish Commission steamer *Albatross.*
With an appendix by Waldo L. Schmitt. *The American Neptune* 5: 5–26.

– 1946. The steamer *Albatross. Scientific Monthly* 65: 17–22.

– 1974. One hundred years of Pacific oceanography. In C.B. Miller, ed., *The
biology of the oceanic Pacific,* pp. 137–55. Corvallis, Oregon: Oregon State University Press.

Heilbron, J.L. 1990. Introductory essay. In T. Frängsmyr, J.L. Heilbron, and R.E.
Rider, eds, *The quantifying spirit in the 18th Century,* pp.1–23. Berkeley: University of California Press.

Helland-Hansen, B. 1905. Report on hydrographical investigations in the
Faeroe-Shetland Channel and the northern part of the North Sea in the year
1902. *Fishery Board for Scotland. Report of the North Sea Fishery Investigation Committee (Northern Area)* 1902–03. *British Sessional Papers,* Vol. 14(1), pp. 1–49.

– 1912. Physical oceanography. In J. Murray and J. Hjort, *The depths of the ocean,*
Ch. 5, pp. 210–306. London: Macmillan.

– 1916. Nogen hydrografiske metoder. *Forhandlinger. ved de skandinaviske natur-
forskeres* 16. møte i Kristiania den 10.–15. juli 1916, pp. 357–9.

– 1934. The Sognefjord Section: Oceanographic observations in the north-
ernmost part of the North Sea and the southern part of the Norwegian Sea.
In R.J. Daniel, ed., *James Johnstone memorial volume,* pp. 257–74. Liverpool:
University Press of Liverpool.

Helland-Hansen, B., and F. Nansen. 1909a. Die jährlichen Schwankungen
der Wassermassen im Norwegischen Nordmeer in ihrer Beziehung zu den
Schwankungen der meteorologischen Verhältnisse, der Ernteerträge und der
Fischereiergebnisse in Norwegen. *Internationale Revue der gesamten Hydrobiologie* 2(3): 337–61.

– 1909b. The Norwegian Sea: Its physical oceanography based upon the Norwegian researches 1900–1904. *Report on Norwegian Fishery and Marine Investigations* 2(2), xx + 390pp; Supplement, i - xii.

– 1917. *Temperatur-Schwankungen des Nordatlantischen Ozeans und in der Atmosphäre: Einleitende Studien über die Ursachen der klimatologischen Schwankungen.* Kristiania: In Kommission bei Jacob Dybwad. VIII + 341pp.

– 1920. Temperature variations in the North Atlantic Ocean and in the atmosphere. *Smithsonian Miscellaneous Collections* 70(4), viii + 408pp.

– 1926. *The eastern North Atlantic. Norsk Videnskaps-Akademi i Oslo. Geografiske Publikasjoner* IV(2), 76pp.

Helland-Hansen, B., and J.S. Worm-Müller. 1940. Nansen, Fridtjof. *Norsk Biografisk leksikon* IX: 599–640.

Herdman, W.A. 1922. H.S.H. Prince Albert of Monaco. *Nature* 110 (2752): 156–8.

Herschel. J.F.W. 1861. *Physical geography.* From the Encyclopaedia Britannica. Edinburgh: Adam and Charles Black. viii + 441pp.

– 1871. 'Sir J. Herschel on ocean currents.' *Nature* 4: 71.

Hertz, H. 1900. *The Principles of mechanics, presented in a new form.* Preface by H. von Helmholtz. Authorized English translation by D.E. Jones and J.T. Whalley. London: Macmillan. xx + 274pp.

Hesse, M. 1961. *Forces and fields: The concept of action at a distance in the history of physics.* London: Thomas Nelson. x + 318pp.

Hesselberg, T.1923. Bjerknes, Vilhelm Friman Koren. *Norsk Biografisk Leksikon* I: 584–8.

– 1940. Mohn, Henrik. *Norsk Biografisk Leksikon* IX: 290–5.

Hettner, A. 1915. *Englands Weltherrschaft und der Krieg.* Leipzig: B.G. Teubner. 269S.

Hjort, J. 1914. Fluctuations in the great fisheries of northern Europe viewed in the light of biological research. *Conseil International pour l'Exploration de la Mer, Rapports et Procès-Verbaux* 20: 1–228.

– 1915. Investigations into the natural history of the herring of the Atlantic waters of Canada, 1914. Preliminary Report No.1. *Ottawa. Department of the Naval Service. Supplement to the Fifth Annual Report of the Department of the Naval Service for 1915,* 38pp.

– 1919. Introduction to the Canadian Fisheries Expedition, 1914–1915. In: Canada, Department of the Naval Service, *Canadian Fisheries Expedition, 1914–1915: Investigations in the Gulf of St. Lawrence and Atlantic waters of Canada under the direction of Dr. Johan Hjort, head of the expedition, director of fisheries for Norway,* pp. xi–xxviii. Ottawa: King's Printer.

– 1923. Hydrographical and biological investigations in North Atlantic waters. In *Festskrift tillägnad professor Otto Pettersson den 12 februari 1923: Sven Otto Pettersson en havets arbetare,* pp. 22–48. Helsingfors: Holger Schildts Tryckeri.

– 1927. Georg Ossian Sars (1837–1927). *Journal du Conseil International pour l'Exploration de la Mer* 2(2): 111–12.

Hopkins, T. 1852. On the causes of the great currents of the ocean. *Memoirs of the Literary and Philosophical Society of Manchester*, Series 2, 10: 1–16.

Hornberger, T. (ed.).1979. *A goodly gallerye: William Fulke's Book of Meteors (1563)*. Philadelphia: American Philosophical Society. 121pp.

Howse,D. 1980. *Greenwich time and the discovery of the longitude*. Oxford: Oxford University Press. xviii + 254pp.

Hubbard, J. 1993. An independent progress: The development of marine biology on the Atlantic coast of Canada, 1898–1939. Ph.D. diss., University of Toronto. v + 446pp.

– 2006. *A science on the scales. The rise of Canadian Atlantic fisheries biology, 1898–1939*. Toronto: University of Toronto Press. x + 351pp.

Hubendick, E. 1950. Fredrik Gustaf Ekman. *Svenskt Biografiskt Lexikon* 13: 117–23.

Humboldt, A. von. 1814. *Voyage aux régions équinoxiales du nouveau continent, fait en 1799–1804 par Al. de Humboldt et A. Bonpland. Part. I. Relation historique, 1.* Paris: F. Schoell.

– 1837. Der Perustrom. In H. Berghaus, *Allgemeine Länder und Volkerkunde. Erster Band*, pp. 575–83. Stuttgart.

– 1845. *Kosmos: Entwurf einer physischen Weltbeschreibung. Erster Band*. Stuttgart u. Tübingen: J.G. Gotta'schen Verlag. xvi + 493pp.

– 1975 (1850). *Views of nature: or contemplations of the sublime phenomena of creation; with scientific illustrations*. Translated from the German by E.C. Otté and Henry G. Bohn. Reprint of the 1850 edition published by H.G. Bohn. New York: Arno Press. xxx + 452pp.

– 1997 (1845). *Cosmos: A sketch of the physical description of the universe. Volume 1.* Translated by E.C. Otté. Introduction by Nicolaas A. Rupke. Baltimore: Johns Hopkins University Press. xlii + 375pp.

Huntsman, A.G. 1919. Some quantitative and qualitative plankton studies of the eastern Canadian plankton. In: Canada, Department of the Naval Service, *Canadian Fisheries Expedition, 1914–1915: Investigations in the Gulf of St. Lawrence and the Atlantic waters of Canada under the direction of Dr. Johan Hjort, leader of the expedition, director of fisheries for Norway*, pp. 404–85. Ottawa: King's Printer.

– 1924a. *The ocean around Newfoundland*. Newfoundland Ministry of Marine and Fisheries, Saint John's Fisheries Research Laboratory, 19pp.

– 1924b. Oceanography. In *Handbook of Canada*, Issued by the Local Committee on the occasion of the meeting of the British Association for the Advancement of Science at Toronto, August 1924. Toronto: University of Toronto Press.

- 1925. The ocean around Newfoundland. *The Canadian Fisherman,* January. 4pp.
- 1927a. The oceanography of the Canadian Atlantic. American Geophysical Union, Section of Oceanography. *Washington. Bulletin of the National Research Council* No. 6. *Transactions of the American Geophysical Union* 8: 204–6.
- 1927b. *The Passamaquoddy Bay power project and its effect on the fisheries.* St John, NB, 45pp.
- 1930. *Biological Board of Canada. Report of the Atlantic Biological Station for 1930.* Toronto: mimeographed, ca. 5pp. + appendices.
- 1931a. Appendix No. 31. Survey of the fishery resources of the Fundy area. In *Biological Board of Canada. Report of the Atlantic Biological Station for 1931,* by A.G. Huntsman. Toronto: mimeographed, unpaged.
- 1931b. *Biological Board of Canada. Report of the Atlantic Biological Station for 1931.* Toronto: mimeographed, 5pp + appendices.
- 1931c. Biological and oceanographic conditions in Hudson Bay. 1. Hudson Bay and the determination of fisheries. *Contributions to Canadian Biology and Fisheries* 6, *Series* A, No. 22: 455–62.
- 1945. Edward Ernest Prince. *Canadian Field-Naturalist* 59: 1–3.
Hutchinson, A.H. 1928. A bio-hydrographical investigation of the sea adjacent to the Fraser River mouth. Paper II. Factors affecting the distribution of the phytoplankton. *Transactions of the Royal Society of Canada, Section* V, *Series* 3, 22: 293–310.
- 1929. The economic effect of the Fraser River on the waters of the Strait of Georgia. *Biological Board of Canada. Progress Reports of Pacific Coast Stations* 4: 3–6.
- 1944. Dr. Arthur Willey. *Journal of the Fisheries Research Board of Canada* 6 (3): 207–8.
Hutchinson, A.H., and C.C. Lucas. 1931. The epithalassa of the Strait of Georgia: Salinity, temperature, pH and phytoplankton. *Canadian Journal of Research* 5: 231–84.
Hutchinson, A.H., C.C. Lucas, and M. McPhail. 1929. Seasonal variations in the chemical and physical properties of the waters of the Strait of Georgia in relation to phytoplankton. *Transactions of the Royal Society of Canada, Section* V, Series 3, 23: 177–83.
- 1930. An oceanographic survey of the Strait of Georgia. In *Contributions to marine biology,* pp. 87–90. Stanford, CA: Stanford University Press.
Hylleraas, E. 1957. Bjørn Helland-Hansen. *Fra Fysikkens Verden* 19: 97–100.
ICES (International Council for the Exploration of the Sea). 1906. Hydrographical Section, February–March 1906. *Conseil international pour l'Exploration de la Mer. Rapports et Procès-verbaux des Réunions* 6C: 30–6.

Idrac, P. 1934. Berget (Alphonse). *Larousse mensuel* 9, No. 329, p. 738.

Imbrie, J., and K.P. Imbrie. 1979. *Ice ages: Solving the mystery*. Hillside, NJ: Enslow. 224pp.

Ince, S. 1987. Some early attempts at theory formation in fluid mechanics. In E.R. Landa and S. Ince, eds. *The history of hydrology*, pp. 35–8. American Geophysical Union, History of Geophysics 3.

International Council for the Exploration of the Sea (ICES). 1906. Hydrographical Section, February–March 1906. *Conseil international pour l'Exploration de la Mer. Rapports et Procès-verbaux des Réunions* 6C: 30–6.

Irons, J.C. 1896. *Autobiographical sketch of James Croll, LL.D., F.R.S., etc. with memoir of his life and work*. London: Edward Stanford. 553pp.

Iselin, C. O'D. 1927. A study of the northern part of the Labrador Current. *Transactions of the American Geophysical Union* No. 61: 217–22.

– 1929. Recent work on the dynamic oceanography of the North Atlantic. *Transactions of the American Geophysical Union* 10: 82–9.

– 1930. A report on the coastal waters of Labrador, based on the explorations of the 'Chance' during the summer of 1926. *Proceedings of the American Academy of Arts and Sciences* 66(1): 1–37.

– 1933a. The development of our conception of the Gulf Stream system. *Transactions of the American Geophysical Union* 14: 226–31.

– 1933b. Some phases of modern deep-sea oceanography: With a description of some of the equipment and methods of the newly formed Woods Hole Oceanographic Institution. *Report of the Smithsonian Institution for 1932*, pp. 251–67.

– 1936. A study of the circulation of the western North Atlantic. *Papers in Physical Oceanography and Meteorology* 4(4), 101pp.

– 1937. The new plans for the cooperative investigation of the North Atlantic's circulation. *Transactions of the American Geophysical Union* 18: 222–3.

– 1938a. The influence of fluctuations in the major ocean currents on the climate and the fisheries. *The Collecting Net* 13(7): 1–4.

– 1938b. Problems in the oceanography of the North Atlantic. *Nature* 14 (3574): 772.

– 1938c. A promising theory concerning the causes and results of long-period variations in the strength of the Gulf Stream system. *Transactions of the American Geophysical Union* 19: 243–4.

– 1939a. The influence of vertical and lateral turbulence on the characteristics of the waters at mid-depths. *Transactions of the American Geophysical Union* 20: 414–17.

– 1939b. Some physical factors which may influence the productivity of New England's coastal waters. *Journal of Marine Research* 2(1): 74–85.

– 1940. Preliminary report on long-period variations in the transport of the Gulf Stream system. *Papers in Physical Oceanography and Meteorology* 8(1): 1–40.

– 1942. IV. Eleventh annual report of the director for the year 1940. *Woods Hole Oceanographic Institution. Report for the year 1940*, pp. 1–30.

Iselin, C. O'D., and F.C. Fuglister. 1948. Some recent developments in the study of the Gulf Stream. *Journal of Marine Research* 7: 317–29.

Jacobsen, J.P. 1913. Beitrag zur Hydrographie der Dänischen Gewässer. *Meddelser fra Kommissionen for Havundersøgelser, Serie: Hydrografi* 1 (2): 1–94.

Jahn, I. 2001. Alexander von Humboldt's cosmical view on nature and his researches shortly before and shortly after his departure from Spain. *Estudios de Historia das Ciencias e das Técnicas. Actos do VII Congreso de la Sociedad Española de Historia de las Ciencias y de las Técnicas*, I: 31–9.

Jamieson, S. 2005. Introduction. In J. Thoulet, *A voyage to Newfoundland*, pp. xiv–xxxiv. Translated from the French and edited by Scott Jamieson. Montreal and Kingston: McGill-Queen's University Press.

Janson, O. 1907. *Meeresforschung und Meeresleben*. Aus Natur und Geisteswelt. Sammlung wissenschaftlich-gemeinverständlichen Darstellungen. 30. Bandchen. Leipzig: B.G. Teubner. iv + 148pp.

Jarrell, R.A. 1985. Science. *The Canadian Encyclopedia* III: 1653–55.

Jewell, R. 1984. The meteorological judgement of Vilhelm Bjerknes. *Social Research* 51: 783–807.

Jilek, A. 1857. *Lehrbuch der Oceanographie zum Gebrauche der k.k. Marine-Akademie*. Wien: Kaiserlich-Königlichen Hof- und Staatsdruckerei. x + 298S.

J.M.R. 1892. Johnston, Alexander Keith. *Dictionary of National Biography* 30: 54–5.

Jochum, M., and R. Murtugudde. 2006. *Physical oceanography: Developments since 1950*. New York: Springer. xii + 250pp.

Johnston, A.K. 1848. *The physical atlas: A series of maps and notes illustrating the geographical distribution of natural phenomena*. Edinburgh: Blackwood. 94pp.+ 30 maps.

Johnstone, K. 1977. *The aquatic explorers: A history of the Fisheries Research Board of Canada*. Toronto: University of Toronto Press. xv + 342pp.

Jones, I., and J. Jones. 1992. *Oceanography in the days of sail*. Sydney: Hale and Iremonger. 288pp.

Jones, J.E., and I.S.F. Jones. 2002. The western boundary current in the Pacific: The development of our oceanographic knowledge. In K.R. Benson and P.F. Rehbock, eds, *Oceanographic history: The Pacific and beyond*, pp. 86–95. Seattle: University of Washington Press.

Jörberg, L. 1973. The industrial revolution in the Nordic countries. In C.M. Cipolla, ed., *The Fontana economic history of Europe: The emergence of industrial societies. Part II*, pp. 375–485. Glasgow: Fontana/Collins.

J[oubin], L. 1934. Le professeur Alphonse Berget. *Bulletin de la Société d'océano-graphie de France* 14(75): 1328.

Jungnickel, C., and R. McCormmach. 1986. *Intellectual mastery of nature: Theoretical physics from Ohm to Einstein. Volume 2. The now mighty theoretical physics, 1870–1925.* Chicago: University of Chicago Press. xxv + 435pp.

Kangro, H. 1973. Kircher, Athanasius. *Dictionary of Scientific Biography* 7: 374–8.

Kargon, R., and P. Achinstein, eds. 1987. *Kelvin's Baltimore Lectures and modern theoretical physics.* Cambridge, MA: MIT Press. vii + 547pp.

Kawai, H. 1998. A brief history of the recognition of the Kuroshio. *Progress in Oceanography* 41: 505–78.

Kayser, J. 1873. *Physik des Meeres: Für gebildete Leser.* Paderborn: Ferdinand Schöningh. x + 359S.

Kerz, W. 1979. Die Entwicklung der Geophysik zur eigenständen Wissenschaft. *Gauss-Gesellschaft E.V. Göttingen Mitteilungen* Nr 16: 41–54.

Kevles, D.J. 1971. 'Into hostile political camps': The reorganization of international science in World War I. *Isis* 62: 47–60.

Kiilerich, A.B. 1939. The Godthaab Expedition 1928: A theoretical treatment of the hydrographical observations. *Meddelelser om Grønland* 78(5), 149pp.

Kircher, A. 1665. *Mundus subterraneus.* Amsterdam [2 books, 346 and 487 pp.]

Kjellen, R. 1914. *Die Grossmächte der Gegenwart.* Leipzig: B.G. Teubner. 208S.

Klein, M.J. 1972. Mechanical explanation at the end of the nineteenth century. *Centaurus* 17: 58–82.

Knight, D.M. 1974. Introduction. In *The Admiralty manual of scientific enquiry. Prepared for the use of officers in Her Majesty's Navy; and travellers in general.* Ed. J.F.W. Herschel. 2nd ed (1851). Reprint. Folkestone, Kent: Dawson. xi + 504pp.

Knudsen, M., ed. 1901. *Hydrographical tables.* Copenhagen: G.E.C. Gadd; London: Williams and Norgate. v + 63pp.

Kobe, G. 1934. Der hydrographische Aufbau und die dadurch bedingten Strömungen im Skagerrak. *Veröffentlichungen des Instituts für Meereskunde an der Universität Berlin.* Neue Folge. A. *Geographisch-naturwissenschaftliche Reihe* 26: 1–62.

Kofoid, C.A. 1910. *The biological stations of Europe.* United States Bureau of Education Bulletin No. 4 (440). Washington, DC: Government Printing Office. xiii + 360pp.

Kohl, J.G. 1868. *Geschichte des Golfstroms und seiner Erforschung von den ältesten Zeiten bis auf den grossen amerikanischen Bürgerkrieg. Eine Monographie zur Geschichte der Oceane und der geographischen Entdeckungen.* Bremen: Müller. xv + 224pp.

Kohlmann, R. 1905. *Beiträge zur Kenntnis der Strömungen des westlichen Ostsee.* Inaugural-Dissertation zur Erlangung der Doctorwürde der hohen philosophischen Facultät der Königlichen Christian-Albrechts-Universität zu Kiel. Kiel: Schmidt and Klaunig. 50pp.

Kortum, G. 1983. Ferdinand von Richthofen und die Kunde vom Meere. *Schriften der Naturwissenschaftlichen Verein für Schleswig-Holstein* 53: 1–32.

– 1987. Berlins Bedeutung für die Entwicklung der geographischen Meereskunde: Das Vermächtnis der Georgenstrasse. *Berliner geographische Studien* 25: 133–56.

– 1990. An unpublished manuscript of Alexander von Humboldt on the Gulf Stream. In W. Lenz and M. Deacon, eds, *Ocean sciences: Their history and relation to man.* Deutsche Hydrographische Zeitschrift, Ergänzungsheft, Reihe B, 22: 122–30.

– 1993a. Otto Krümmel an der Deutsche Seewarte, Hamburg. In: G. Wegner, ed., *Meeresforschung in Hamburg: Von Vorgestern bis Übermorgen,* pp. 63–76. Deutsche Hydrographische Zeitschrift, Ergänzungsheft, Reihe B, Nr. 25.

– 1993b. Überfahrten in die Neue Welt, Die Atlantikquerungen von Christoph Columbus (1492) und Alexander von Humboldt (1799) im ozeanographiegeschichtliche Vergleich. *Zeitschrift für geologischen Wissenschaft* 21(5/6): 605–16.

– 1994a. Alexander von Humboldts Besuch auf Helgoland 1790 und die frühe Entwicklung der Meeresbiologie in Deutschland. *Schriften der Naturwissenschaftlichen Verein für Schleswig-Holstein* 64: 111–33.

– 1994b. Alexander von Humboldts Forschungsfahrt auf dem Kaspischen Meer 1829. *Deutsche Gesellschaft für Meeresforschung – Mitteilungen* 3/94: 3–9.

Kortum, G., and A. Lehmann. 1997. A. v. Humboldts Forschungsfahrt auf der Ostsee im Sommer 1834. *Schriften der Naturwissenchaftlichen Verein für Schleswig-Holstein* 67: 45–58.

Kröber, G. 1965. *L. Euler. Briefe an eine deutsche Prinzessin.* Leipzig.

Krümmel, O. 1877. *Die aequatorialen Meeresströmungen des Atlantischen Oceans und das allgemeine System der Meerescirculation.* Leipzig: Duncker und Humblot. 52S.

– 1879. *Versuch einer vergleichenden Morphologie der Meeresräume.* Leipzig: Duncker und Humblot. xii + 110S.

– 1886. *Der Ozean: Eine Einführung in die allgemeine Meereskunde.* Das Wissen der Gegenwart 52. Leipzig and Prague: G. Frentag und F. Tempsky. viii + 242S.

– 1907. *Handbuch der Ozeanographie. Band I. Die räumlich, chemischen und physikalischen Verhältnisse des Meeres.* Stuttgart: J. Engelhorn. 526pp.

– 1911. *Handbuch der Ozeanographie. Band II. Die Bewegungsformen des Meeres.* Stuttgart: J. Engelhorn. 766pp.

Krusenstern, A.J. 1813. *Voyage around the world, in the years 1803, 1804, 1805 and 1806, by order of His Imperial Majesty Alexander I, on board the ships Nadeshda and Neva.* Translated by R.B. Hoppner. 2 vols. London.

Kullenberg, B. 1954. Vagn Walfrid Ekman 1874–1954. *Journal du Conseil International pour l'Exploration de la Mer* 20: 140–3.

Kutzbach, G. 1970. Buchanan, John Young. *Dictionary of Scientific Biography* 2: 557–8.

– 1979. *The thermal theory of cyclones: A history of meteorological thought in the nineteenth century.* Historical Monographs Series, American Meteorological Society. Boston: American Meteorological Society. 255pp.

Lacombe, H. 1965. *Cours d'océanographie physique (Théories de la circulation générale: Houles et vagues).* Paris: Gauthier-Villars. 392pp.

Lacroix, A., G. Durand-Viel, L. Fage, and C. Vallaux. 1946. Hommages au Docteur Jules Richard, directeur du Musée océanographique de Monaco. *Monaco. Bulletin de l'Institut océanographique* 43 (892), 24pp.

Lamb, H. 1879. *A Treatise on the mathematical theory of the motion of fluids.* Cambridge: Cambridge University Press. x + 258pp.

Lankester, E.R. 1885. Dr. Carpenter, C.B., F.R.S. *Nature* 33: 83–5.

Laqueur, W. 1974. *Weimar: A cultural history, 1918–1933.* London: Weidenfeld and Nicolson. xi + 308pp.

Latour, B. 1987. *Science in action: How to follow scientists and engineers through Society.* Cambridge, MA: Harvard University Press. viii + 274pp.

Lauzier, L. 1950. Multiple ship survey of the Gulf Stream, Operation Cabot. In H.B. Hachey, *Canadian Joint Committee on Oceanography, Annual Report – 1950,* pp. 14–15. Ottawa. 34pp + appendices.

– 1953. The St Lawrence spring run-off and summer salinities in the Magdalen Shallows. *Journal of the Fisheries Research Board of Canada* 10: 146–7.

Lauzier, L., and R.W. Trites. 1959. The deep waters in the Laurentian Channel. *Journal of the Fisheries Research Board of Canada* 15: 1247–57.

Lea, E. 1919. Report on 'age and growth of the herring in Canadian waters.' In: Canada, Department of the Naval Service, *Canadian Fisheries Expedition, 1914–1915: Investigations in the Gulf of St. Lawrence and the Atlantic waters of Canada under the direction of Dr. Johan Hjort, leader of the expedition, director of fisheries for Norway,* pp. 75–171. Ottawa: King's Printer.

Leary, W.M. 1999. *Under ice: Waldo Lyon and the development of the Arctic submarine.* College Station, TX: Texas A & M University Press. xxviii + 303pp.

Le Danois, E. 1924. Étude hydrologique de l'Atlantique-Nord. *Paris. Annales de l'Institut océanographique,* N.S., 1(1): 1–52.

– 1937a. Hydrographie et hydrologie de l'Atlantique Nord. *Paris. Annales de l'Institut océanographique,* N.S., 16(4): 453–66.

– 1937b. Hydrographie et hydrologie des bancs de Terre-Neuve. *Paris. Annales de l'Institut océanographique,* N.S., 16(4): 466–79.

Lehmann, H. 1965. Grund, Alfred Johannes. *Neue Deutsche Biographie* 7: 216–17.

Leighly, J. 1963. Introduction. In: J. Leighly, ed., *The physical geography of the sea and its meteorology,* by Matthew Fontaine Maury, pp. ix–xxx. Cambridge, MA: Harvard University Press.

– 1968. Matthew Fontaine Maury in his time. *Monaco. Bulletin de l'Institut océanographique, No. spécial 2 (Premier Congrès internationale d'histoire de l'océanographie)* 1: 147–59.

Leim, A.H., and H.B. Hachey. 1935. A transgression of marginal waters over the Scotian Shelf. *Transactions of the American Fisheries Society* 65: 279–83.

Lenz, E. 1830. Ueber das Wasser des Weltmeeres in verschiedenen Tiefen in Rücksicht auf die Temperatur und den Salzgehalt. *Annalen der Physik und Chemie* 20: 73–140.

– 1831. Physikalische Beobachtungen angestellt auf einer Reise um die Welt unter dem Commando des Capitains Otto von Kotzebue in den Jahren 1823, 1824, 1825 und 1826. *Mémoires de l'Académie Impériale des Sciences de St-Petersbourg, 6me Série. Sciences Mathématiques, Physiques et Naturelles* 1: 221–341.

– 1832. On the temperature and saltness of the waters of the ocean at different depths. *Edinburgh Journal of Science, 2nd Series,* 6: 341–4.

– 1845 [1847]. Bemerkungen über die Temperatur des Weltmeeres in verschiedenen Tiefen. *Bulletin de la Classe physico-mathématique de l'Académie impériale des sciences de St. Pétersbourg* 5 (1845–6): 65–74.

– 1847. Bericht über die ozeanischen Temperaturen in verschiedenen Tiefen. *Bulletin de la Classe d'histoire et philosophie de l'Académie impériale des sciences de St. Pétersbourg,* 3, Supplement, pp.11–12.

– 1848. Bermerkungen über die Temperatur des Weltmeeres in verschieden Tiefen. *Annalen der Physik und Chemie,* Ergänzungsband 72: 615–26.

Lenz, W. 1986. Gerhard Schott 1866–1961. *Deutsche Gesellschaft für Meeresforschung, DGM-Mitteilungen* 1/86: 19–22.

Levere, T.H. 1993. *Science and the Canadian Arctic: A century of exploration, 1818–1918.* Cambridge: Cambridge University Press. xii + 438pp.

Levings, C. 2000. Tully, John Patrick 'Jack.' In D. Francis, ed., *Encyclopedia of British Columbia,* p. 721. Vancouver: Harbour Publishing.

Lewis, J.M. 1996. C.-G. Rossby: Geostrophic adjustment as an outgrowth of modelling the Gulf Stream. *Bulletin of the American Meteorological Society* 77(11): 2711–28.

Lezhneva, O.A. 1973. Lenz, Emil Khristianovich (Heinrich Friedrich Emil). *Dictionary of Scientific Biography* 8: 187–9.

Liljequist, G. 1993. *High latitudes: A history of Swedish polar travels and research.* Swedish Polar Research Secretariat and Streiffert Forlag AB. 607pp.

Limburg, P.R. 1979. *Oceanographic institutions.* Amsterdam: Elsevier. 265pp.

Lindberg, D.C. 1992. *The beginnings of Western science: The European scientific tradi-*

tion in philosophical, religious, and institutional context, 600 B.C. to A.D. 1450. Chicago: University of Chicago Press. xvii + 455pp.

Longard, J.R. 1993. *Knots, volts and decibels – An informal history of the Naval Research Establishment, 1940–1967.* Dartmouth, NS: Defence Research Establishment Atlantic. 113pp.

Lucas, C.C. 1929. Further oceanographic studies of the sea adjacent to the Fraser River mouth. *Transactions of the Royal Society of Canada, Series 3, Section V,* 23: 29–58.

Lucas, C.C., and A.H. Hutchinson. 1927. A bio-hydrographical investigation of the sea adjacent to the Fraser River mouth. *Transactions of the Royal Society of Canada, Series 3, Section V,* 21: 485–512.

Lüdecke, C. 1995. Die Deutsche Polarforschung seit der Jahrhundertwende und der Einfluss Erich von Drygalskis. *Berichte zur Polarforschung* 158: i–xiii, 1–340, A1–A72.

Lutjeharms, J.R.E., W.P.M. de Ruiter, and R.G. Peterson. 1992. Interbasin exchange and the Agulhas retroflection: The development of some oceanographic concepts. *Deep-Sea Research* 39: 1791–807.

Luyten, J., and N. Hogg, eds. 1992. A tribute to Henry Stommel. *Oceanus,* Special Issue, 132pp.

Luyten, J.R., and W.J. Schmitz, Jr. 1992. Henry Melson Stommel. *Deep-Sea Research* 39 (7/8A): I–II.

Mackenzie, O.F. 1943. Tribute to the late J.J. Cowie. *Fisheries Research Board of Canada, Progress Reports (Atlantic),* No. 34.

MacLeod, R.M. 1995. 'Kriegsgeologen and practical men': Military geology and modern memory 1914–1918. *British Journal for the History of Science* 28: 427–50.

Maffioli, C.S. 1994. *Out of Galileo: The science of waters, 1628–1718.* Rotterdam: Erasmus. xx + 509pp.

Mahoney, M.S. 1974. Mariotte, Edme. *Dictionary of Scientific Biography* 9: 114–22.

Mamayev, O.I. 1987. Björn Helland-Hansen, 1877–1957 (on the 110th anniversary of his birth). *Oceanology* 27(5): 664–7.

Marcet, A. 1819. On the specific gravity and temperature of sea waters, in different parts of the ocean, and in particular seas; with some account of their saline contents. *Philosophical Transactions of the Royal Society of London* 109: 161–208.

Markham, C.R. 1895. *Major James Rennell and the rise of modern geography.* New York: Macmillan. vii + 232pp.

Marmer, H.A. 1926. *Coastal currents along the Pacific coast of the United States.* U.S. Coast and Geodetic Survey, Special Publication 121, iv + 80pp.

Marsigli, L.F. 1681. *Osservazioni Intorno al Bosforo Tracio Overo Canale di Con-*

stantinopli Rappresentate in Lettera alla Sacra Real Maesta di Cristina Regina di Svezia. Roma. (Reprinted, 1935. *Bolletina di Pesca, Piscicultura e Idrobiologia* 11: 734–58.)

– 1999 (1725). *Histoire physique de la mer.* Edizione Fotostatico con Versione Inglese. Bologna: Museo di Fisica dell'Università di Bologna. ix + 546pp.

Martire de Anghiera, P. 1516. *De Rebus Oceanis et Orbe Novo Decades Tres.* Decade 3, Book 6. Alcala.

– 1555. *The decades of the New Worlde or West Indies.* Compiled by Richard Eden. 1996 facsimile edition. Ann Arbor, MI: University Microfilms. Unpaged + 233pp.

Matthäus, W. 1967. Der Ozeanograph Johann Gottfried Otto Krümmel (1854– 1912). *Wissenschaftliche Zeitschrift der Universität Rostock – 16 Jahrgang 1967. Mathematisch-Naturwissenschaftliche Reihe,* 9/10: 1219–24.

Matthews, D.J. 1914. Report by Mr. D.J. Matthews. In *Report on the work carried out by the SS 'Scotia,' 1913: Ice observation, meteorology and oceanography in the North Atlantic Ocean,* pp. 4–47. London: HMSO.

Maurice, H.G. 1923. Foreword. In *Festskrift Tillägnad Professor Otto Pettersson den 12 Februari 1923. Sven Otto Pettersson en Havets Arbetare,* pp. 9–12. Helsingfors: Holger Schildts Tryckeri.

– 1948. Prof. Johan Hjort, For. Mem. R.S. *Nature* 162: 765–6.

Maury, M.F. 1844. Remarks on the Gulf Stream and currents of the sea. *American Journal of Science and Arts* 47: 161–81.

– 1963 (1861). *The physical geography of the sea and its meteorology.* Edited by John Leighly. Cambridge, MA: Harvard University Press. xxx + 432pp.

Mavor, J.W. 1920a. Circulation of water in the Bay of Fundy and the Gulf of Maine. *Transactions of the American Fisheries Society* 50: 334–44.

– 1920b. Drift bottles indicating a superficial circulation in the Gulf of Maine. *Science* 52: 442–3.

– 1922. The circulation of water in the Bay of Fundy. I. Introduction and drift bottle experiments. *Contributions to Canadian Biology and Fisheries (New Series)* 1(8): 101–24.

– 1923. The circulation of water in the Bay of Fundy. Part II. The distribution of temperature, salinity and density in 1919 and the movements of water which they indicate in the Bay of Fundy. *Contributions to Canadian Biology and Fisheries (New Series)* 1(18): 353–75.

McCartney, M., B. Beardsley, and H. Bryden, eds. 1982. Cold wind two gyres: A tribute to Val Worthington. *Journal of Marine Research* 40, Supplement, xxii + 860pp.

McConnell, A. 1982. *No sea too deep: The history of oceanographic instruments.* Bristol: Adam Hilger. xii + 162pp.

– 1990. The art of submarine cable laying: Its contribution to physical oceanography. In W. Lenz and M. Deacon, eds, *Ocean sciences: Their history and relation to man*, pp. 467–73. *Deutsche Hydrographische Zeitschrift*, Ergänzungsheft Reihe B. Nr 22.

– 1999. Introduction. In L.F. Marsigli, *Histoire physique de la mer*, pp. 3–28. Edizione Fotostatico con Versione Inglese, a Cura di Giorgio Dragoni. Bologna: Museo di Fisica dell'Università di Bologna.

McCormmach, R. 1970. H.A. Lorentz and the electromagnetic view of nature. *Isis* 61: 457–97.

– 1972. Hertz, Heinrich. *Dictionary of Scientific Biography* 6: 340–50.

– 1982. *Night thoughts of a classical physicist*. Cambridge, MA: Harvard University Press. 219pp.

McEwen, G.F. 1910. Preliminary report on the hydrographic work carried on by the Marine Biological Association of San Diego. *University of California Publications in Zoology* 6: 189–204.

– 1912. The distribution of ocean temperatures along the west coast of North America deduced from Ekman's theory of the upwelling of cold water from the adjacent ocean depths. *Internationale Revue der gesamten Hydrobiologie und Hydrographie* 5: 243–86.

– 1915. Oceanic circulation and temperature off the Pacific coast. In *Nature and science on the Pacific Coast*, pp. 133–40. San Francisco: Elder.

– 1919. Ocean temperatures, their relation to solar radiation and oceanic circulation: Quantitative comparisons of certain empirical results with those deduced by principles and methods of mathematical physics. *Miscellaneous Studies in Agriculture and Biology. Semicentennial Publication of the University of California 1868–1918*, pp. 335–431.

– 1921a. The science of oceanography. *Special Publications of the Bernice P. Bishop Museum* No. 7, pp. 597–607.

– 1921b. The status of oceanographic studies of the Pacific. *Special Publications of the Bernice P. Bishop Museum* No. 7, pp. 487–97.

– 1922. Suggestions relative to the application of mathematical methods to certain basic problems of dynamical oceanography. *Transactions of the American Geophysical Union* 2: 787–81.

– 1924. A mathematical theory of the temperature distribution in water due to solar radiation, evaporation, and convection. *Transactions of the American Geophysical Union* 4: 139–40.

– 1927. Recent progress in the dynamical oceanography of the northeastern part of the North Pacific Ocean. *Transactions of the American Geophysical Union* 8: 222–35.

– 1929a. Application of the Bjerknes dynamical theory of oceanic circulation to the North Pacific off southern California and Cape Mendocino. *Proceedings of the Third Pan-Pacific Science Congress, Tokyo 1926*, 1, 1p. (abstract)

- 1929b. A mathematical theory of the vertical distribution of temperature and salinity in water under the action of radiation, conduction, evaporation and mixing due to the resulting convection. *Bulletin of the Scripps Institution of Oceanography, Technical Series,* 2: 197–306.
- 1929c. Ocean surface drift in the Pacific coastal belt off North America. *Proceedings of the Third Pan-Pacific Science Congress, Tokyo 1926,* pp. 191–8.
- 1929d. Qualitative review of a theory of vertical temperature gradients in relation to radiation, evaporation, and convection. *Proceedings of the Third Pan-Pacific Science Congress, Tokyo 1926,* p. 199.
- 1930a. Movement of ocean water between Ocean Cape and Cape Chiniak, Alaska, in December, 1927, and January, 1928. *Proceedings of the Fourth Pan-Pacific Science Congress, Java 1929,* 2b: 1031.
- 1930b. A survey of some of the methods used in modern dynamical oceanography. In *Contributions to marine biology: Lectures and symposia given at the Hopkins Marine Laboratory, December 20–21, 1929, at the midwinter meeting of the Western Society of Naturalists,* pp. 57–68. Stanford, CA: Stanford University Press.
- 1932. A summary of basic principles underlying modern methods of dynamical oceanography. In *Physics of the earth – V. Oceanography,* pp. 310–57. Bulletin of the National Research Council No. 85. Washington, DC,
- 1937a. Calculations of the rate of vertical displacement and turbulence in the San Diego region of the Pacific at monthly intervals. *International Association of Physical Oceanography, Procès-Verbaux* No. 2: 129–31.
- 1937b. Calculation of the velocity of horizontal surface currents in the eastern North Pacific. *International Association of Physical Oceanography, Procès-Verbaux* No. 2: 132–4.
- 1937c. The University and the Pacific. *California Monthly* (March), pp. 9–11, 37–9.
- 1938. Some energy relations between the sea surface and the atmosphere. *Journal of Marine Research* 1: 217–38.
McEwen, G.F., T.G. Thompson, and R. Van Cleve. 1930. Hydrographic sections and calculated currents in the Gulf of Alaska, 1927 and 1928. *Report of the International Fisheries Commission appointed under the treaty between the United States and Great Britain for the preservation of the Northern Pacific halibut fishery,* No. 4: 1–36.
McGowan, J.A. 2004. Sverdrup's biology. *Oceanography* 17(2): 106–12.
McKenzie, R.A., and H.B. Hachey. 1939. The cod fishery and the water temperature on the Scotian Shelf. *Fisheries Research Board of Canada. Progress Reports of Atlantic Coast Stations,* No. 24, pp. 6–8.
McLellan, H.J. 1957. On the distinctiveness and origin of the slope water off

the Scotian Shelf and its easterly flow south of the Grand Banks. *Journal of the Fisheries Research Board* of Canada 14: 213–39.

McLellan, H.J., L. Lauzier, and W.B. Bailey. 1953. The slope water off the Scotian Shelf. *Journal of the Fisheries Research Board of Canada* 10: 155–76.

McMurrich, J.P. 1884. Science in Canada. *The Week* 1(49): 776–7.

Menaché, M. 1952. Détermination indirecte de la densité de l'eau de mer: Dosage de la chlorinité par la méthode de Mohr-Knudsen. (Bibliographie). *Circulaires du Centre de recherches et d'études océanographiques, renseignements techniques et bibliographiques*, No. 7, 25pp.

Merriman, D. 1972. Hjort, Johan. *Dictionary of Scientific Biography* 6: 441–2.

Merz, A. 1910. Über die Bedeutung 24stündiger Beobachtungen für die Ozeanographie. *Internationale Revue der gesamtem Hydrobiologie und Hydrographie* 3: 44–9.

– 1911. Die Sprungschichte der Seen. *Mitteilungen des Vereins der Geographen an der Universität Leipzig* I: 79–81.

– 1912. Berliner Seenstudien und Meeresforschung. *Zeitschrift der Gesellschaft für Erdkunde zu Berlin*, 1913: 166–79.

– 1913. Einleitung. Die Nordseearbeiten des Instituts und die Methode vielstündiger Beobachtungen. In F. Wendicke, *Hydrographische und biologische Untersuchungen auf den deutschen Feuerschiffen der Nordsee 1910/11. Ausgeführt vom Institut für Meereskunde in Berlin und von der Biologischen Anstalt auf Helgoland. Die hydrographischen Ergebnisse*, pp. iii–xii. Veröffentlichungen des Instituts für Meereskunde an der Universität Berlin. Neue Folge. A. Geographisch-naturwissenschaftliche Reihe 3: xv + 124S.

– 1915. Neue Anschauungen über das nordatlantische Stromsystem. *Zeitschrift der Gesellschaft für Erdkunde zu Berlin*, 1915: 111–22.

– 1922a. Meereskunde, Wirtschaft und Staat. In *Meereskunde: Sammlung volkstümlicher Vorträge zum Verstandnis der nationalen Bedeutung von Meer und Seewesen*, pp. 1–36. Vierzehnter Jahrgang, Heft 1. Berlin: Institut für Meereskunde.

– 1922b. Temperaturschichtung und Vertikalzirkulation im Südatlantischen Ozean nach den 'Challenger' – und 'Gazelle' – Beobachtungen. *Zeitschrift der Gesellschaft für Erdkunde zu Berlin*, 1922: 288–98.

– 1925. Die Deutsche Atlantische Expedition auf dem Vermessungs- und Forschungsschiff 'Meteor.' I. Bericht. *Sitzungsberichte der Preussischen Akademie der Wissenschaften* 21: 562–86.

Merz, A., and G. Wüst. 1922. Die atlantische Vertikalzirkulation. *Zeitschrift der Gesellschaft für Erdkunde zu Berlin* 1922: 1–35.

– 1923a. Die atlantische Vertikalzirkulation: Eine Erwiderung an Professor O. Petterson. *Annalen der Hydrographie und maritimen Meteorologie* 51: 149–50.

– 1923b. Die atlantische Vertikalzirkulation. 3. Beitrag. *Zeitschrift der Gesellschaft für Erdkunde zu Berlin*, 1923: 132–44.

Merz, J.T. 1903. *A history of European thought in the nineteenth century. Volume II.* Edinburgh: William Blackwood and Sons. xiii + 807pp.

Meyer, H.C. 1955. *Mitteleuropa in German thought and action, 1815–1945.* The Hague: Martinus Nijhoff. xv + 378pp.

Middleton, W.E.K. 1979. *Physics at the National Research Council of Canada, 1929–1952.* Waterloo, ON: Wilfrid Laurier University Press. viii + 238pp.

Mill, H.R. 1922. H.S.H. the Prince of Monaco. *Geographical Journal* 60(3): 236–7.

Mills, E.L. 1978. Edward Forbes, John Gwyn Jeffreys, and British dredging before the *Challenger* Expedition. *Journal of the Society for the Bibliography of Natural History* 8: 507–36.

– 1983. Problems of deep-sea biology: an historical perspective. In G.T. Rowe, ed, *The sea, volume 8: Deep-sea biology*, ch. 1, pp. 1–79. New York: John Wiley and Sons.

– 1989. *Biological oceanography: An early history, 1870–1960.* Ithaca, NY: Cornell University Press. xvii + 378pp.

– 1990. Useful in many capacities: An early career in American physical oceanography. *Historical Studies in the Physical and Biological Sciences* 20(2): 265–311.

– 1991. The oceanography of the Pacific: George F. McEwen, H.U. Sverdrup and the origin of physical oceanography on the west coast of North America. *Annals of Science* 48: 241–66.

– 1993. *The Scripps Institution: Origin of a habitat for ocean science.* La Jolla, CA: Scripps Institution of Oceanography. 29pp.

– 1994 (1995). Bringing oceanography into the Canadian university classroom. *Scientia Canadensis* 18(1): 3–21.

– 1997. 'Physische Meereskunde': From geography to physical oceanography in the Institut für Meereskunde, Berlin, 1900–1935. *Historisch-Meereskundliches Jahrbuch* 4: 45–70.

– 2001a. Canadian marine science research in Arctic and northern waters. *History of Oceanography (Newsletter)* 13: 3–6.

– 2001b. Enlightened natural history or the beginnings of oceanic science? *Annals of Science* 58: 403–8.

– 2001c. Exploring a space for science: The marine laboratory as observatory. *Estudios de Historia des Ciencias e des Técnicas. Actas VII Congreso de la Sociedad Española de la Historia de las Ciencias e des Técnicas* I: 51–7.

– 2002. Pacific waters and the POG: The origin of physical oceanography on the West Coast of Canada. In K.R. Benson and P.F. Rehbock, eds, *Oceano-*

graphic history: The Pacific and beyond, pp. 303–15. Seattle: University of Washington Press.

– 2004. Mathematics in Neptune's garden: Making the physics of the sea quantitative, 1876–1900. In H.M. Rozwadowski and D.K. van Keuren, eds, *The machine in Neptune's Garden: Historical perspectives on technology and the marine environment,* pp. 39–63. Sagamore Beach, MA: Science History Publications.

– 2005. From *Discovery* to discovery: The hydrology of the Southern Ocean, 1885–1937. *Archives of Natural History* 32(2): 246–64.

– 2007. Creating a global ocean conveyor: George Deacon and *The Hydrology of the Southern Ocean.* In K.R. Benson and H.M. Rozwadowski, eds, *Extremes: Oceanography's polar adventures,* pp. 107–32. Sagamore Beach, MA: Science History Publications.

Mills, E.L., and J. Carpine-Lancre. 1992. The Oceanographic Museum of Monaco. In E. Mann-Borgese, ed., *Ocean frontiers: Explorations by oceanographers on five continents,* pp. 120–35. New York: Harry N. Abrams.

Milner, M. 1988. Inshore ASW: The Canadian experience in home waters. In W.A.B. Douglas, ed., *The RCN in transition, 1910–1985,* pp. 143–58, 371–3. Vancouver: University of British Columbia Press.

– 1999. *Canada's navy: The first century.* Toronto: University of Toronto Press. xiii + 356pp.

Mohn, H. 1875. *Gründzuge der Meteorologie.* Berlin: Dietrich Reimer.

– 1880. Die Norwegischen Nordmeer-Expedition: Resultate der Lothungen- und Tiefseetemperatur-Beobachtungen. *Petermanns Geographische Mitteilungen,* Ergänzungsband XIV, 63: 1–24.

– 1883. X. Meteorology. *The Norwegian North-Atlantic Expedition, 1876–1878.* Christiania: Grøndahl and Søn.

– 1885. Die Strömungen des Europäischen Nordmeeres. *Petermanns Geographische Mitteilungen,* Ergänzungsband XVIII, 79: 1–20.

– 1887. The North Ocean: Its depths, temperature and circulation. *The Norwegian North-Atlantic Expedition, 1876–1878 [Den Norske Nordhavs-expedition, 1876–1878]* II (2): 212pp. Christiania: Grøndahl and Søn.

Monmonier, M. 1999. *Air apparent: How meteorologists learned to map, predict and dramatize weather.* Chicago: University of Chicago Press. xiv + 309pp.

Montgomery, R.B. 1937. Fluctuations in monthly sea level on eastern U.S. coast as related to dynamics of Western North Atlantic Ocean. *Journal of Marine Research* 1: 165–85.

– 1938. Circulation in upper layers of southern North Atlantic deduced with use of isentropic analysis. *Papers in Physical Oceanography and Meteorology* 6(2): 1–55.

– 1981. Notes related to Stommel's early years in Woods Hole. In B.A. Warren and C. Wunsch, eds, *Evolution of physical oceanography: Scientific surveys in honor of Henry Stommel*, pp. xxiv–xxvi. Cambridge, MA: MIT Press.

Mosby, H. 1958. Bjørn Helland-Hansen. 1877–1957. *Journal du Conseil International pour l'Exploration de la Mer* 23: 321–3.

– 1959. Minnetale over Professor Bjørn Helland-Hansen holdt i den mat.-naturv. Klasses møte den 28 de februar 1958. *Det Norske Videnskaps-Akademi I Oslo. Årbok* 1958, pp. 37–43.

– 1961. Fridtjof Nansen: Oceanographer. In: P. Vogt, ed., *Fridtjof Nansen: Explorer – scientist – humanitarian*, pp. 106–18. Oslo: Dreyers Forlag.

Mounce, I. 1922. Effect of marked changes in specific gravity upon the amount of phytoplankton in Departure Bay waters. *Contributions to Canadian Biology and Fisheries, New Series*, 1(6): 81–94.

Mulligan, J.F. 1987. The influence of Hermann von Helmholtz on Heinrich Hertz's contributions to physics. *American Journal of Physics* 55(8): 711–19.

Munk, W.H. 1950. On the wind-driven ocean circulation. *Journal of Meteorology* 7: 79–93.

– 1955. The circulation of the oceans. *Scientific American* 193: 96–104.

– 1980. Affairs of the sea. *Annual Review of Earth and Planetary Sciences* 8: 1–16.

– 1992. *The ocean* 'bible': Reminiscences. *Oceanography* 5(3): 155–7.

Munk, W., and D. Day. 2003. Harald U. Sverdrup, 1888–1957. In R.L. Fisher, E.D. Goldberg, and C.S. Cox, eds, *Coming of age: Scripps Institution of Oceanography: A centennial volume, 1903–2003*, pp. 177–201. University of California at San Diego. Scripps Institution of Oceanography.

Murray, J., and J. Hjort. 1912. *The depths of the ocean: A general account of the modern science of oceanography based largely on the scientific researches of the Norwegian steamer 'Michael Sars' in the North Atlantic*. London: Macmillan. xx + 821pp.

Nansen, F. 1901. Some oceanographical results of the expedition with the *Michael Sars* in the summer of 1900. Preliminary report. *Nyt Magazin for Naturvidenskab* 39(2): 129–61.

– 1902. The oceanography of the North Polar Basin. *The Norwegian North Polar Expedition 1893–1896, Scientific Results* III: 1–427.

– 1906. Northern waters: Captain Roald Amundsen's oceanographic observations in the Arctic seas in 1901. With a discussion of the formation of the bottom waters of the northern seas. Christiania. *Videnskabs-Selskabets Skrifter I. Mathematisk-Naturvidenskabelig Klasse* 1906, No. 3, pp. 1–145.

– 1912. Das Bodenwasser und die Abkühlung des Meeres. *Internationale Revue der gesamten Hydrobiologie und Hydrographie* 5(1): 42 pp.

– 1913. The waters of the northeastern North Atlantic: Investigations made during the cruise of the Frithjof, of the Norwegian Royal Navy, in July 1910. *Internationale Revue der gesamten Hydrobiologie und Hydrographie* 4: 1–139.

Naylor, R. 2007. Galileo's tidal theory. *Isis* 98(1): 1–22.

Neatby, L.H. 1973. *Discovery in Russian and Siberian waters*. Athens, OH: Ohio University Press. vii + 226pp.

Needler, A.W.H. 1958. Biological Station Nanaimo, B.C., 1908–1958. *Journal of the Fisheries Research Board of Canada* 15: 759–87.

– 1975. Archibald Gowanloch Huntsman: 1883–1973. *Proceedings of the Royal Society of Canada, Series IV*, 13: 67–9.

– 1985. Prince, Edward Ernest. *The Canadian Encyclopedia* III: 1474.

Newitt, M. 1986. Prince Henry and the origins of Portuguese expansion. In: M. Newitt, ed., *The first Portuguese colonial empire*, pp. 9–35. Exeter: Exeter University Press.

Nierenberg, W.A. 1996. Harald Ulrik Sverdrup, 1888–1957. *Washington, D.C. National Academy of Sciences, Biographical Memoirs* 69: 1–38.

Nordenskiöld, A.E. 1973. *Facsimile-atlas to the early history of cartography with reproductions of the most important maps printed in the XV and XVI centuries. With a new Introduction by J.B. Post*. New York: Dover. x + 141pp + maps I-LI.

Nordgaard, O. 1918. *Michael og Ossian Sars*. Kristiania: Steenske Forlag. 96pp.

Nordgaard, O., and W. Kielhau. 1934. Hjort, Johan. *Norsk Biografisk Leksikon* IV: 139–44.

Nornvall, F. 1999. Reasons for marine science: Arguments for hydrographic research in Sweden. *Historisch-Meereskundliches Jahrbuch* 6: 35–58.

North American Council on Fishery Investigations (NACFI). 1932. *Proceedings 1921–1930. No. 1*. Ottawa: F.A. Acland. 53pp.

Nye, M.J. 1986. *Science in the provinces: Scientific communities and provincial leadership in France, 1860–1930*. Berkeley: University of California Press. xi + 328pp.

Oceanic Observations of the Pacific. 1955. *The NORPAC atlas*. Berkeley: University of California Press. 11pp. + 123 charts.

Oettingen, A.J. von, ed. 1904. *J.C. Poggendorff's biographisch-literarisches Handwörterbuch zur Geschichte der exakten Wissenschaften. Vierter Band*, I. Abteilung (A-L). Leipzig: J.A. Barth.

Oldroyd, D.R. 1996. *Thinking about the earth. A history of ideas in geology*. Cambridge, MA: Harvard University Press. xxx + 410pp.

Olson, R. 1975. *Scottish philosophy and British physics, 1750–1880: A study in the foundations of the Victorian scientific style*. Princeton, NJ: Princeton University Press. vii + 349pp.

Open University Course Team. 2001. *Ocean circulation*. 2nd. edn. Oxford: Butterworth-Heinemann. 286pp.

Oreskes, N., and R. Rainger. 2000. Science and security before the atomic bomb: The loyalty case of Harald U. Sverdrup. *Studies in the History and Philosophy of Modern Physics* 31(3): 309–69.

Ortelius, A. 1570. *Theatrum orbis terrarum*. Antwerp: Gillis Coppans van Diest. X, [16], 53 [55], 70 maps.

Otto, J.F. 1800. *Versuch einer physischen Erdbeschreibung. Nach den neuesten Beobachtungen und Entdeckungen. Hydrographie. System einer allgemeinen Hydrographie des Erdbodens.* Berlin: G.C. Nauck. vi + 662pp.

Paffen, K., and G. Kortum. 1984. Die Geographie des Meeres: Disciplingeschichtliche Entwicklung seit 1650 und heutiger methodischer Stand. *Kieler geographische Schriften* 60: xiv + 293pp.

Pantiulin, A. 2002. Water masses: Birth of the idea. *ICES Marine Science Symposia* 215: 100–3.

Parr, A.E. 1937. Report on hydrographic observations at a series of anchor stations across the Straits of Florida. *Bulletin of the Bingham Oceanographic Collection* 6(3): 1–62.

Parrot, G.F. 1833. Expériences de forte compression sur divers corps. *Mémoires de l'Académie impériale des sciences de St-Pétersbourg* 2: 495–630.

Parry, J.H. 1963. *The age of reconnaissance.* London: Weidenfeld and Nicolson. xv + 365pp.

– 1966. *The establishment of the European hegemony: 1415–1715. Trade and exploration in the age of the Renaissance.* New York: Harper and Row. 176pp.

– 1971. *Trade and dominion: The European overseas empires in the eighteenth century.* London: Weidenfeld and Nicolson. xvii + 409pp.

– 1974. *The discovery of the sea.* New York: Dial Press. xv + 302pp.

Paul, H.W. 1985. *From knowledge to power: The rise of the science empire in France, 1860–1939.* Cambridge: Cambridge University Press. ix + 415pp.

Pauly, P.J. 1987. *Controlling life: Jacques Loeb and the engineering ideal in biology.* New York: Oxford University Press. iv + 252pp.

– 2000. *Biologists and the promise of American life: From Meriwether Lewis to Alfred Kinsey.* Princeton, NJ: Princeton University Press. xvi + 313pp.

Pearce, A.F. 1980. Early observations and historical notes on the Agulhas Current circulation. *Transactions of the Royal Society of South Africa* 44: 205–12.

Pedersen, O. 1974. Mohn, Henrik. *Dictionary of Scientific Biography* 9: 442–3.

Pedlosky, J. 2006. A history of thermocline theory. In M. Jochum and R. Murtugudde, eds, *Physical oceanography: Developments since 1950,* pp. 139–52. New York: Springer.

Penck, A. 1910. Das Institut und Museum für Meereskunde an der Königl. Friedrich-Wilhelms-Universität in Berlin. In M. Lenz, ed., *Geschichte der Universität Berlin, Band III,* pp. 1–5. Halle: Waisenhauses.

– 1912. Das Museum und Institut für Meereskunde in Berlin. *Mitteilungen der K.K. Geographischen Gesellschaft in Wien* 7/8: 413–33.

– 1925. Die Deutsche Atlantische Expedition. *Zeitschrift der Gesellschaft für Erdkunde zu Berlin* 1925: 243.

– 1926. Alfred Merz. *Zeitschrift der Gesellschaft für Erdkunde zu Berlin* 1926: 81–103.

Petermann, A. 1871. Der Golfstrom und Standpunkt der thermometrischen Kenntniss des Nord-atlantischen Oceans und Landgebiets im Jahre 1870. *Petermann's Geographische Mitteilungen* 16(6): 201–44.

Peterson, R.G., L. Stramma, and G. Kortum. 1996. Early concepts and charts of ocean circulation. *Progress in Oceanography* 37: 1–115.

Petit, G. 1970. Albert Ier of Monaco (Honoré Charles Grimaldi). *Dictionary of Scientific Biography* 1: 92–3.

Petit-Thouars, A. 1840–55. *Voyage autour du monde sur la frégatte la Vénus, pendant les années 1836–1839*. 10 vols. Paris.

Pettersson, H. 1923. Forteckning över Otto Pettersson's från trycket utgivna skrifter. In *Festskrift Tillägnad Otto Pettersson den 12 Februari 1923, Göteborg*, pp. 99–114. Helsingfors.

– 1950. Fredrik Laurents Ekman. *Svenskt Biografiskt Lexikon* 13: 111–16.

– 1953. *Westward ho with the 'Albatross.'* New York: E.P. Dutton. 218pp.

Pettersson, O. 1894. A review of Swedish hydrographic research in the Baltic and North Seas. *Scottish Geographical Magazine* 10: 281–302, 352–9, 413–27, 449–62, 525–39, 617–35.

– 1896. Ueber die Beziehungen zwischen hydrographischen und meteorologischen Phänomenen. *Meteorologische Zeitschrift* 13: 285–321.

– 1899. Ueber den Einfluss der Eisschmelzung auf die oceanische Cirkulation. *Öfversigt af Konglige Vetenskaps-Akademiens Förhandlingar* 1899, No. 3: 141–66.

– 1900a. Die Wasserzirkulation im Nordatlantischen Ozean. *Petermann's geographische Mitteilungen* 46: 61–5, 81–92.

– 1900b. Über systematische hydrographisch-biologische Erforschung der Meere, Binnenmeere und tieferen Seen Europas. *Verhandlungen des VII. Internationalen Geographen-Kongresses in Berlin, 1899*, pp. 334–42.

– 1904. Über die Wahrscheinlichkeit von periodischen Schwankungen in dem atlantische Strome. Mit einem Anhang: Beschreibung zu dem dynamischen Schnitten von J.W. Sandström. *Svenska Hydrografisk-Biologiska Kommissionens Skrifter* II: 11–40.

– 1907. On the influence of ice-melting upon ocean circulation. *Geographical Journal* 30(3): 273–95.

– 1923. Bemerkungen zu der Abhandlung: Die atlantische Vertikalzirkulation von Professor A. Merz und Dr. G. Wüst. *Annalen der Hydrographie und maritimen Meteorologie* 51: 64–71.

– 1930. Gustaf Ekman. 1852–1930. *Journal du Conseil international pour l'exploration de la mer* 5: 287–9.

Pettersson, O., and G. Ekman. 1899. Redogörelse för de Svenska hydrografiska undersökningarne åren 1896–1899 under ledning af G. Ekman, O. Pettersson och A. Wijkander. V. Ytvattnets tillstånd I Nordsjön och Skagerack *Bihang till Kongliga Svenska Vetenskaps-Akademiens Handlingar* 25(II), No.1, pp. 1–47.

Phillips, N. 1998. Carl-Gustaf Rossby: His times, personality and actions. *Bulletin of the American Meteorological Society* 79: 1097–1112.

Pickard, G.L., and W.J. Emery. 1982. *Descriptive physical oceanography: An introduction.* 4th (SI) edn. New York: Pergamon Press. xiv + 249pp.

Phillips, C.R. 1990. The growth and composition of trade in the early modern world. In J.D.Tracy, ed., *The rise of merchant empires: Long-distance trade in the early modern world, 1350–1750,* pp. 34–101. Cambridge: Cambridge University Press.

Pihl, M. 1972a. Bjerknes, Carl Anton. *Dictionary of Scientific Biography* 2: 166–7.

– 1972b. Bjerknes, Vilhelm Frimann Koren. *Dictionary of Scientific Biography* 2: 167–9.

Poilleux, C.-F. 1924. Le Bureau hydrographique international à Monaco. *Rives d'Azur* 8(191): 1027–33.

Pollard, R., and G. Griffiths. 1993. James Rennell, the father of oceanography. *Ocean Challenge* 4 (1–2): 24–5.

Porter, R. 1980. The terraqueous globe. In: G.S. Rousseau and R. Porter, eds, *The ferment of knowledge: Studies in the historiography of eighteenth-century science,* pp. 285–324. Cambridge, Cambridge University Press.

Porter, T.M. 1994. Making things quantitative. *Science in Context* 7(3): 389–407.

Portier, P. 1936. Le professeur J. Thoulet. *Bulletin de la Société d'océanographie de France* 16(90): 1553–4.

– 1945. Le docteur Jules Richard, directeur du Musée océanographique de Monaco, correspondant de l'Institut de France (1863–1945). *Monaco. Bulletin de l'Institut océanographique* 42(881), 19pp.

Preston, D., and M. Preston. 2004. *A pirate of exquisite mind: Explorer, naturalist, and buccaneer; the life of William Dampier.* New York: Walker and Co. ix + 372pp.

Prestwich, G.A. 1899. *Life and letters of Sir Joseph Prestwich M.A., D.C.L., F.R.S., formerly Professor of Geology in the University of Oxford.* Edinburgh and London: William Blackwood. xvi + 444pp.

Prestwich, J. 1871. Anniversary address. *Quarterly Journal of the Geological Society of London* 27: xxx–lxxv.

– 1874. Tables of temperatures of the sea at various depths below the surface, taken between 1749 and 1868; collated and reduced, with notes and sections. *Proceedings of the Royal Society of London* 22: 462–8.

– 1875. Tables of temperature of the sea at various depths below the surface, reduced and collated from the various observations made between the years 1749 and 1868, discussed. With map and sections. *Philosophical Transactions of the Royal Society of London* 165B: 587–674.

Priesner, C. 1993. Merz, Alfred. *Neue Deutsche Biographie* 17: 196–8.

Prince, E.E. 1897. The fisheries of Canada. In: *Handbook of Canada,* pp. 264–74.

British Association for the Advancement of Science Toronto meeting 1897. Toronto: Publication Committee of the Local Executive.

– 1913. The Biological Board of Canada. In: Canada, Commission of Conservation, *Report of the Fourth Annual Meeting, 1913,* pp. 87–104. Toronto: Warwick Bros. and Rutter.

– 1916a. The herring fishery of Canada. In Canada, Commission of Conservation, *Conservation of fish, birds and game,* pp. 37–46. Toronto: Methodist Book and Publishing House.

– 1916b. Unutilized fisheries resources of Canada. In: Canada, Commission of Conservation, *Conservation of fish, birds and game,* pp.47–60. Toronto: Methodist Book and Publishing House.

– 1919. Preface. In: Canada, Department of the Naval Service, *Canadian Fisheries Expedition, 1914–1915: Investigations in the Gulf of St. Lawrence and Atlantic waters of Canada under the direction of Dr. Johan Hjort, leader of the expedition, director of fisheries for Norway,* pp. v–viii. Ottawa: King's Printer.

– 1924. The fisheries of Canada. In: *Handbook of Canada,* pp.263–73. Issued by the Local Committee on the occasion of the meeting of the British Association for the Advancement of Science at Toronto, August, 1924. Toronto: University of Toronto Press.

Probst, B. 1995. Das Institut und Museum für Meereskunde – eine bewegte Geschichte? In *Aufgetaucht. Das Institut und Museum für Meereskunde im Museum für Verkehr und Technik Berlin,* pp. 11–25. Berlin: Nicolaische Verlagsbuchhandlung.

Proudman, J. 1953. *Dynamical oceanography.* London: Methuen. xii + 409pp.

Purrington, R.D. 1997. *Physics in the nineteenth century.* New Brunswick, NJ: Rutgers University Press. xx + 249pp.

Rabot, C. 1987. Le 75e anniversaire de l'inauguration de l'Institut océanographique. *Acta Geographica,* 3° série, No. 70, pp. 36–7.

Rainger, R. 2003. Adaptation and the importance of local culture: Creating a research school at the Scripps Institution of Oceanography. *Journal of the History of Biology* 36: 461–500.

Raitt, H., and B. Moulton. 1967. *Scripps Institution of Oceanography: First Fifty Years.* San Diego: Ward Ritchie Press. xix + 217pp.

Ramster, J. 1975. Dr. J.N. Carruthers, 24 November 1895 – 8 March 1973. *Journal du Conseil international pour l'Exploration de la Mer* 36 (2): 101–5.

R.E.A. 1890. Hadley, George (1685–1768). *Dictionary of National Biography* 23: 434–5.

Redfield, A.C. 1976. Henry Bryant Bigelow. October 3, 1879 – December 11, 1967. *National Academy of Sciences of the U.S.A., Biographical Memoirs* 48: 50–80.

Reed, G.B. 1948. Obituary of A.T. Cameron. *Journal of the Fisheries Research Board of Canada* 7: 217.

Reichs-Marine-Amt. 1909. *Forschungsreise S.M.S. "Planet" 906/07. Herausgegeben vom Reichs-Marine-Amt. I. Band. Reisebeschreibung.* Berlin: Karl Siegismund. xviii + 104pp.

Rennell, J. 1793. Observations on a current that often prevails to the westward of Scilly; endangering the safety of ships that approach the British Channel. *Philosophical Transactions of the Royal Society of London* 83: 182–200.

– 1832. *Investigation of the currents of the Atlantic Ocean, and of those which prevail between the Indian Ocean and the Atlantic.* London: J.G. and F. Rivington. 359pp.

Revelle, R. 1978. Columbus O'Donnell Iselin, II. (1904–1971). *American Philosophical Society Yearbook,* 1977, pp. 61–71.

– 1980. The Oceanographic and how it grew. In M. Sears and D. Merriman, eds, *Oceanography: The past,* pp. 10–24. New York: Springer-Verlag.

– 1990. Bowie Medal to Walter H. Munk. *Eos* 71(1): 13.

Revelle, R., and W. Munk. 1948. Harald Ulrik Sverdrup – an appreciation. *Journal of Marine Research* 7: 127–38.

Rice, A.L. 1986. *British oceanographic vessels, 1800–1950.* London: Ray Society. 193pp.

Richard, J. 1907. *L'océanographie.* Paris: Vuibert et Nony. vi + 398pp.

– 1910. Les campagnes scientifiques de S.A.S. le Prince Albert Ier de Monaco. *Monaco. Bulletin de l'Institut océanographique* 7(162), 159pp., index bibliographique, pp. i–xxix.

– 1932. Notice biographique sur le baron Jules de Guerne. *Monaco. Bulletin de l'Institut océanographique* 29(590), 11pp.

– 1934a. Caption to 'Le Prince Albert sur la passerelle de la PRINCESSE ALICE.' *Résultats des campagnes scientifiques, accomplies sur son yacht par Albert Ier, Prince souverain de Monaco,* 89: 454–5.

Richard, J. 1934b. Liste générale des stations des campagnes scientifiques du Prince Albert de Monaco, avec neuf cartes des itinéraires et des notes et observations. *Résultats des campagnes scientifiques accomplies sur son yacht par Albert Ier, Prince souverain de Monaco,* 89, 471pp.

Richardson, P.L. 2005. WHOI and the Gulf Stream. http://www.whoi.edu/75th/book/whoi-richardson.pdf. 10pp.

Richie, A. 1998. *Faust's metropolis: A history of Berlin.* London: HarperCollins. xxviii + 1139pp.

Richthofen, F. von. 1904. *Das Meer und die Kunde vom Meer. Rede zur Gedächtnisfeier des Stifters der Berliner Universität König Friedrich Wilhelm III in der Aula am 3. August 1904.* Berlin: Universitäts-Buchdruckerei von Gustav Schade. 45pp.

Ricker, W.E. 1979. Obituary. Neal Marshall Carter 1902–1978. *Journal of the Fisheries Research Board of Canada* 36(5): 595–6.

Ricketts, N.G. 1932. The 'Marion' Expedition to Davis Strait and Baffin Bay. In N.G. Ricketts and P.D. Trask, eds, *The Marion Expedition to Davis Strait and Baffin Bay under direction of the United States Coast Guard, 1928. Scientific Results, Part 1. The bathymetry and sediments of Davis Strait*, pp. 1–52. U.S. Treasury Department. United States Coast Guard Bulletin No. 19.

Riis-Karstensen, E. 1931. The Godthaab Expedition 1928: Report on the expedition. *Meddelelser om Grönland* 78(1): 1–105.

– 1936. The Godthaab Expedition 1928: The hydrographic work and material. *Meddelelser om Grönland* 78(3), 101pp.

Ritchie, G.S. 1967. *The Admiralty Chart: British naval hydrography in the nineteenth century.* London: Hollis and Carter. xi + 388pp.

– 2003. Pre-GEBCO history of deep-sea sounding. In D. Scott, ed., *The history of GEBCO, 1903–2003: The 100-year story of the General Bathymetric Chart of the Oceans*, pp. 7–13. Lemmer, Netherlands: GITC.

Ritter, W.E. 1905a. A general statement of the ideas and present aims and status of the Marine Biological Association of San Diego. *University of California Publications in Zoology* 2: i–xvii.

– 1905b. Organization in scientific research. *Popular Science Monthly* 67: 49–53.

– 1908. The scientific work of the San Diego Marine Biological Station during the year 1908. *Science* 28: 329–33.

– 1912a. The duties to the public of research institutions in pure science. *Popular Science Monthly* 80: 51–7.

– 1912b. The Marine Biological Station of San Diego: Its history, present conditions, achievements, and aims. *University of California Publications in Zoology* 9: 137–248.

– 1913. Scripps Institution for Biological Research. *University of California at Berkeley. Annual Report of the President of the University 1912–1913*, pp. 113–23.

– 1915. The biological laboratories of the Pacific coast. *Popular Science Monthly* 84: 223–32.

– 1916a. The culture value of science. *Science* 44: 261–4.

– 1916b. What the Scripps Institution is trying to do. *Bulletin of the Scripps Institution for Biological Research* (30 December): 19–24.

– 1923. A proposed American institute of oceanography. *Science* 58: 44–5.

Ritter, W.E., and E.W. Bailey. 1928. The organismal conception: Its place in science and its bearing on philosophy. *University of California Publications in Zoology* 31(14): 307–58.

Robertson, A.J. 1907. I. Report on hydrographical investigations in the Faeroe-Shetland Channel and the northern part of the North Sea during the years

1904–1905. In *North Sea Fisheries Investigation Committee. Second report (northern area) on the fishery and hydrographical investigations in the North Sea and adjacent waters, 1904–1905. Part 1. Hydrography,* pp. 1–40. British Sessional Papers, Cd 3358. London: HMSO. xii + 755pp.

Robinson, A., and H. Stommel. 1959. The oceanic thermocline and the associated thermohaline circulation. *Tellus* 11: 295–308.

Röhr, A. 1981. *Bilder aus dem Museum für Meereskunde in Berlin 1906–1945.* Bremerhaven: Deutsches Schiffahrtsmuseum. 72pp.

Roll, H.U. 1987. Georg Wüst 1890–1977. *Deutsche Gesellschaft für Meeresforschung – DGM Mitteilungen* 2: 25–8.

Rollefsen, G. 1966. Norwegian fisheries research. *Fiskeridirektoratets Skrifter, Serie Havundersøkelser* 14(1): 1–36.

– 1970. Paul Bjerkan (1874–1968). *Journal du Conseil international pour l'Exploration de la Mer* 33 (2): 124–5.

Rollet de l'Isle, M. 1951. Étude historique sur les ingénieurs hydrographes et le Service hydrographique de la marine (1814–1914). *Annales hydrographiques,* 4ᵉ série, 1 bis [=1950], 378pp.

Ronan, C.A. 1968. Edmond Halley and early geophysics. *Geophysical Journal of the Royal Astronomical Society* 15: 241–8.

– 1972. Halley, Edmond. *Dictionary of Scientific Biography* 6: 67–72.

Rosenman, H. 1987. The historical background. In J.S. Dumont d'Urville, *An account in two volumes of two voyages to the south seas. Volume 1. 'Astrolabe,' 1826–1829,* pp. xxi–xl. Translated and edited by Helen Rosenman. Honolulu: University of Hawaii Press.

Rossby, C.-G. 1938. On the mutual adjustment of pressure and velocity distributions in certain simple current systems, II. *Journal of Marine Research* 1: 239–63.

Rouch, J. 1922. *Manuel d'océanographie physique.* Paris: Masson. 229pp.

– 1943–8. *Traité d'océanographie physique. 1. Sondages 2. L'eau de mer 3. Les mouvements de la mer.* Paris: Payot. 256 + 349 + 413pp.

– 1948. Le docteur Jules Richard, directeur du Musée océanographique de Monaco (1863–1945). *Riviera scientifique (Bulletin de l'Association des naturalistes de Nice et des Alpes-Maritimes)* 35: 1–10.

– 1958. Un océanographe de Douai, Jules de Guerne. *Comité des travaux historiques et scientifiques, Bulletin de la Section de géographie,* 70 [=1957]: 97–110.

Rouse, H. 1971. Du Buat, Pierre-Georges-Louis. *Dictionary of Scientific Biography* 4: 207–8.

– 1987. The origins of fluid mechanics. *Proceedings of the American Society of Civil Engineers. Journal of Engineering Mechanics* 113(1): 66–71.

Rouse, H., and S. Ince. 1957. *History of hydraulics.* Iowa City: State University of Iowa / Iowa Institute of Hydraulic Research. xii+269pp.

Rozwadowski, H. 2002. *The sea knows no boundaries: A century of science under ICES.* Seattle: University of Washington Press. ix + 448pp.

Rubin, M. 1982. James Cook's scientific programme in the Southern Ocean, 1772–75. *Polar Record* 21: 33–49.

Rudwick, M.J.S. 1972. *The meaning of fossils: Episodes in the history of palaeontology.* New York: American Elsevier. iii + 287pp.

Rupke, N. 1997. Introduction to the 1997 edition. In A. von Humboldt, *Cosmos: A sketch of the physical description of the universe. Volume 1,* pp. vii–xlii. Translated by E.C. Otté. Introduction by Nicolaas A. Rupke. Baltimore: Johns Hopkins University Press.

Russell, E.S. 1948. Prof. Johan Hjort, For. Mem. R.S. *Nature* 162: 764–5.

Russell, P.E. 1984. *Prince Henry the Navigator: The rise and fall of a culture hero. Taylorian Special Lecture, 10 November 1983.* Oxford: Clarendon Press. 30pp.

– 2000. *Prince Henry 'the Navigator.' A Life.* New Haven: Yale University Press. xvi + 448pp.

Ruud, J. 1948. Johan Hjort in memoriam. *Norsk Hvalfangst-Tidende (The Norwegian Whaling Gazette)* 37(11): 441–51.

Ryan, P.R. 1984. Henry Stommel. 'Apprentice' oceanographer. *Oceanus* 27(1): 55–9.

Saint-Pierre, J. 1994. *Les chercheurs de la mer: Les débuts de la recherche en océanographie et en biologie des pêches du Saint-Laurent.* Québec: Institut Québécois de la recherche sur la culture. 256pp.

Sandström, J.W. 1904. Beschreibung zu den dynamischen Schnitten in Pl. XIII–XVI. *Svenska Hydrografisk-Biologiska Kommissionens Skrifter* II: 38–40, pls. IX–XVII.

– 1908. Dynamische Versuche mit Meerwasser. *Annalen der Hydrographie und maritimen Meteorologie* XXXVI Jahrgang, pp. 6–23.

– 1919. The hydrodynamics of Canadian Atlantic waters. In Canada, Department of the Naval Service, *Canadian Fisheries Expedition, 1914–1915: Investigations in the Gulf of St. Lawrence and Atlantic waters of Canada under the direction of Dr. Johan Hjort, leader of the expedition, director of fisheries for Norway,* Ottawa: King's Printer.

– 1923. Deux théorèmes fondamentaux de la dynamique de la mer: Traité elementaire expérimental. *Svenska Hydrografisk-Biologiska Kommissionens Skrifter. Hydrografi.* VII: 1–6.

Sandström, J.W., and B. Helland-Hansen. 1903. Über die Berechnung von Meeresströmungen. *Report on Norwegian Fishery and Marine Investigations* 2(4): 1–43.

– 1905. On the mathematical investigation of ocean currents (transl. D'Arcy Wentworth Thompson). *Report of the North Sea Fishery Investigation Committee (Northern Area), 1902–1903*, pp. 135–63. British Sessional Papers, Vol. 14.

Sargent, B. 1983. Walter H. Munk: Unifier of ocean fields. *Oceanus* 26(4): 57–62.

Sargent, P. 1979. *The sea acorn: Scripps Institution of Oceanography, the people and the place, 1936–1942.* San Diego: privately printed.

Schlee, S. 1978. *On almost any wind: The saga of the oceanographic research vessel 'Atlantis.'* Ithaca, NY: Cornell University Press. 301pp.

– 1980. The R/V *Atlantis* and her first oceanographic institution. In M. Sears and D. Merriman, eds, *Oceanography: The past*, pp. 49–56. New York: Springer-Verlag.

Schmelck, L. 1882. X. Chemistry. I. On the solid matter in sea-water. II. On oceanic deposits. *The Norwegian North Atlantic Expedition 1876–1878.* Christiania: Grøndahl and Søn.

Schmitt, W.L. 1948. C. McLean Fraser: An appreciation. June 1, 1872 – December 26, 1946. *Allan Hancock Pacific Expeditions* 4: i–xv.

Schneer, C.J. 1984. *The evolution of physical science: Major ideas from the earliest times to the present.* Lanham, MD: University Press of America. xvii + 398pp.

Schott, G. 1902. Oceanographie und maritime Meteorologie im Auftrage des Reichs-Marine-Amtes. *Wissenschaftliche Ergebnisse der Deutschen Tiefsee-Expedition auf dem Dampfer "Valdivia" 1898–1899*, I, xii + 403pp.

– 1903. *Physische Meereskunde.* Sammlung Göschen. Leipzig: G.J. Göschen'sche Verlagshandlung. 162pp.

– 1923. Merz, A.: Meereskunde, Wirtschaft und Staat. *Annalen der Hydrographie und maritimen Meteorologie* 51: 53.

– 1924. Dr. Wilhelm Brennecke †. *Annalen der Hydrographie und maritimen Meteorologie* 52: 49–50.

– 1942. *Geographie des Atlantischen Ozeans.* Hamburg: Verlag von C. Boysen. xvi + 438pp.

Schott, W. 1987. *Early German oceanographic institutions, expeditions and oceanographers.* Hamburg: Deutsches Hydrographisches Institut. 50pp.

Schulz, B. 1936. Zur Vollendung des 70. Lebensjahres von Gerhard Schott. *Annalen der Hydrographie und maritimen Meteorologie* 8: 329–35.

Schumacher, A. 1959. Konteradmiral a.D. Dr. h.c. Fritz Spiess. *Der Seewart* 20: 117.

Schwach, V. 2000. *Havet, Fisken og Vitenskapen: Fra Fiskeriundersøkelser til Havforskningsinstitutt 1860–2000.* Oslo: John Grieg. 405pp.

Scott, D. ed. 2003. *The history of GEBCO, 1903–2003: The 100-year story of the General Bathymetric Chart of the Oceans.* Lemmer, Netherlands: GITC bv. vii + 140pp.

Scott, J.F. 1973. Maclaurin, Colin. *Dictionary of Scientific Biography* 8: 609–12.

Shinn, T. 1979. The French science faculty system, 1808–1914: Institutional change and research potential in mathematics and the physical sciences. *Historical Studies in the Physical Sciences* 10: 271–332.

Shor, E.N. 1978. *Scripps Institution of Oceanography: Probing the oceans, 1936 to 1976.* San Diego: Tofua Press. x + 502pp

– 1981. How the Scripps Institution came to San Diego. *Journal of San Diego History* 27: 161–73.

Siegel, D.M. 1991. *Innovation in Maxwell's electromagnetic theory: Molecular vortices, displacement current, and light.* Cambridge: Cambridge University Press. x + 225pp.

Sivertsen, E. 1968. Michael Sars, a pioneer in marine biology, with some aspects from the early history of biological oceanography in Norway. *Monaco. Bulletin de l'Institut océanographique, No. spécial 2 (Premier Congrès internationale d'histoire de l'océanographie)* 2: 439–52.

Smed, J. 1992. Early discussion and tests of the validity of Knudsen's hydrographical tables. *Historisch- Meereskundliches Jahrbuch* 1: 77–86.

– 2005. The Central Laboratory of the International Council for the Exploration of the Sea and its successors. *Earth Sciences History* 24(2): 225–46.

Smith, C. 1998. *The science of energy: A cultural history of energy physics in Victorian Britain.* Chicago: University of Chicago Press. xi + 404pp.

Smith, C., and M.N. Wise. 1989. *Energy and empire: A biographical study of Lord Kelvin.* Cambridge: Cambridge University Press. xxvi + 866pp.

Smith, E.H. 1926a. Oceanography for the Ice Patrol. *Transactions of the American Geophysical Union* 7: 106–12.

– 1926b. A practical method for determining ocean currents. *Bulletin of the United States Coast Guard* 14, 50pp.

– 1927. Oceanographic investigations of the International Ice Patrol. *American Geophysical Union, Section of Oceanography. Bulletin of the National Research Council* 61: 212–17.

– 1930. Arctic ice: With reference to its distribution in the North Atlantic Ocean. Ph.D diss., Harvard University. 263pp.

– 1931.Arctic ice, with special reference to its distribution in the North Atlantic Ocean. *The Marion Expedition to Davis Strait and Baffin Bay, 1928. Scientific Results Part 3. United States Coast Guard Bulletin,* No. 19.

Smith, E.H., F.M. Soule, and O. Mosby. 1937. The 'Marion' and 'General Green' expeditions to Davis Strait and the Labrador Sea under the direction of the U.S. Coast Guard, 1928–1931–1933–1934–1935. *United States Coast Guard Bulletin* 19, 259pp.

Sobel, D. 1995. *Longitude: The true story of a lone genius who solved the greatest scientific problem of his time.* New York: Walker and Co. viii + 184pp.

Soffientino, B., and M.E.Q. Pilson. 2005. The Bosporus Strait, a special place in the history of oceanography. *Oceanography* 18(2): 16–23.

Solemdal, P. 1997. The three cavaliers: A discussion from the Golden Age of Norwegian marine research. In: R.C. Chambers and E.L. Trippel, eds, *Early life history and recruitment in fish populations,* pp. 551–65. London: Chapman and Hall.

Sorrenson, R. 1996. The ship as a scientific instrument in the eighteenth century. *Osiris, 2nd Series,* 11: 221–36.

Spiess, F. 1926. I. Die Aufgaben und bisherigen Arbeiten der Expedition. Die Deutsche Atlantische Expedition auf dem Vermessungs- und Forschungsschiff 'Meteor.' *Annalen der Hydrographie und maritimen Meteorologie* LIV Jahrgang (1926), III: 1–19.

– 1928. *Die Meteor-Fahrt: Forschungen und Ergebnisse der Deutschen Atlantischen Expedition, 1925–1927.* Berlin: Dietrich Reimer. 376pp.

– 1935. Henrik Mohn. Zur hundertsten Wiederkehr seines Geburtstages. *Annalen der Hydrographie,* LXIII Jahrgang (1935), V, pp. 181–2.

– 1985. *The Meteor Expedition: Scientific Results of the German Atlantic Expedition, 1925–1927* (English translation edited byW.J. Emery). New Delhi: Amerind. xvi + 429pp.

Spilhaus, A.F. 1938. A bathythermograph. *Journal of Marine Research* 1: 95–100.

Spilhaus, A.F., and A.R. Miller. 1948. The sea sampler. *Journal of Marine Research* 7: 370–85.

Spratt, T.A.B. 1871. On the undercurrent theory of the ocean, as it is propounded by recent explorers. *Proceedings of the Royal Society of London* 19: 528–56.

Stahlberg, W. 1929. *Das Institut und Museum für Meereskunde an der Friedrich Wilhelms-Universität in Berlin.* Hamburg: Druckerei- Gesellschaft Hattung and Co. 16pp.

Stern, F. 1999. *Einstein's German World.* Princeton, NJ: Princeton University Press. ix + 335pp.

Stevenson, E.L., ed. 1932. *Geography of Claudius Ptolemy, based upon Greek and Latin manuscripts and important late fifteenth and early sixteenth century printed editions.* New York: New York Public Library. 167pp.

Stommel, H. 1948. The westward intensification of wind-driven ocean currents. *Transactions of the American Geophysical Union* 29(2): 202–6.

– 1950a. The Gulf Stream: A brief history of the ideas concerning its cause. *Scientific Monthly* 70(4): 242–53.

– 1950b. Note on the deep circulation of the Atlantic Ocean. *Journal of Meteorology* 7: 245–6.

– 1951. An elementary explanation of why ocean currents are strongest in the west. *American Meteorological Society Bulletin* 32: 21–33.

- 1954. Circulation in the North Atlantic Ocean. *Nature* 173: 886.
- 1955. The anatomy of the Atlantic. *Scientific American* 192: 30–5.
- 1957a. The abyssal circulation of the ocean. *Nature* 180: 733–4.
- 1957b. A survey of ocean current theory. *Deep-Sea Research* 4: 149–84.
- 1958a. The abyssal circulation. *Deep-Sea Research* 5: 80–2.
- 1958b. The circulation of the abyss. *Scientific American* 199(1): 85–90.
- 1958c. *The Gulf Stream: A physical and dynamical description.* Berkeley and Los Angeles: University of California Press. xiii + 202pp.
- 1965. *The Gulf Stream: A physical and dynamical description.* 2nd ed. Berkeley and Los Angeles: University of California Press. xiii + 248pp.
- (with Luis Capurro and Mary Swallow). 1980. *The ocean and William Leighton Jordan, Esq.* Privately printed. Ca. 12pp. (Reproduced in Luyten and Hogg 1992.)
- 1987. *A view of the sea: A discussion between a chief engineer and an oceanographer about the machinery of the ocean circulation.* Princeton, NJ: Princeton University Press. xiv + 165pp.
- 1989. Why we are oceanographers. *Oceanography* 2(2): 48–54.
- 1994. Columbus O'Donnell Iselin. September 25, 1904 – January 5, 1971. *National Academy of Sciences of the U.S.A., Biographical Memoirs* 64: 165–86.
Stommel, H., and A.B. Arons. 1960a. On the abyssal circulation of the world ocean – I. Stationary planetary flow patterns on a sphere. *Deep-Sea Research* 6(2): 140–54.
- 1960b. On the abyssal circulation of the world ocean – II. An idealized model of the circulation pattern and amplitude in ocean basins. *Deep-Sea Research* 6: 217–33.
- 1972. On the abyssal circulation of the world ocean – V. The influence of bottom slope on the broadening of inertial boundary currents. *Deep-Sea Research* 19: 707–18.
Stommel, H., A.B. Arons, and A.J. Faller. 1958. Some examples of stationary planetary flow patterns in unbounded basins. *Tellus* 10: 179–87.
Stoye, J. 1994. *Marsigli's Europe, 1680–1730: The life and times of Luigi Ferdinando Marsigli, soldier and virtuoso.* New Haven: Yale University Press. xii + 356pp.
Stuardo, J.R. 2004. *Alexander von Humboldt y el Inicio de la Biología Marina y la Oceanografía en el Mar del Sur.* Concepción, Chile: Editorial Universidad de Concepción. 116pp.
Stutzbach-Michelsen, M. 1989. Early German oceanographic expeditions. In B. Watermann, ed., *Exposition on historical aspects of marine research in Germany. Deutsche Hydrographische Zeitschrift, Ergänzungsheft, Reihe B*, No. 21, pp. 9–45.
Sullivan, W. 1992. Henry Stommel, 71, theoretician influential in ocean current study. *New York Times*, Tuesday, 21 January.

Sumner, F.B. 1945. *The life history of an American naturalist.* Lancaster, PA: Jacques Cattell Press.

Svansson, A. 1965. Some hydrographic problems of the Skagerrak. *Progress in Oceanography* 3: 355–72.

– 1996. Ekman, Vagn Walfrid. In E.J. Dasch, ed., *Encyclopedia of earth sciences, Vol. 1,* pp.256–7. New York: Macmillan.

– 1999. Herring and hydrography: Otto Pettersson and his ideas of the behaviour of the period herring. In B. Andersson, ed., *Swedish and International fisheries: Papers presented at the conference in Göteborg 1998-11-22,* pp. 22–36. Rapport från Ekonomisk-Historiska Institutionen vid Göteborgs Universitet 13. Göteborg.

– 2006. *Otto Pettersson: Oceanografen, Kemisten, Uppfinnaren.* Göteborg: Tre Böcker Förlag. 376pp.

Sverdrup, H.U. 1926. Scientific work of the 'Maud' expedition, 1922–1925. *Scientific Monthly* 22: 400–10.

– 1927. Scientific work of the 'Maud' expedition, 1922–1925. *Smithsonian Institution Report for 1926,* pp. 219–33.

– 1928. Review of the present state of work within physical oceanography in Europe. Scripps Institution of Oceanography Archives. SIO Office of the Director (Sverdrup), Records, 1936–1948, #82–56, Box 1, Folder 2, 'Sverdrup – Lectures and Manuscripts folder 1,' 9pp.

– 1931. The origin of the deep-water of the Pacific Ocean as indicated by the oceanographic work of the 'Carnegie.' *Gerlands Beiträge zur Geophysik* 29: 95–105.

– 1934. Helland-Hansen, Bjørn. *Norsk biografisk leksikon* VI: 10–13.

– 1938a. The fluid problems of ocean circulation. National Meeting of the American Association of Mechanical Engineers, March 25, 1938. Scripps Institution of Oceanography Archives. SIO Office of the Director (Sverdrup), Records, 1936–1948, #82–56, Box 1, Folder 2. 10pp.

– 1938b. On the process of upwelling. *Journal of Marine Research* 1: 155–64.

– 1938c. Research within physical oceanography and submarine geology at the Scripps Institution of Oceanography during April 1937 to April 1938. *Transactions of the American Geophysical Union* 19: 238–42.

– 1939a. *Physics and geophysics. With special reference to problems in physical oceanography. Faculty Research Lecture at the University of California at Los Angeles delivered March 23, 1938.* Berkeley: University of California Press. 23pp.

– 1939b. Research within physical oceanography and submarine geology at the Scripps Institution of Oceanography during April 1938 to April 1939. *Transactions of the American Geophysical Union* 20: 422–7.

– 1940. Research in oceanography. California scientists study the ocean. *California Monthly,* December 1940, 3pp.

– 1947. Wind-driven currents in a baroclinic ocean: With application to the equatorial currents of the eastern Pacific. *Proceedings of the National Academy of Sciences* 33: 318–26.

– 1948. Research in geophysics: Seminar address at the SIO on January 30, 1948. Scripps Institution of Oceanography Archives. SIO Office of the Director (Sverdrup), Records, 1936–1948, #82–56, Box 1, Folder 3, 14pp.

Sverdrup, H.U., M.W. Johnson, and R.H. Fleming. 1942. *The Oceans: Their physics, chemistry, and general biology.* Englewood Cliffs, NJ: Prentice-Hall. x + 1087pp.

Swallow, J.C. 1955. A neutral buoyancy float for measuring deep currents. *Deep-Sea Research* 3: 74–81.

Swallow, J.C., and L.V. Worthington. 1957. Measurement of deep currents in the western North Atlantic. *Nature* 179: 1183–4.

– 1961. An observation of a deep countercurrent in the Western North Atlantic. *Deep-Sea Research* 8: 1–19.

Tabata, S. 1987a. John Patrick Tully 1906–1987. *Atmosphere-Ocean* 25(4): 355–7.

– 1987b. Obituary. John Patrick Tully 1906–1987. *Canadian Journal of Fisheries and Aquatic Sciences* 44: 1674–5.

Tait, J.B. 1958a. Bjørn Helland-Hansen, Hon. FRSE. *Yearbook of the Royal Society of Edinburgh 1956–1957,* pp. 26–29.

– 1958b. Prof. Bjørn Helland-Hansen. *Nature* 181(4607): 453–4.

Tanner, Z.L. 1897. Deep-sea exploration: A general description of the steamer *Albatross,* her appliances and methods. *Bulletin of the United States Fish Commission for 1896* 16: 257–428.

Taylor, G.I. Internal waves and turbulence in a fluid of variable density. *Conseil permanent pour l'Exploration de la Mer, Rapports et Procès-Verbaux des Réunions* 76: 35–42.

Thomas, K.B. 1971. Carpenter, William Benjamin. *Dictionary of Scientific Biography* 3: 87–9.

Thompson, B. (Count Rumford). 1798. The propagation of heat in fluids. In: *Essays political, economical, and philosophical. Volume 2,* pp. 199–313. London: Cadell and Davies.

Thompson, D.W. 1905. Introduction. In *Fishery Board for Scotland. Report on fishery and hydrographical investigations in the North Sea and adjacent waters conducted for the Fishery Board for Scotland in conjunction with the International Council for the Exploration of the Sea under the supervision of D'Arcy Wentworth Thompson, C.B., Scientific Member of the Board, 1902–1903,* pp. iii–vi. British Sessional Papers 14, pp. i–vi, 1–618.

– 1941. Dr. Otto Pettersson. *Nature* 147: 701–2.

– 1947. Otto Pettersson. 1848–1941. *Journal du Conseil international pour l'exploration de la Mer* 15: 121–5.

Thompson, J.H., and A. Seager. 1985. *Canada, 1922–1939: Decades of discord.* Toronto: McClelland and Stewart. xiv + 438pp.

Thomson, C.W. 1870. Letter from Prof. Wyville Thomson to the Rev. A.M. Norman on the successful dredging of H.M.S. 'Porcupine.' *Report of the British Association for the Advancement of Science, Exeter, 1869,* pp. 115–16.

– 1873. *The depths of the sea: An account of the general results of the dredging cruises of the H.M.S.S. "Porcupine" and "Lightning" during the summers of 1868, 1869, and 1870, under the scientific direction of Dr. Carpenter, F.R.S., J. Gwyn Jeffreys, F.R.S., and Dr. Wyville Thomson, F.R.S.* London: Macmillan. xx + 527pp.

– 1877. *The voyage of the 'Challenger': The Atlantic: A preliminary account of the general results of the exploring voyage of H.M.S. 'Challenger' during the year 1873 and the early part of the year 1876.* Volume 1. London: Macmillan. xxx + 424pp.

– 1878. Section E. Geography. Opening address by the President, Prof. Sir C. Wyville Thomson, F.R.S. *Nature* 18: 448–52.

Thone, F.E.A., and E.W. Bailey. 1927. William Emerson Ritter: Builder. *Scientific Monthly* 24: 256–62.

Thorade, H. 1909. Über die kalifornische Meeresströmungen: Oberflächentemperaturen und Strömungen, an der Westküste Nordamerikas. *Annalen der Hydrographie und maritimen Meteorologie* 37: 17–34, 63–76.

– 1935. Henrik Mohn und die Entwicklung der Meereskunde. *Annalen der Hydrographie und maritimen Meteorologie* LXIII Jahrgang, V: 182–6.

– 1944. A. Defant sechzig Jahre alt. *Die Naturwissenschaften,* 32 Jahrgang, Heft 27/39: 165–6.

Thoulet, J. 1887. Observations faites à Terre-Neuve à bord de la frégate 'La Clorinde' pendant la campagne de 1886. *Revue maritime et coloniale* 93: 398–430.

– 1889a. De l'état des études d'océanographie en Norvège et en Écosse: Rapport sur une mission du ministère de l'Instruction publique. *Archives des missions scientifiques et littéraires, 3ᵉ Série,* 15: 187–240.

– 1889b. Les études océanographiques en Norvège et en Écosse. *Revue scientifique (Revue rose)* 43(18): 554–8.

– 1890a. Les eaux abyssales. *Revue générale des sciences pures et appliquées* 1(16): 500–8.

– 1890b. *Océanographie (statique).* Paris: Librairie militaire de L. Baudoin et Cie. x + 492pp.

– 1890c. De quelques objections à la théorie de la circulation verticale profonde dans l'océan. *Paris. Comptes rendus hebdomadaires des séances de l'Académie des sciences* 110(7): 324–6.

– 1890d. Sur la circulation verticale profonde océanique. *Paris. Comptes rendus hebdomadaires des séances de l'Académie des sciences* 110(25): 1350–2.

– 1891a. Considérations sur les eaux abyssales. *Paris. Comptes rendus hebdomadaires des séances de l'Académie des sciences* 112(20): 1144–6.

– 1891b. Le sol sous-marin et les eaux abyssales. *Revue générale des sciences pures et appliquées* 2(10): 326–30.

– 1892. Sur l'immobilité des eaux océaniques profondes. *Paris. Comptes rendus hebdomadaires des séances de l'Académie des sciences* 114(20): 1143–4.

– 1893. Les courants de la mer (1). *Revue scientifique (Revue rose)* 51(9): 257–66.

– 1895a. *Guide d'océanographie pratique.* Encyclopédie scientifique des aide-mémoire 124B. Paris: Gauthier-Villars. 224pp.

– 1895b. Quelques considérations générales sur l'étude des courants marins. *Annales de géographie* 4: 257–70.

– 1896. *Océanographie (dynamique), première partie.* Paris: Librairie militaire de L. Baudoin et Cie. 131pp.

– 1899. Oceanography. *Annual Report of the Smithsonian Institution for 1898*, pp. 407–25.

– 1901. L'étude internationale de l'océan. *Revue scientifique (Revue rose)*, 4ᵉ série, 15(7): 193–8.

– 1902a. Étude des échantillons d'eaux et de fonds récoltés pendant la campagne du yacht *Princesse-Alice* dans l'Atlantique Nord en 1901. *Résultats des campagnes scientifques, accomplies sur son yacht par Albert Iᵉʳ, Prince souverain de Monaco* 22, 76pp.

– 1902b. La circulation océanique. *Revue scientifique (Revue rose)*, 4ᵉ série, 17(18): 545–55.

– 1902c. Sur une série verticale de densités d'eaux marines en Mediterranée. *Paris. Comptes rendus hebdomadaires des séances de l'Académie des sciences* 134(24): 1459–60.

– 1903. Étude de la circulation marine. *Paris. Comptes rendus hebdomadaires des séances de l'Académie des sciences* 137(1): 97–8.

– 1904a. Mesure des courants marins au moyen de l'analyse physique et chimique d'échantillons d'eaux récoltés en séries. *Monaco. Bulletin de l'Institut océanographique* 1(12), 8pp.

– 1904b. *L'océan, ses lois et ses problèmes.* Paris: Librairie Hachette. viii + 397pp.

– 1904c. Méthode physique et chimique de reconnaissance et de mesure des courants sous-marines profonds. *Paris. Comptes rendus hebdomadaires des séances de l'Académie des sciences* 138(8): 527–9.

– 1905a. Analyses d'eaux de mer récoltées à bord de la *Princesse-Alice* en 1902 et 1903 et considérations générales sur la circulation océanique. *Résultats des campagnes scientifiques, accomplies sur son yacht par Albert Iᵉʳ, Prince souverain de Monaco* 29: 81–112.

– 1905b. Cours d'océanographie fondé à Paris par S.A.S. le Prince de Monaco.

Leçons faites par M. Thoulet. *Monaco. Bulletin du Musée océanographique* 2(34): 1–10.

– 1905c. La méthode en océanographie. *Revue scientifique (Revue rose)*, 5ᵉ série, 4(2): 33–9.

– 1908. *Instruments et opérations d'océanographie pratique*. Paris: Librairie militaire R. Chapelot et cie. vi + 186pp.

– 1921a. Circulation océanique; densités *in situ* et indices de réfraction. *Monaco. Bulletin de l'Institut océanographique* 18(394), 26pp.

– 1921b. La circulation océanique et la densité des eaux. *Paris. Comptes rendus hebdomadaires des séances de l'Académie des sciences* 172(14): 861–3.

– 1921c. Sur la circulation océanique profonde. *Monaco. Bulletin de l'Institut océanographique*, 18(404), 10pp.

– 1922. *L'océanographie*. Paris: Gauthier-Villars. ix + 287pp.

– 1924a. La circulation océanique. *Paris. Annales de l'Institut océanographique, nouvelle série*, 1(2): 53–76.

– 1924b. La densité des eaux marines: Son rôle dans l'étude de la circulation océanique et la pêche maritime. *Paris. Comptes rendus hebdomadaires des séances de l'Académie des sciences* 178(10): 858–60.

– 1924c. Études locales de la circulation des eaux dans l'Océan Atlantique. *Paris. Comptes rendus hebdomadaires des séances de l'Académie des sciences* 179(1): 61–3.

– 1924d. Sur la circulation océanique. *Paris. Comptes rendus hebdomadaires des séances de l'Académie des sciences* 178(3): 335–7.

– 1924e. Sur une couche particulière au sein de la masse des eaux océaniques. *Paris. Comptes rendus hebdomadaires des séances de l'Académie des sciences* 178(20): 1621–3.

– 1925a. Le calme des eaux abyssales de l'océan. *Revue scientifique (Revue rose)* 63(19): 648–51.

– 1925b. Contributions à l'étude de la circulation océanique. *Paris. Annales de l'Institut océanographique, nouvelle série*, 2(4): 409–25.

– 1925c. Découverte et mesure des courants marins profonds. *Monaco. Bulletin de l'Institut océanographique* 22(463), 7pp.

– 1925d. Reconnaissance et mesure des courants océaniques profonds. *Paris. Comptes rendus hebdomadaires des séances de l'Académie des sciences* 181(17): 561–3.

– 1926a. Essai d'une densimétrie des océans. *Paris. Annales de l'Institut océanographique, nouvelle série*, 3(3): 137–60.

– 1926b. Relations entre la composition des sédiments sous-marins et les conditions des eaux superficielles. *Monaco. Bulletin de l'Institut océanographique* 23(470), 28pp.

– 1927a. Densimétrie et volcanicité abyssale dans le Pacifique. *Paris. Annales de l'Institut océanographique, nouvelle série*, 4(2): 25–45.

– 1927b. Étude densimétrique dans le Pacifique. *Paris. Annales de l'Institut océanographique, nouvelle série*, 4(7): 261–81.

– 1927c. Sur une double circulation superficielle et abyssale de l'océan. *Paris. Comptes rendus hebdomadaires des séances de l'Académie des sciences* 185(17): 865–6.

– 1930a. Étude densimétrique des eaux océaniques abyssales. *Paris. Annales de l'Institut océanographique, nouvelle série*, 8(3): 185–226.

– 1930b. Études océanographiques. I. Volcanicité abyssale. II. Le courant Kurosio-Oyasio. *Paris. Annales de l'Institut océanographique, nouvelle série*, 7(2): 25–48.

– 1933. Notes d'océanographie. *Monaco. Bulletin de l'Institut océanographique* 30(634), 16pp.

– 2005. *A voyage to Newfoundland*. Translated from the French and edited by Scott Jamieson. Montreal and Kingston: McGill-Queen's University Press. xxxiv + 195pp.

Thoulet, J., and A. Chevallier. 1906. Sur la circulation océanique. *Paris. Comptes rendus hebdomadaires des séances de l'Académie des sciences* 142(4): 245–6.

Thrower, N.J.W. 1969. Edmond Halley and thematic geo-cartography. In N.J.W. Thrower and C.J. Glacken, *The terraqueous globe: The history of geography and cartography*, pp. 3–43. Los Angeles: William Andrews Clark Memorial Library, University of California. iv + 75pp.

Tibbetts, G.R. 1971. *Arab navigation in the Indian Ocean before the coming of the Portuguese*. London: Royal Asiatic Society of Great Britain and Ireland. xxvi + 614pp.

Tizard, T.H., H.N. Moseley, J.Y. Buchanan, and J. Murray. 1885. Narrative of the cruise of H.M.S. Challenger. *Report on the scientific results of the voyage of H.M.S. Challenger during the years 1873–76, Narrative* 1(2): 948–1003.

Tomczak, M., and J.S. Godfrey. 1994. *Regional oceanography: An introduction*. Oxford: Pergamon. vii + 422pp.

Tornøe, H. 1880. Chemistry. I. On the air in sea-water. II. On the carbonic acid in sea-water. III. On the amount of salt in the water of the Norwegian Sea. *The Norwegian North Atlantic expedition, 1876–1878*. Christiania: Grøndahl and Søn.

Tracy, J.D., ed. 1990. *The rise of merchant empires: Long-distance trade in the early modern world, 1350–1750*. Cambridge: Cambridge University Press. ix + 442pp.

Trites, R.W. 1956. The oceanography of Chatham Sound, British Columbia. *Journal of the Fisheries Research Board of Canada* 13(3): 385–434.

Tsimplis, M.N., and H.L. Bryden. 2000. Estimation of the transports through the Strait of Gibraltar. *Deep-Sea Research* I, 47(12): 2219–42.

Tuan, Yi-Fu. 1968. *The hydrologic cycle and the wisdom of God.* Toronto: University of Toronto Press. xiii + 160pp.

Tully, J.H. 1988. John Patrick Tully 1906–1987. *Transactions of the Royal Society of Canada,* Series V, 3: 206–8.

Tully, J.P. 1933. A preliminary oceanographical survey in Nootka Sound. *Biological Board of Canada. Pacific Biological Station. PBS Summary Reports. Investigations in Fisheries and Oceanography,* No. 66, 1p.

– 1935. Oceanographic investigations No. 2. *Biological Board of Canada. Pacific Biological Station. PBS Summary Reports. Investigations in Sea Fisheries Research.* 1p.

– 1936a. Ocean current survey. *Biological Board of Canada. Summary Reports of the Pacific Biological Station for 1936,* No. 56. 2pp.

– 1936b. Ocean currents. *Biological Board of Canada. Progress Reports of Pacific Coast Stations,* No. 30, pp. 16–19.

– 1936c. Outline of procedure at sea in dynamic current survey. *Biological Board of Canada. Manuscript Reports of Biological Stations,* No. 219, 6pp.

– 1936d. Weather and the ocean. *Biological Board of Canada. Pacific Biological Station, Progress Reports,* No. 26, pp. 5–10.

– 1937a. Coastal current investigations. *Fisheries Research Board of Canada. Summary Reports of the Pacific Biological Station for 1937,* No. 52, 3pp.

– 1937b. Gradient current surveys. *Fisheries Research Board of Canada. Summary Reports of the Pacific Biological Station for 1937,* No. 54.

– 1937c. Gradient currents. *Fisheries Research Board of Canada. Progress Reports of the Pacific Biological Station,* No. 32, pp. 13–14.

– 1937d. Oceanographic program. 1937. *Fisheries Research Board of Canada. Pacific Biological Station, Nanaimo, B.C. Manuscript,* 18pp + appendices, chart.

– 1937e. Oceanography of Nootka Sound. *Journal of the Biological Board of Canada* 3(1): 43–69.

– 1937f. Report on dynamic studies off the Canadian Pacific Coast, 1936. *Transactions of the American Geophysical Union, 18th Annual Meeting,* pp. 228–31.

– 1937g. Why is the water along the east coast of Vancouver Island so cold? *Fisheries Research Board of Canada. Progress Reports of Pacific Coast Stations,* 34, pp. 13–15.

– 1938a. Hydrographical investigations. *Fisheries Research Board of Canada. Summary Reports of the Pacific Biological Station for 1938,* No. 60, 2pp.

– 1938b. Some relations between meteorology and coast gradient currents off the Pacific Coast of North America. *Transactions of the American Geophysical Union, 19th Annual Meeting,* pp. 176–83.

– 1939a. Alberni Inlet investigation. *Fisheries Research Board of Canada. Summary Reports of the Pacific Biological Station for 1939*, No. 84, 2pp.

– 1939b. The program in oceanography. *Fisheries Research Board of Canada. Summary Reports of the Pacific Biological Station for 1939*, No. 75, 3pp.

– 1942 Surface non-tidal currents in the approaches to Juan de Fuca Strait. *Journal of the Fisheries Research Board of Canada* 5(4): 398–409.

– 1949. Oceanography and prediction of pulp mill pollution in Alberni Inlet. *Fisheries Research Board of Canada Bulletin*, No. 83, 169pp.

– 1951. Oceanography on the Pacific Coast of Canada. *Fisheries Research Board of Canada. Pacific Biological Station, Nanaimo B.C. Pacific Oceanographic Work Lecture.* File N 7–4–1, April 20, 1951. Manuscript, 10pp. + figures.

– 1956. Norpac. *Proceedings of the Hawaiian Academy of Science, Thirty-first Annual Meeting,* 1955–6, 1p.

– 1958. Canadian Pacific oceanography since 1953. *Proceedings of the Ninth Pacific Science Congress* 16: 6–13.

Tully, J.P., and A.J. Dodimead. 1956. Pacific salmon water? *Fisheries Research Board of Canada. Progress Reports of Pacific Coast Stations*, 107, pp. 28–32.

Turrell, B., and M. Angel. 1995. A centenary of hydrographic work in the Faeroe-Shetland Channel. *Ocean Challenge* 6(1): 2–3.

Ulrich, J. 1986. Johann Gottfried Otto Krümmel 1854–1912. *Geographers Biobibliographical Studies* 10: 99–104.

Ulrich, J., and G. Kortum. 1997. Otto Krümmel (1854–1912). Geograph und Wegbereiter der modernen Ozeanographie. *Kieler Geographische Schriften* 93, vii + 310pp.

Vachon, A.H. 1918. Hydrography in Passamaquoddy Bay and vicinity, New Brunswick. *Contributions to Canadian Biology and Fisheries (New Series)* 15: 295–328.

Vallaux, C. 1927. L'expédition scientifique du 'Meteor' au sud de l'Atlantique et dans l'océan Austral (1925–1926). Premiers résultats. *Paris. Annales de l'Institut océanographique, nouvelle série*, 4(1): 1–24.

– 1936. Notice sur Julien-Olivier Thoulet (1843–1936). *Monaco. Bulletin de l'Institut océanographique* 33(702), 28pp.

– 1937a. La circulation océanique de surface et la circulation profonde. *Paris. Annales de l'Institut océanographique, nouvelle série*, 16(4): 323–33.

– 1937b. Nouvelles recherches américaines sur la circulation de l'Atlantique Nord. *Monaco. Bulletin de l'Institut océanographique* 34(723), 16pp.

– 1938. Recherches du *Discovery II* sur la dynamique de l'océan Austral. *Monaco. Bulletin de l'Institut océanographique* 35(751), 15pp.

van der Wee, H. 1990. Structural changes in European long-distance trade, and particularly in the re-export trade from south to north, 1350–1750. In

J.D. Tracy, ed., *The rise of merchant empires: Long-distance trade in the early modern world, 1350–1750*, pp. 14–33. Cambridge: Cambridge University Press.

Varenius (Varen), B. 1650. *Geographia Generalis, in qua Affectiones Generales Telluris Explicantur.* Amsterdam. 748pp.

Vaughan, T.W. 1926. Scripps Institution of Oceanography, University of California, Berkeley. *Annual Report to the President of the University, 1924–25 and 1925–26*, pp. 61–4.

Vaughan, T.W., et al. 1937. *International aspects of oceanography: Oceanographic data and provisions for oceanographic research.* Washington, DC: National Academy of Sciences. xvii + 225pp.

Veronis, G. 1981. A theoretical model of Henry Stommel. In B.A. Warren and C. Wunsch, eds, *Evolution of physical oceanography: Scientific surveys in honor of Henry Stommel*, pp. xix–xxiii. Cambridge, MA: MIT Press.

– 1992. Henry Melson Stommel. 27 September, 1920 – 17 January, 1992. *Journal of Marine Research* 50: i–viii.

Wagner, H. 1885. Karl Zöppritz. *Verhandlungen der Gesellschaft für Erdkunde zu Berlin* 12: 298–304.

Waitz, J.S. 1755. Undersökning om orsaken, hvarföre vattnet i Atlantiska hafvet altid strömar in uti Medelhafvet, genom Sundet vid Gibraltar. *Kongliga Svenska Vetenskaps-Akademiens Handlingar* 16: 27–50.

Waldie, R.J., L.A.E. Doe et al. 1950. Oceanographic discovery. *Fisheries Research Board of Canada. Progress Reports of Pacific Coast Stations*, 84, pp. 59–63.

Walker, J.M. 1991a. Farthest north: Dead water and the Ekman spiral. Part 1. An audacious venture. *Weather* 46(4): 103–7.

– 1991b. Farthest north: Dead water and the Ekman spiral. Part 2. Invisible waves and a new direction in current theory. *Weather* 46(6): 158–64.

Wallace, W.J. 1974. *The development of the chlorinity/salinity concept in oceanography.* Amsterdam: Elsevier. xii + 227pp.

Walther, J. 1893. *Allgemeine Meereskunde.* Leipzig: J.J. Weber. xvi + 296pp.

Warren, B.A. 1966. Medieval Arab references to the seasonally reversing currents of the North Indian Ocean. *Deep-Sea Research* 13: 167–71.

– 1981. Deep circulation of the world ocean. In B.A. Warren and C. Wunsch, eds, *Evolution of physical oceanography: Scientific surveys in honor of Henry Stommel*, pp. 6–41. Cambridge, MA: MIT Press.

– 1987. Ancient and medieval records of the monsoon winds and currents of the Indian Ocean. In J.S. Fein and P.L. Stephens, eds, *Monsoons*, pp. 137–58. New York: John Wiley.

– 1992. Physical oceanography in *The Oceans. Oceanography* 5: 157–9.

– 2006. Historical introduction: Oceanography of the general circulation to the middle of the twentieth century. In M. Jochum and R. Murtugudde, eds, *Physical oceanography: Developments since 1950*, pp. 1–14. New York: Springer.

Warren, B.A., and C. Wunsch, eds. 1981. *Evolution of physical oceanography: Scientific surveys in honor of Henry Stommel.* Cambridge, MA: MIT Press. xxxiv + 623pp.

Watermann, B., ed. 1989. *Exposition on historical aspects of marine research in Germany.* Deutsche Hydrographische Zeitschrift, Ergänzungsheft, Reihe B, Nr 21, 130pp.

Wegemann, G. 1899. Die Oberflächenströmungen des Nordatlantischen Ozeans nördlich von 50⁰ N-Br. *Aus dem Archiv der Deutsche Seewarte* 22(4): 1–27.

– 1900. *Die Oberflächen-Strömungen des Nordatlantischen Ozeans von 50⁰ N-Br.* Inaugural-Dissertation zur Erlangung der Doktorwürde der hohen philosophischen Fakultät der Christian-Albrechts-Universität in Kiel. Altona. 27pp.

– 1915. Krümmel, J.G. Otto. *Biographischer Jahrbuch und deutscher Nekrolog* 17 (1912): 200–6.

Weir, G.E. 2001. *An ocean in common: American naval officers, scientists, and the ocean environment.* College Station, TX: Texas A & M University Press. xx + 404pp.

Weisz, G. 1983. *The emergence of modern universities in France, 1863–1914.* Princeton, NJ: Princeton University Press. xiii + 397pp.

Welander, P. 1971. Ekman, Vagn Walfrid. *Dictionary of Scientific Biography* 4: 344–5.

Wells, N.C., W.J. Gould, and A.E.S. Kemp. 1996. The role of ocean circulation in the changing climate. In C.P. Summerhayes and S.A. Thorpe, eds, *Oceanography: An illustrated guide*, pp. 41–58. New York and Toronto: John Wiley and Sons.

Wendicke, F. 1912. *Meereskundliche Untersuchungen in der Deutschen Bucht.* Inaugural-Dissertation zur Erlangung der Doktorwürde genehmigt von der Philosophischen-Fakultät der Friedrich-Wilhelms-Universität zu Berlin. 71pp.

Went, A.E.J. 1972. Seventy years agrowing: A history of the International Council for the Exploration of the Sea 1902–1972. *Conseil international pour l'Exploration de la Mer, Rapports et Procès-Verbaux des Réunions* 165: 1–252.

Wharton, W.J.L. 1873. Observations on the currents and undercurrents of the Dardanelles and Bosphorus, made by Commander W.J.L. Wharton of H.M. Surveying-Ship *Shearwater,* between the months of June and October, 1872. *Proceedings of the Royal Society of London* 21: 387–93.

– 1886. *Report on the currents of the Dardanelles and Bosporus.* London: Lord Commissioners of the Admiralty.

Whitfield, P. 1994. *The image of the world: Twenty centuries of world maps*. London: The British Library. vii + 144pp.

– 1996. *The charting of the oceans: Ten centuries of maritime maps*. Rohnert Park, CA: Pomegranate. 136pp.

Wille, C. 1882a. The apparatus and how used. *The Norwegian North Atlantic Expedition. 1876–1878*. Christiania: Gröndahl and Sön.

– 1882b. 1. Historical account. *The Norwegian North Atlantic Expedition, 1876–1878*. Christiania: Gröndahl and Sön.

Williams, F.L. 1963. *Matthew Fontaine Maury: Scientist of the sea*. New Brunswick, NJ: Rutgers University Press. xx + 720pp.

Williams, G. 1966. *The expansion of Europe in the eighteenth century: Overseas rivalry, discovery and exploitation*. London: Blandford. x + 309pp.

– 1997. *The Great South Sea: English voyages and encounters, 1570–1750*. New Haven: Yale University Press. xv + 300pp.

– 1999. *The prize of all the oceans: The triumph and tragedy of Anson's voyage around the world*. New York: HarperCollins. xxi + 264pp.

Wissemann, W. 1906. Die Oberflächenströmungen des Schwarzen Meeres. *Annalen der Hydrographie und maritimen Meteorologie* 34: 162–79.

Witte, E. 1878. *Über Meeresströmungen*. Pless: A. Kramer. 45pp.

Wolf, A. 1950. *A history of science, technology and philosophy in the 16th and 17th centuries*. Volume 1. Second edition prepared by Douglas McKie. London: Macmillan. xvi + 349pp.

Woods Hole Oceanographic Institution. 2001. In Memoriam: Arnold B. Arons. *Obituary. Woods Hole Oceanographic Institution*. http://www.whoi.edu/mr/obit.do?id=774

Woodward, H.W. 1893. Eminent living geologists. No. 8. Professor Joseph Prestwich, D.C.L., F.R.S., F.G.S., Assoc, Inst. C.E., Corr. Inst. France, etc., etc. *Geological Magazine, New Series*, Decade III, X: 241–6.

Wüst, G. 1920. Die Verdunstung auf dem Meere. *Veröffentlichungen des Instituts für Meereskunde an der Universität Berlin. Neue Folge. A. Geographisch-naturwissenschaftliche Reihe* 6: 1–95.

– 1924. Florida- und Antillenstrom: Eine hydrodynamische Untersuchung. *Veröffentlichungen des Instituts für Meeeskunde an der Universität Berlin. Neue Folge. A. Geographisch-naturwissenschaftliche Reihe* 12: 1–48.

– 1928a. Ozeanographische Methoden und Instrumente der Deutschen Atlantischen Expedition. *Zeitschrift der Gesellschaft für Erdkunde zu Berlin*, Ergänzungsheft III: 66–83.

– 1928b. Der Ursprung der Atlantischen Tiefenwässer (aus den Ergebnissen der Deutschen Atlantischen Expedition). *Zeitschrift der Gesellschaft für Erdkunde zu Berlin*, Jubiläums-Sonderband 1928: 506–34.

– 1935. Schichtung und Zirkulation des Atlantischen Ozeans. Die Stratosphäre. *Deutsche Atlantische Expedition, 1925–1927. Wissenschaftliche Ergebnisse* 6(1) 2, 288pp.

– 1936. Kuroshio und Golfstrom: Eine vergleichende hydrodynamische Untersuchung. *Veröffentlichungen der Instituts für Meereskunde and der Universität Berlin. Neue Folge* A. *Geographisch-naturwissenschaftliche Reihe* 29, 69pp.

– 1938. Bodentemperatur und Bodenstrom in der Atlantischen, Indischen und Pazifischen Tiefsee. *Gerland's Beiträge zur Geophysik* 54(1): 1–8.

– 1951. Über die Fernwirkungen Antarktischen und Nordatlantischen Wassermassen in den Tiefen des Weltmeeres. *Naturwissenschaftliche Rundschau* 3: 97–108.

– 1959. Alexander von Humboldts Stellung in der Geschichte der Ozeanographie. In J.H. Schultze, ed., *Alexander von Humboldt: Studien zu seiner universalen Geisteshaltung,* pp. 90–104. Berlin: Walter de Gruyter.

– 1964a. Albert Defant achtzig Jahre alt. *Die Naturwissenschaften* 51(13): 301–2.

– 1964b. Albert Defant zum 80. Geburtstag. *Beiträge zur Physik der Atmosphäre* 37(2): 59–68.

– 1968. History of investigations of the longitudinal deep-sea circulation (1800–1922). *Monaco. Bulletin de l'Institut océanographique, No. Spécial* 2 (*Premier Congrès international d'histoire de l'océanographie*) 1: 109–20.

Wunsch, C. 1997. Henry Stommel. September 27, 1920 – January 17, 1992. *National Academy of Sciences of the U.S.A. Biographical Memoirs* 72: 331–50.

– 2002. What is the thermohaline circulation? *Science* 298: 1179–81.

Wunsch, C., and R. Ferrari. 2004. Vertical mixing, energy, and the general circulation of the oceans. *Annual Review of Fluid Mechanics* 36: 281–314.

Zaslow, M. 1971. *The opening of the Canadian North, 1870–1914.* Toronto: McClelland and Stewart. xii + 339pp.

– 1975. *Reading the rocks: The story of the Geological Survey of Canada, 1842–1972.* Toronto: Macmillan. 599pp.

– 1988. *The northward expansion of Canada, 1914–1967.* Toronto: University of Toronto Press. xv + 421pp.

Zeller, S. 1987. *Inventing Canada: Early Victorian science and the idea of a transcontinental nation.* Toronto: University of Toronto Press. vii + 356pp.

Zimmerman, D. 1988. The Royal Canadian Navy and the National Research Council, 1939–45. *Canadian Historical Review* 69: 203–21.

– 1989. *The great naval battle of Ottawa.* Toronto: University of Toronto Press. xiv + 209pp.

Zöppritz, K. 1878a. Hydrodynamische probleme in Beziehung zur Theorie der Meeresströmungen. *Annalen der Physik und Chemie, N.F.,* 3: 582–607.

– 1878b. Hydrodynamic problems in reference to the theory of ocean currents. *Philosophical Magazine, Series 5*, 6: 192–211.

– 1879a. Hydrodynamische Probleme in Beziehung zur Theorie der Meer-esströmungen. *Annalen der Physik und Chemie, N.F.*, 6: 1–13, 599–611.

– 1879b. Zöppritz on ocean currents [Introduction by James Croll]. *Nature* 19: 202–4.

INDEX